SYNTHETIC FUELS

ISPRA
COURSES

ON ENERGY SYSTEMS AND TECHNOLOGY

A series devoted to the publication of courses and educational seminars given at the Joint Research Centre, Ispra Establishment, as part of its education and training program. Published for the Commission of the European Communities, Directorate-General Information Market and Innovation.

Already published:

ENERGY STORAGE AND TRANSPORTATION

PROSPECTS FOR NEW TECHNOLOGIES

Edited by G. BEGHI

1981, x + 497 pp. ISBN 90-277-1166-6

THERMAL ENERGY STORAGE

Edited by G. BEGHI

1982, vi + 505 pp. ISBN 90-277-1428-2

PERFORMANCE OF SOLAR ENERGY CONVERTERS: THERMAL COLLECTORS AND PHOTOVOLTAIC CELLS

Edited by G. BEGHI

1983, viii + 527 pp. ISBN 90-277-1545-9

NATIONAL ENERGY PLANNING AND MANAGEMENT IN DEVELOPING COUNTRIES

Edited by H. NEU and D. BAIN

1983, xiii + 442 pp. ISBN 90-277-1589-0

SYNTHETIC FUELS

Lectures of a Course held at the Joint Research Centre, Ispra, Italy, May 7-11, 1984

Edited by

G. E. BEGHI
Joint Research Centre, Ispra, Italy

D. REIDEL PUBLISHING COMPANY
A MEMBER OF THE KLUWER ACADEMIC PUBLISHERS GROUP
DORDRECHT / BOSTON / LANCASTER

Library of Congress Cataloging in Publication Data
Main entry under title:

Synthetic fuels.

Includes index.
1. Synthetic fuels–Congresses. I. Beghi, G. II. Commission of the European Communities. Joint Research Centre. Ispra Establishment.
TP360.S9474 1985 666'.66 85-8204
ISBN 90-277-2016-9

CIP

Commission of the European Communities Joint Research Centre Ispra (Varese), Italy

Publication arrangements by
Commission of the European Communities
Directorate-General Information Market and Innovation, Luxembourg

EUR 9832

Published by D. Reidel Publishing Company
P.O. Box 17, 3300 AA Dordrecht, Holland

Sold and distributed in the U.S.A. and Canada
by Kluwer Academic Publishers,
190 Old Derby Street, Hingham, MA 02043, U.S.A.

In all other countries, sold and distributed
by Kluwer Academic Publishers Group,
P.O. Box 322, 3300 AH Dordrecht, Holland

Printed in The Netherlands

TABLE OF CONTENTS

INTRODUCTION TO THE ISPRA COURSE

G. Beghi
Commission of the European Communities
Joint Research Centre - Ispra Establishment
21020 Ispra (Varese)
Italy

The main characteristic of any primary source is the degree of ease with which its energy can be converted and transported. From this point of view, petroleum-based fuels are most convenient and they are indispensable in our present society. Attention has been focussed in recent years on possible ways of obtaining liquid and gaseous fuels from alternative sources. These fuels which can be produced from non-petroleum sources, but with the same attractive characteristics of easy conversion and transportability, are called synthetic fuels, or synfuels.

The importance of the transport of energy in the energy systems, and of the synthetic fuels, from this point of view, is the motivation for a course to be held at the Joint Research Centre Ispra which has been active in the field of energy problems since its beginning. These problems are the main objectives of the research and development programme of the JRC including various activities, from experimental work to theoretical studies, together with different actions for the dissemination of information.

The Course gives a review of the present state of synfuel production technologies from the two main primary energy sources considered to-day: coal and biomass.

Coal is an important primary energy source, because of availability and reserves, but a conversion is useful. It may have a significant sulphur content and it produces dust when burned. When it is converted to gas or liquid fuels, combustion produces substantially less air pollution. Furthermore, moving gas or oil through a pipeline is cheaper than transporting coal.

Biomass is considered a sources of synthetic fuels for other reasons: biomass resources are renewable and may provide a valuable alternative source to petroleum. Some of the processes for converting biomass are well-known, for instance fermentation to produce alcohols suitable for car engines.

While basic technologies for most of these conversion processes are already available, and it is technically feasible to replace petroleum-based fuels, synfuels are not yet competitive

G. E. Beghi (ed.), Synthetic Fuels, 1–2.

and there are problems of a different nature hindering their penetration and wider application in the energy system.

Nevertheless in the long term synthetic fuels can be expected to make a contribution to the energy requirements and in some cases transition phases have already begun.

The main conversion processes using coal and biomass as primary energy sources are described, for the production of both gaseous fuels (low energy gas, methane, etc.) and liquid fuels (alcohols, synthetic hydrocarbons, etc.).

In addition to the state of the art of the production processes, the Course gives information on practical experience in the use of synfuels, particularly in automobile engines and on some aspects of environmental pollution.

The prospects of synthetic fuels depend strongly on local situations: the first element is economic competitivity but aspects other than the purely economic can be important, such as the availability of primary sources, needs and the energy policy.

The Course does not aim to arrive at conclusions, the goal is to give updated and reliable data, which will enable each participant to make his own evaluations and assessments for his particular situation. It is an opportunity to collect information on technologies and their potential development and to discuss these with experts; these discussions on the current status and problems could be helpful in the evaluation of the prospects of synthetic fuels in the energy system and of their practical applications.

The Ispra Course is organized in cooperation with KFA-Jülich, Project Management for Energy Research (PLE), Federal Republic of Germany and thanks are expressed to Dr. N. Stump. We should like to acknowledge the contribution of other services of the Commission of the European Communities: Directorate Alternative Energy Sources, Directorate General XII - Science and Research (Dr. G. Imarisio) and Directorate Coal, Directorate General XVII Energy (Dr. S. Furfari)

It should be mentioned that in addition to the contributions of the experts in each area, the discussions among participants and lecturers are also important in ensuring the satisfactory outcome of the Course.

GENERAL STRATEGY FOR ENERGY IN THE EUROPEAN COMMUNITY

Dr. Ing. S. Furfari
Commission of the European Communities
Directorate-General for Energy
Coal Directorate
200 rue de la Loi
1049 Brussels
Belgium

1.ENERGY SITUATION IN THE COMMUNITY

1.1 Consumption

Since 1979, energy consumption in the Community has decreased each year to the extent that in 1983 the lowest level of demand since 1975 was reached. This trend of falling consumption must be placed in the economic context of virtually zero growth in the Community's real gross domestic product in 1982, a fall in the index of industrial production and an unemployment rate equivalent to about 10% of the total labour force. Since 1979, there has been a dissociation between energy consumption and real economic growth. Figure 1 shows that setting the ratio of energy consumption to the gross domestic product (GDP) at the value of 100 for the year 1979, it consequently falls to 94 in 1980, 91 in 1981, 88 in 1982 and 86 in 1983 (the forecast for 1984 is also 86). The parallel index for oil consumption per unit of GDP (1979 = 100) moves to 91 in 1980, 83 in 1981, 79 in 1982 and 75 in 1983 (the forecast for 1984 is 74). A study by the Commission services of the reasons for changes in final energy use in the periods 1973-1979 and 1979-1981 in 7 of the Community countries suggests that a large part of the changes can be ascribed to genuine increases in energy efficiency rather than simply to changes in the level of economic activity and in economic and industrial structures.

Table I and Figure 2 show that the largest falls in demand have been registered in the oil sector. The share of oil in the gross inland consumption of energy was 52.3% in 1980 but fell to 47.1% in 1983. A long way has been covered since 1973 when the figure was 61%.

G. E. Beghi (ed.), Synthetic Fuels, 3–33.

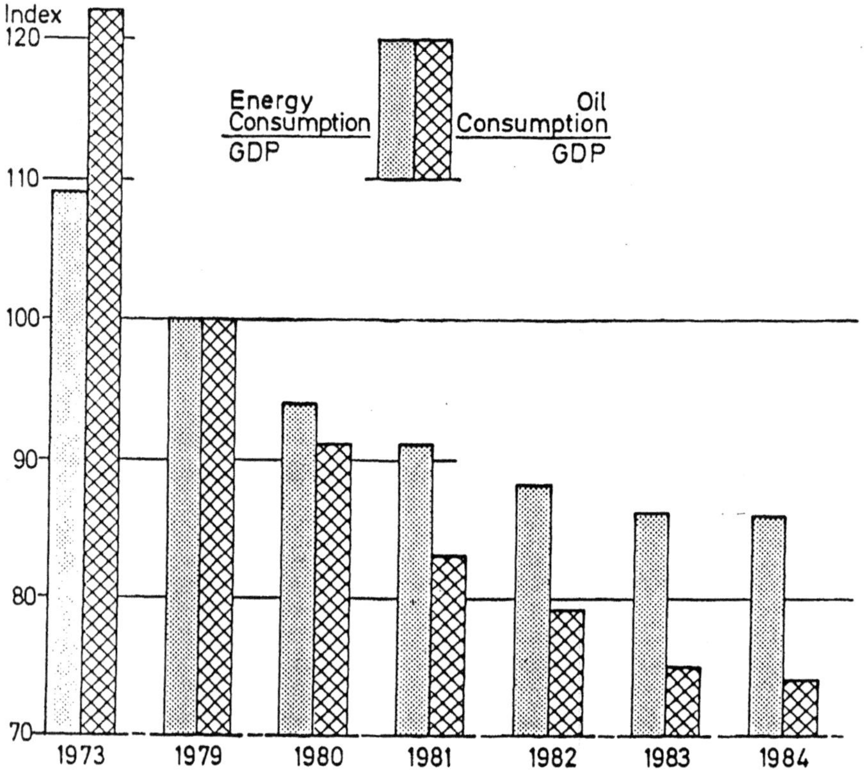

Fig. 1 Ratio of: $\frac{\text{Energy Consumption}}{\text{Gross Domestic Product}}$ compared to the Ratio of : $\frac{\text{Oil Consumption}}{\text{Gross Domestic Product}}$ (Ratio settled to 100 in 1979)

Table I
Gross inland consumption of energy in the Community

	1980 Mtoe	1980 %	1981 Mtoe	1981 %	1982 Mtoe	1982 %	1983* Mtoe	1983* %
Hard coal and equiv.	189	20.1	186	20.5	{		{	
Lignite and equivalent	33	3.5	34	3.7	216	24.5	207	23.7
Crude oil and equiv.	494	52.3	452	49.7	430	48.7	410	47.0
Natural gas	163	18.0	166	18.2	158	17.9	162	18.6
Nuclear energy	43	4.5	57	6.2	64	7.2	78	8.9
Hydro & others	15	1.6	16	1.7	14	1.6	14	1.6
Total**	944	100	910	100	883	100	873	100

*Forecasts
**Including other fuels

The other main features of the changing primary energy balance at Community level are:

- a decline in solid fuel use as a whole since 1979 (principally as a result of the contraction of output in the steel industry) but stabilisation in solid fuel use in power stations. Overall solid fuel use is now around 7% below its 1973 level;

- a doubling in the contribution of nuclear power since 1979 and a quadrupling since 1973, nuclear now accounting for the equivalent of 8.9% of total primary energy demand (78mtoe) and for 27% of the inputs to electricity supply. The result has been a marked divergence in the growth rates of nuclear and solid fuels as substitutes for hydrocarbons in electricity generation.

  a decline in the use of natural gas since 1979, reversing the high rates of growth of gas demand throughout the 1970s. The fall in gas use was particularly marked in 1982 and since then there has been a limited recovery. But gas demand is still 6% below its 1979 level.

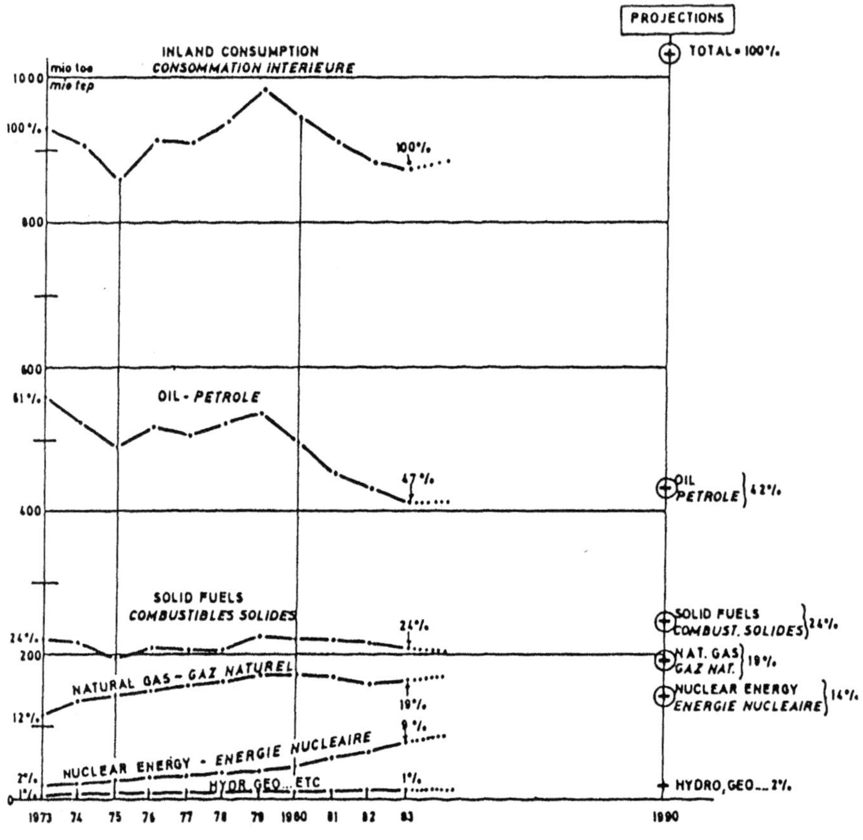

FIGURE 2

EUR-10: Main Trends in the Energy Balance

Principales tendances du bilan énergétique

1973–1990

1.2 Supply

For the fourth successive year indigenous energy production in the Community has increased. Net energy imports have therefore fallen even more quickly than energy demand (down to less than 42% of gross primary energy demand compared with 55% in 1979 and 64% in 1973). (Table II and Figure 3).

Over the same period there has also been an important increase in indigenous oil supply of over 41 mtoe or 46%, essentially from the North Sea, which has helped to reduce net oil imports to 59% of their 1979 level. In 1973 practically all Community oil was imported. In 1983 the Community was producing 2.6 mbd or over 30% of its own oil requirements, mainly from the North Sea.

Table II
Energy supply in the Community (Mtoe)

a.= production
b.= net imports

	1980		1981		1982		1983*	
	a.	b.	a.	b.	a.	b.	a.	b.
Solid fuels	185	47	187	42	182	46	174	37
Oil	91	438	102	358	118	326	130	287
Natural gas	129	41	125	43	116	44	119	47
Primary electricity	57	1	70	2	78	2	92	2
TOTAL	462	527	484	445	494	418	515	373
Share (%)	46.7	53.3	52.1	47.9	54.2	45.8	58.0	42.

In 1973 nuclear provided 7.5% of the inputs to electricity production and met less than 2% of inland energy demand. By 1979 the figures had risen to 13.3% and 3.8% respectively. In the past four years there has been a "quantum leap". In 1983 nuclear provided over 27% of the inputs to electricity and met nearly 9% of inland energy demand.

Natural gas production in the Community in 1983 was around the same level as in 1973, having declined from its peak during the mid-late 1970s

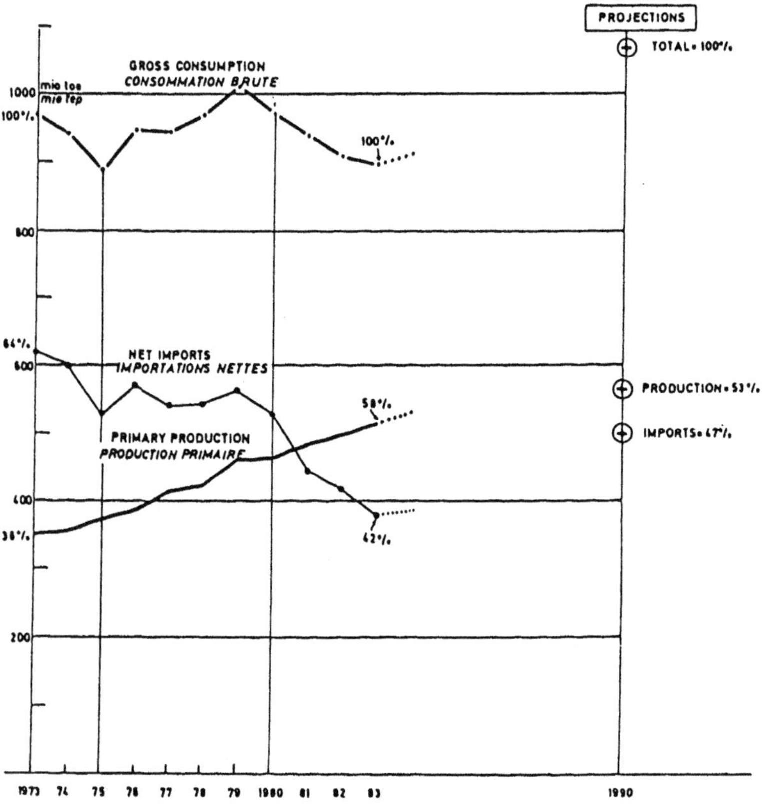

FIGURE 3

EUR-10: ENERGY: Gross Consumption, Primary Production and Net Imports

Consommation brute d'énergie et approvisionnements

1973—1990

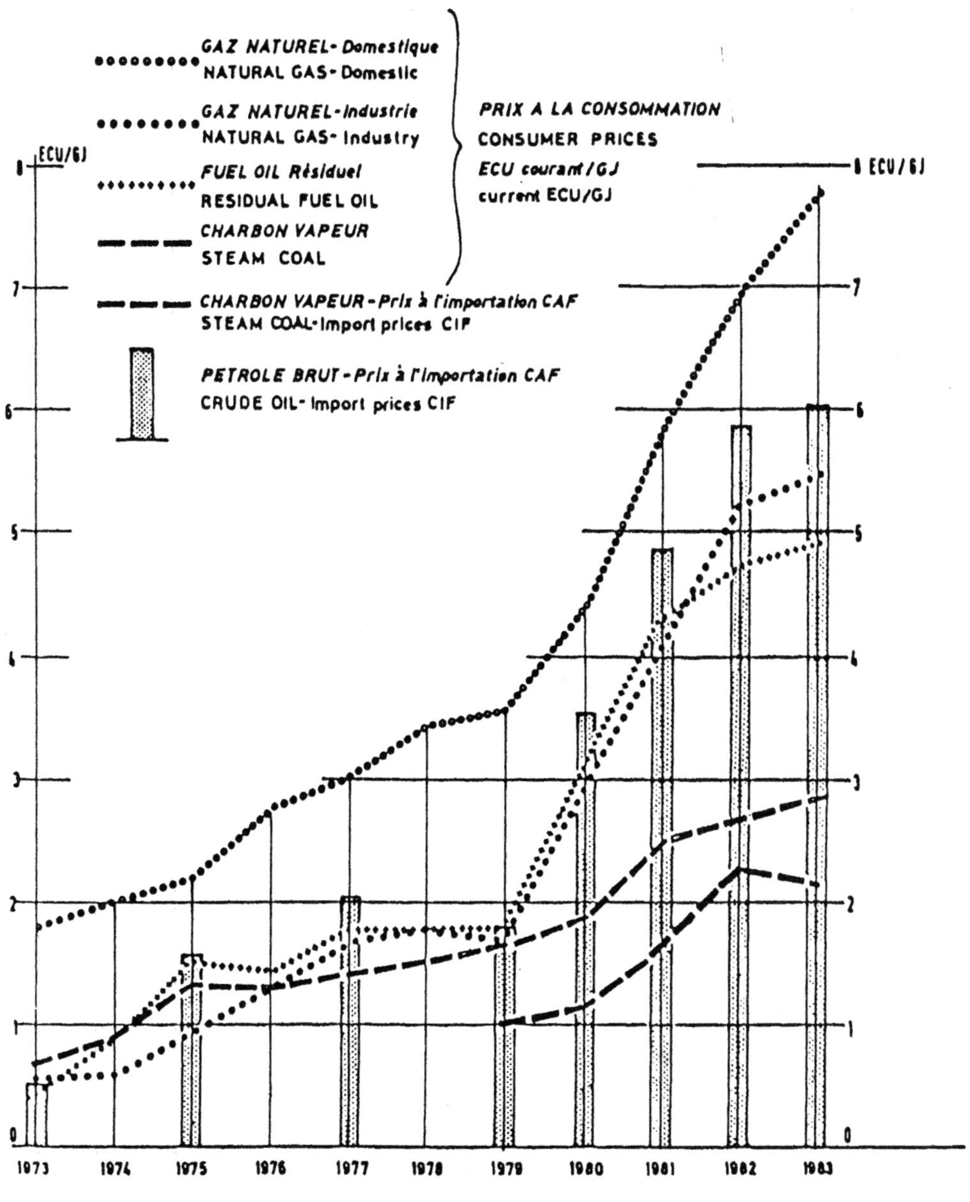

FIGURE 4

EUR-10 : Energy Prices - *Prix de l'énergie*

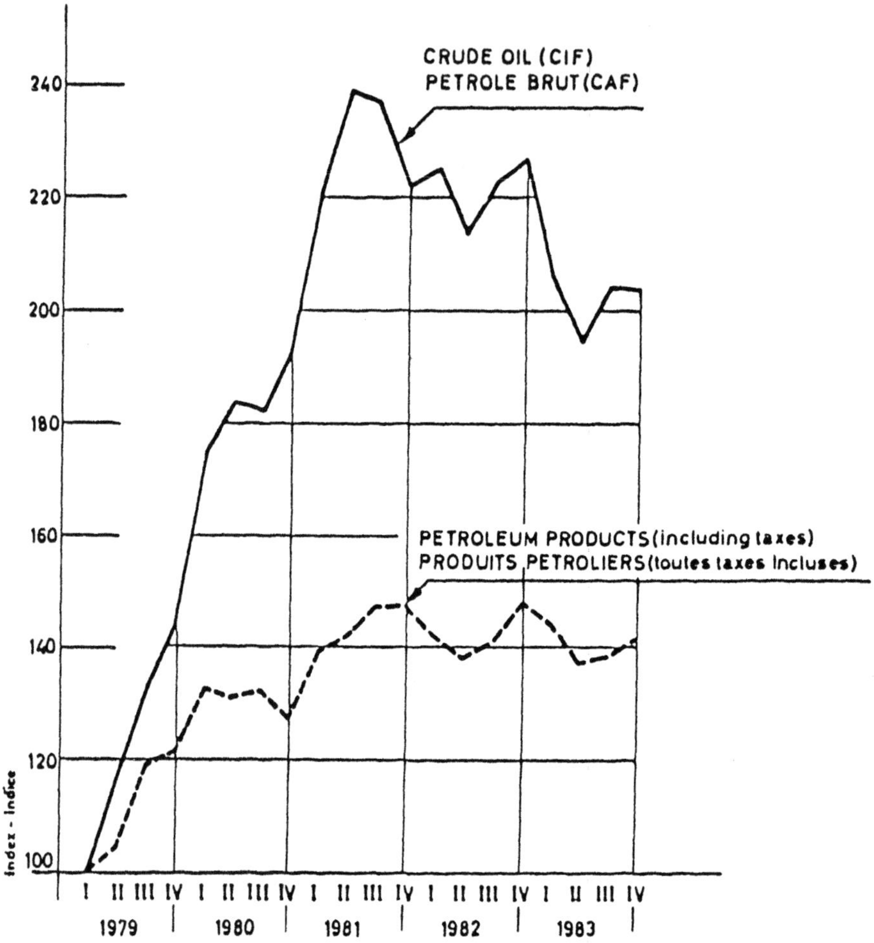

FIGURE 5

Real Oil Price Indices*
Indices des prix réels du pétrole*

* Calcuted as follow: The weighted sum of consumer oil prices or crude oil imports prices in current ECU for 4 member States (F, D, I, U.K.) is deflated by the weighted sum of the ECU deflators for the same 4 member States
NB: Weights are based on the 1982 structure of consumption/imports

Calculés comme suit: la somme pondérée des prix au consommateur ou des prix du brut importé en ECU courants pour 4 pays membres (F, D, I, U.K.) est déflatée par la somme pondérée des déflateurs de l'ECU des mêmes pays.
N.B.: La structure de la consommation ou des importations 1982 est le pondérateur

The share of solid fuels and nuclear as inputs to electricity generation is up to around 74%, compared with 61% in 1979 and less than 54% in 1973. Moreover the share of solid fuels, nuclear and hydro and other minor sources together has risen to 79%, while that of oil and natural gas has fallen to 21%, compared to 42% in 1973. The main figures are set out in Tables III and IV which also include figures for 1973 and the forecasts for 1990 discussed later.
As far as the introduction of new and renewable energies is concerned, progress so far has been very slow. Their total contribution to the Community's inland energy consumption in 1979 was 14 Mtoe or 1.4%. In 1983 the figure was still 14 Mtoe, equivalent to 1.6% of a lower total consumption. In each of these years, moreover, virtually the whole of this contribution came from hydro electricity, which provides nearly 5% of total electricity supplies in the Community as a whole and around 11% in France and Italy. These figures make no allowance for the limited use of, for example, solar collectors and heat pumps in the Community (whose contribution is reflected only in reduced energy demand and not in the supply figures themselves) and they therefore understate the real contribution of new and renewable energies and new energy technologies. But even allowing for the statistical imperfections, it is clear that new and renewable energies will find an increasingly important place in the Community's total energy supplies only at a later stage and then only as a results of sustained efforts in R, D&D and in the commercialisation of new technologies.

1.3 Energy price trends.

The most significant developments in energy price trends are as follows: (figures 4 and 5)

(i) since 1979 there has been a sharp divergence in the trends in prices for oil, on the one hand, and coal on the other. The price which the Community has to pay in national currencies for its crude oil imports is now 3 times as high as in January 1979 in nominal terms and twice as high in real terms, despite a fall of 25% in the average dollar price of imported crude from the peak in early 1981. The price of key oil products (pre- and post-tax) has continued to increase in nominal ECU terms since 1980. On average the nominal 1983 prices of residual fuel oil and gas oil (ex-VAT) and heating oil (tax paid) were over 2.5 times those of 1979. Steam coal prices, on the other hand, increased by only 1.5 times. The price advantage of steam coal over competing oil products widened significantly during 1983;

(ii) the real prices of the main oil products in the four largest Community Member States are now generally higher than in January 1980 although somewhat below their peaks of 1981.

(iii) natural gas prices have risen sharply since 1981 in many Community countries. The available figures point to a marked change in the price competitivity of natural gas versus oil products in a number of Member States. In France, Belgium and the Netherlands the price of natural gas to small-and medium-sized industrial consumers increased by over 25% in 1982 alone at a time when the rise in the cost of competing oil products had slowed down to 5-10%. The same trend is understood to have occurred in Germany. During much of 1983 too, the price of natural gas continued to rise more quickly than that of oil products in all countries for which comparable figures exist, and notably Italy. One important factor appears to be the mechanisms in gas pricing contracts linking gas prices, with some delay, to crude oil and oil product prices;

(iv) ever since 1973 there has been a marked divergence between trends in electricity prices, on the one hand, and those for oil and gas on the other. In the last three years industrial electricity prices have continued to rise much more slowly than prices of competing oil products.

The effects of these absolute and relative price changes on the level and pattern of energy demand in the past 3-4 years have been significant, underlining the importance of energy pricing to both operating decisions (fuel-saving and fuel-switching with existing plant) and energy investment decisions (the introduction of new or replacement plant which is more fuel efficient and/or involves substitution of oil for other products).

2. ENERGY PROGRAMMES OF THE MEMBER STATES OF THE EUROPEAN COMMUNITY

2.1 Belgium

Belgium has made steady progress in improving her energy situation since 1973, reducing dependence on oil from 62% in 1973 to close to 50% in 1982(table V). The main element has been the nuclear power programme which has driven down oil demand in the electricity sector. In 1982 nuclear supplied some 33% of primary energy inputs to Belgian electricity production, the second highest percentage in the Community (after France). Belgium expects to reduce her dependence on oil still further as nuclear output more than doubles during the rest of this decade, with oil dependence projected to fall to 41% by 1990.

TABLE III: MAIN ENERGY INDICATORS - EUROPEAN COMMUNITY (related to 1990 objectives)

	1973	1979	1982	1983 (est)	1990 (proj)
I. Share of oil in gross energy consumption	62.1%	55.7%	50.1%	48.2%	43%
II. Fuel inputs to electricity production[6]. Shares taken by:					
Solid fuels	46.1%	47.4%	48.7%	46.6%	41%
Nuclear	7.5%	13.3%	22.8%	27.5%	40%
Hydro and others	4.5%	4.9%	4.9%	4.8%	4%
Natural gas	10.0%	10.4%	7.1%	8.0%	5%
Oil	31.9%	24.0%	16.5%	13.1%	10%
III. Supply dependence on:[7]					
Imported oil	61.6%	48.2%	35.9%	32.0%	33%
Imported solid fuels	2.0%	3.3%	5.1%	4.1%	6%
Imported natural gas	0.4%	3.6%	4.9%	5.2%	7.5%
IV. Share of imports in[8] gross consumption of:					
Energy	64.0%	55.2%	46.1%	41.6%	47%
Oil	97.8%	86.5%	71.7%	66.1%	77%
Solid fuels	8.6%	15.1%	21.4%	17.9%	27%
Natural gas	3.5%	21.0%	28.0%	29.0%	41%
	1973/63	1975/73	1979/75	1982/79	1990/85
V. Average annual growth rates:					
G D P	+4.7%	+0.3%	+3.6%	+0.4%	+2.4%
Inland energy consumption[9]	+4.6%	-3.8%	+3.4%	-3.6%	+1.6%
Final energy consumption	+4.8%	-4.0%	+3.3%	-4.4%	+1.0%
VI. Energy coefficient[10]	0.98	-	0.94	-	0.67
	1973	1979	1982	1983	1990
VII. Energy ratio[11]	833.0	760.6	673.3	659.4	669
VIII. Intensity of final energy[12] demand	573.8	521.9	455.6	n.a.	439.5

---

6. Basic data expressed in oil equivalent terms
7. Respective shares of imported oil, imported natural gas or imported solid fuels in gross energy consumption.
8. Share of each imported energy source, in gross consumption of that source.
9. Non energy use included
10. Energy coefficient = ratio between growth rates of inland energy consumption and GDP
11. Energy ratio = volume of inland energy consumption per unit of GDP, expressed in real terms, at 1975 prices (in kg of oil equivalent per 10^3 ECU)
12. Intensity of final energy demand = volume of final energy demand per unit of GDP, expressed in real terms at 1975 prices (kg per ECU)

TABLE IV MAIN TRENDS IN THE COMMUNITY ENERGY BALANCE 1973-1990

MTOE

	1973	1979	1982	1983 (estimated)	1990 (projected)
DEMAND					
Gross Primary Energy Demand	968	1012	908	896,5	1061
Inland Energy Demand					
(total):	931	985	883	872,5	1032
Oil	564	537	430	410	430
Solid Fuels	222	223	216	207	245
Natural Gas	116	172	158	162	194
SUPPLY					
Indigenous Production					
(total):	351	458	494	516	565
Oil	13	89	118	130	106
Hard Coal	171	149	151)	174	144
Lignite and peat	26	31	31)		35
Natural Gas	112	137	116	119	114
Nuclear	18	37	64	78	144
Hydro and Geothermal	9	12	12)	14	13
Other Renewables	1	1,5	1,5)		11
Net Imports					
(total):	620	559	418	373	496
Oil	596	487	326	287	353
Solid Fuels	19	34	46	37	65
Natural Gas	4	36	44	47	80

Sourçes: Statistical Office of the European Communities and submissions by Member States (July 1983 - January 1984)

TABLE V VARIATIONS AMONG MEMBER STATES

(i) Oil Dependence
(The share of oil in primary energy demand)

	1973	1982	1990 (projected)
Belgium	62	51	41
Denmark	89	66	52
Germany	56	44,5	41
Greece	81	73	57
France	72	53	36
Ireland	79	56	49
Italy	79	68	57
Luxembourg	37	35	32
Netherlands	56	48	52
United Kingdom	50	40	38.5
EUR-10	62	50	43

(ii) Oil and Natural Gas Use in Electricity Generation
(% of total primary energy inputs)

	1973		1982		1990 (projected)	
	Oil	Natural Gas	Oil	Natural Gas	Oil	Natural Gas
Belgium	52,5	24	25	4	8	2
Denmark	61	-	8	-	9	-
Germany	14	11	5	10	3	10
Greece	51	-	30	-	7	-
France	44	6	10	2	2	1
Ireland	62,5	-	27	48	36	3
Italy	73	4	59	8	46	8
Luxembourg	24	9	19	-	13	7
Netherlands	13	78	23	45	14	35
United Kingdom	25	1	11	1	5	-
EUR-10	32	10	16	7	10	5

Until recently less attention has been given to energy savings policies than to management of the supply side. The industrial restructuring of the past few years, and especially the decline in importance of the heavy energy-intensive industries, together with a policy of relatively high energy prices, has led to a considerable improvement in the intensity of energy use. But the unrealised potential for energy saving is large and the outlook projected by the Belgian government seems less than satisfactory in this respect. Recently, however, new emphasis has been given to energy savings policy, backed up by Government finance, which should encourage greater progress.

Other areas where important issues arise are:

(i) the Belgian coal industry which is the highest-cost producer in the Community;

(ii) natural gas demand. At a time when demand expectations were high Belgium contracted for increased supplies of natural gas over the coming years, including Algerian LNG, which are more than the market in the medium term now seems likely to require. Excess quantities are being burned in power stations to replace fuel oil.

In addition, final decisions have yet to be taken on the electricity ordering programme for the early-mid 1990s, but these are understood to be imminent.

The conclusion last year of a long-awaited Parliamentary debate on energy should provide a helpful impetus to future policy decisions in these and other areas.

2.2 Denmark

Denmark has effected an impressive switch away from oil over the past ten years by conversion from oil to imported coal in the power-station field, vigorous energy saving policies,realistic energy pricing and high energy taxation, and development of the tradition of collective heat arrangements. Dependence on oil fell from nearly 89% in 1973 to less than 66% in 1982 (table V); coal in 1982 provided nearly 92% of the inputs to electricity-generation compared with less than 40% ten years ago; the reduction in energy intensity has been marked, notably in the residential and industrial sectors. Further consolidation of progress away from oil is planned, bringing projected oil dependence down to 52% by 1990.

The high dependence of the electricity system on imported coal demands a correspondingly high level of "risk-spreading" through geographical diversification of sources of imports. The Danes have made considerable efforts in this direction.

Important decisions have also been taken to develop Denmark's offshore hydrocarbon resources. Danish offshore oil now meets 15% of Danish oil requirements and by 1990 this could be as high as 40%. Natural gas from the Danish sector of the North Sea is planned to play an increasingly important rôle in the domestic and industrial heat supply system later in the decade and also to be a limited source of exports.

Energy planning in Denmark has produced highly successful results. The Danish experience, notably in energy conservation and district heating, should be of considerable interest to other Community members. The expertise developed in the planning field now needs to be brought fully to bear to help minimise the high costs involved in the major task of introducing natural gas into the energy system.

Early decisions are also desirable on the place (if any) of nuclear energy in the electricity system in the 1990s. Reports are expected to be submitted to the government this year on nuclear safety and waste disposal. Rapid follow-up action is needed to facilitate long-term planning of the electricity supply system.

2.3 Germany

During the past ten years Germany has consolidated an already relatively diversified structure of energy supply and demand by significantly greater recourse to nuclear energy, continuing commitments by the electricity industry to the German hard coal industry, more widespread use of natural gas and important improvements in energy intensity. The process of change has been encouraged by a generally market-based approach to pricing and vigorous policies to improve energy efficiency. During the remainder of this decade there should be a further major shift towards nuclear power as new stations come on stream, helping to reduce projected oil dependence to around 41% compared with 44.5% in 1982 (table V).

There are four important issues for the future:

- the problems of the German hard coal industry which now has to adjust to a new and less buoyant market outlook. The speed of phasing out excess capacity, a process already under way, will need to be a continuing focus of policy attention;

- the electricity ordering programme for the 1990s, and notably the respective rôles for nuclear and coal;

- the costs of the electricity system. The commitments by the electricity industry to burn large quantities of German coal, entered into a time when demand expectations were higher than today, inevitably raise the average costs of German electricity production in more depressed market conditions. The new more stringent environmental legislation will also impact on electricity costs (as well as coal-based district heating);

- the continuing need to reduce oil demand outside the electricity generating sector. On some scenarios oil consumption could increase in volume terms. These scenarios may well be pessimistic, but the outlook clearly needs to be kept under careful review and further opportunities for reducing oil dependence vigorously sought. It is noteworthy in this context that the German government has ceased to give financial support to energy conservation in the domestic sector where, despite a good deal of progress, there appears to remain considerable potential for reducing oil demand.

2.4 Greece

During the 1970s Greece made useful progress in reducing dependence on oil in electricity generation by some conversion to solid fuels. But in other sectors oil use increased. Energy intensity in industry has fallen since 1973 but is still high compared with the Community average.

In 1982 Greece depended on oil for 73% of energy supplies (table V). Current plans aim to ensure a very major improvement in this situation during the remainder of this decade and beyond, bringing oil's share in primary energy demand down to 57% by 1990. This will require in particular, a very heavy capital expenditure programme to develop domestic lignite resources and lignite-based power generating capacity; attention to renewable energies; and hydrocarbon exploration.

Key questions for the future are:

- whether adequate financial resources and technical and managerial expertise will be mobilised to implement this ambitious programme. In this respect it is desirable that Greece should encourage more widespread recourse to external experience and resources;

- how far the burden of the investment programme and the long-term cost of the electricity system could be reduced by greater recourse to imports of hard coal than currently planned. The aim of the Greek authorities is to develop their indigenous resources in an optimal fashion. It is essential in this context that costs are fully assessed.

- whether or not nuclear should play a rôle in the power station mix in the 1990s. Early decisions will be required if there is to be any reasonable expectation of a nuclear contribution by the late 1990s.

- the adequacy of policies outside the electricity generating sector. Plans to improve oil substitution in industry and the domestic sector are limited and energy saving policy is relatively weak. The weaknesses are accentuated by pricing policy.

- the importance of adapting the Greek oil monopoly to Community rules as early as possible and in any case by the end of 1985.

France

The clear commitment to nuclear power in France, combined with vigorous efforts in the field of energy saving, has brought radical changes in the French energy economy since 1973. Oil dependence has fallen from 72% to less than 53% (table V). Nuclear now accounts for over half of the inputs to power generation, and coal for over one-quarter; energy intensity has fallen significantly. The process of substitution for oil is likely to continue and indeed to accelerate during the rest of this decade as available nuclear capacity doubles, giving France probably the highest share of dependence on nuclear energy for electricity generation (around 85%) of any country in the world.

Together with Denmark, France can claim some of the most impressive achievements in the engineering of structural change in the energy sector. But the changes in the energy situation of the past few years and a more conservative view of likely rates of economic growth are posing difficult problems, notably in the electricity sector and in the coal industry. There is also uncertainty about the future rôle of natural gas. One option for the electricity industry for the longer-term is a sustained policy of net exports of nuclear electricity to other Member States: this is obviously of considerable interest in a Community context. In the coal industry there has been a good deal of uncertainty about the direction of future policy, but recent decisions imply a gradual reduction in output during the remainder of the decade. On some scenarios there could be a significant potential excess supply of natural gas later in the 1980s. Important decisions will be required in each of these areas in the next few years.

France has ambitious objectives for the contribution of new and renewables and for further improvements in energy efficiency, to be promoted in both cases by a special coordinating agency, the Agence Française pour la Maîtrise de l'Energie (AFME). These programmes are supported in part by investment aids from a special fund financed by a tax on motor fuel. The operations of the AFME and the financing arrangements should be of particular interest to other Community members where less vigorous programmes are in force. It is notable that in contrast to some other Member States with a good record on energy saving, France shows no signs of reducing financial incentives in this field and apparently sees further potential still to be exploited.

2.6 Ireland

Ireland has made considerable progress in reducing dependence on oil. Key elements have been, first and foremost, the development of the Kinsale Head gas field which has allowed natural gas to substitute for oil on a large scale in the electricity generating sector; the development of indigenous peat; and the switch to solid fuels (mainly imported coal) for domestic heating. Dependence on oil has fallen from 79% in 1973 to less than 56% in 1982 (table V). Current forecasts assume further improvements in this situation but at a slower pace than in the last ten years. The major change will be the shift from natural gas to imported coal in power stations, although oil use in power stations is expected to increase in volume terms. Oil and gas were discovered in the Celtic Sea Basin in autumn 1983, which could change substantially the longer-term energy outlook.

Other main issues for the future are:

- energy saving policy, where the rôle of regulatory measures and financial incentives has been particularly weak;

- the development of premium markets for natural gas. Agreement has been reached on the construction of a gas pipeline for supplying Northern Ireland with natural gas from Kinsale. This will provide an export outlet to help reduce gas use in power generation. Natural gas from Kinsale is now also available in Dublin. But continuing efforts are needed to ensure more widespread access to gas by domestic consumers.

2.7 Italy

Progress in Italy so far has been disappointing in a number of respects. Oil dependence has fallen since 1973 but proportionately less than in all Member States apart from Greece (which is at a rather lower level of development) and Luxembourg (where oil use was already relatively low). Italy still depended on oil for over two-thirds of her energy supplies in 1982 and solid fuels, nuclear, hydro and other renewables played together a smaller rôle in electricity generation than in almost all other Community countries.

There has been some improvement in the relationship between energy and GDP and the intensity of final energy demand in industry fell by 27% between 1973 and 1982. But these improvements have been heavily influenced by structural change in the economy. Conservation policy as such has been limited. Progress has been made in recent years in developing a more realistic approach to electricity and oil product pricing than in the 1970s but much still remains to be done.

While specific geographical and economic factors help to explain the limited progress that has been made, the lack until 1981 of a clear political and parliamentary consensus on energy policy prevented rapid and sustained action.

In its last report on Member States' programmes, the Commission welcomed the 1981 Italian Energy Plan as a major step forward, offering the prospect of a radical improvement in the situation over the coming years. Discussion of the Plan appeared to demonstrate a general recognition of the need for a more coherent energy policy and, specifically, progress in the fields of electricity generation and energy saving. But progress in implementing many aspects of the 1981 Plan has been slow and many of the objectives have now been scaled-down or re-timed. This applies particularly to the power station programme for the longer-term where some projected investments have been put back by several years. The major Energy Conservation Law of 1982 has only been partially implemented and as yet there has been no commitment of expenditure. Problems have also arisen in the introduction of natural gas into the Italian energy system because of the fall in gas demand and difficulties in distribution to premium markets; there has been consequent pressure to burn natural gas in power stations.

These elements taken together suggest considerable uncertainty about the rate of future progress in oil substitution and more efficient energy use. Even if current energy projections are realised, nearly half of Italian electricity will still be produced from oil in 1990 and Italy will remain dependent on oil for 57% of her energy supplies. The outturn could be even less satisfactory. It is essential that the planned switch to nuclear and coal in the power-generating sector should not be further delayed by continued problems over siting or the interim use of natural gas in power stations; and that the provisions of the Energy Conservation Law are fully enacted. There is a major risk that, unless the outstanding logjams are broken, Italy's energy position vis-à-vis the other larger Member States will further deteriorate between now and 1990.

2.8 Luxembourg

Luxembourg's small size and the domination of the economy by the steel industry have set a special framework for the development of energy policy. Oil dependence was relatively low in 1973 and it has since been reduced still further through the diversification of inputs (electricity, natural gas) to heat-raising in the steel industry and in the domestic sector. Practically all Luxembourg's energy is imported and the planning of future imports of gas and electricity from neighbouring countries will continue to be a key element in energy policy. Other important areas of policy for the future are:

- the need to introduce a more comprehensive energy saving policy (policies in the domestic sector and transport have been particularly limited so far);

- further progress in bringing the post-tax prices of motor fuels more into line with those in other Community countries;

- improving and extending the natural gas distribution network so as to give wider access by domestic consumers to gas as an alternative to oil for heating.

2.9 Netherlands

The energy economy of the Netherlands is dominated by natural gas, and gas revenues (like those from oil in the United Kingdom) are a major element in the financing of the government budget. The availability of large supplies of indigenous natural gas helped in itself to reduce Dutch dependence on oil during the 1970s. But it has also affected the pace at which substitution by other non-oil fuels has occurred, notably in the electricity sector.

In 1982 the Netherlands was still dependent on oil for nearly a quarter of the inputs to electricity-generation. A programme of conversion from oil to coal in electricity generation is underway and the projections indicate that by 1990 solid fuels could provide over 40% of the inputs to electricity. In the meantime, however, natural gas is expected to substitute increasingly for oil in this sector under new arrangements between Gasunie and the electricity industry. Even if the programme of conversion is completed total inland oil demand will increase by 15% between 1982 and 1990 and oil dependence will increase. The completion of the conversion programme is therefore of major importance.

The pace of restructuring on the demand side has been faster. The Netherlands has had one of the most wide-ranging energy conservation programmes in the Community, with a high level of government budgetary expenditure. The government is now reexamining aspects of this programme against a background of budgetary stringency and a general policy of "deregulation". Account must be taken in this review of the constraints on investment in efficient energy use in a generally depressed investment climate.

Other policy issues of particular interest in a Community context are:

- gas export policy. The Dutch government has indicated a change in gas export policy which could release larger quantities of Dutch gas for sale in other Member States in the 1990s.

- nuclear policy. A long Public Debate on (Nuclear) Energy has reached a turning point with the recent publication of the final report by its Steering Committee. It is important for the planning of the electricity supply system in the 1990s that early decisions should now be taken by the government about the future rôle of nuclear power.

2.10 United Kingdom

The United Kingdom has enjoyed a uniquely favourable set of circumstances in the energy field since 1973 in comparison with other Member States - an already diversified energy supply structure, the advent of North Sea oil, and powerful tools of government intervention. The harnessing of North Sea oil resources is one of the major success stories of the 1970s. In 1973 the United Kingdom depended on imported oil for nearly 50% of her energy demand. In 1982 she was a sizeable net exporter of oil.

A large share of oil exports from the UK has gone to other Community members. But access to North Sea oil and gas has reduced pressures to develop a more integrated gas market, in particular, with continental Europe. A limited cross-Channel electricity link exists with France; but there is no direct gas link with continental Europe. It could be of interest both to the UK and to other Community members to encourage better links with the continent, exploiting the scope for cost savings all round and improving energy security.

A second effect of access to large-scale indigenous resources has perhaps been to circumscribe the rôle of energy saving policy. The United Kingdom has made progress in reducing energy intensity but a slower pace than in many other Community countries, and in the industrial sector the scope for improvement appears to be particularly great. Successive policies of restraint on energy prices in the 1970s undoubtedly discouraged improvements in energy efficiency. During the past four years pricing policy has been put on a sounder footing, and the financial situation of the gas and electricity industries is now much healthier. But there are uncertainties about the future direction of energy pricing policy and specific weaknesses in gas pricing. The recent creation of an Energy Efficiency Office is intended to reflect a new priority for energy savings policy. It has a major job to do in ensuring that the potential for energy saving can be exploited.

The available data suggest that oil dependence in the United Kingdom should have fallen to 38/39% in 1990 compared with 40% in 1982. The main element will be the coming on stream of new nuclear stations. Demand for coal is expected, at best, to stabilise. A difficult process of structural adjustment is now under way in the UK coal industry and will be a continuing focus of policy attention. There are also important decisions to be made in the coming years about the "new generation" nuclear programme in the light of the results of the ongoing enquiry into the construction of a PWR at Sizewell. The results of this enquiry should be acted on in a timely fashion so as to reduce uncertainty for the nuclear and electricity industries.

3. COMMUNITY ENERGY STRATEGY

3.1 Generalities

The new oil market situation and the uncertainties about the future pace of structural change have led the Commission to suggest a policy directed particularly at the following six main areas:

- exploitation of the potential for energy saving and substitution of oil;

- penetration of solid fuels, particularly in the industrial sector;

- new energy technologies (both to expand the potential for rational use of energy and to exploit new and renewable energy sources on a wider scale);

- identification and development of economic Community energy resources;

- the resolution of uncertainties about the development of nuclear power;

- security and flexibility in energy supply.

Although the Commission has suggested that much of the action towards common energy objectives could be pursued satisfactorily at national rather than Community level, it has also stated that it is desirable to exploit the Community dimension whenever it is more economic and more effective to do so for the Community as a whole. The following items show the advantages of such a policy.

- economy of scale;

- pooling of knowledge and experience, both of energy technologies and systems, and of policy measures in specific sectors,

- expansion of the internal market in energy products;

- collective interest of Community Member States in the security provided by competitive energy resources produced and traded within the Community;

- common interest in the flexibility provided by a more integrated infrastructure for Community energy supply;

- collective interest of Member States in protecting their economies against future energy supply and price crises by improved collective contingency management.

The present energy policy objectives of the Community were defined by a Council Resolution of June 1980[1]; which agreed five main guidelines for the Community for 1990, viz.:

- reducing to 0.7 or less the average ratio between the rate of growth in gross primary energy demand and the rate of growth of gross domestic product;
- reducing oil consumption in the Community to a level of about 40% of primary energy consumption;
- covering 70-75% of primary energy requirements for electricity production by means of solid fuels and nuclear energy;
- encouraging the use of renewable energy sources so as to increase their contribution to the Community's energy supplies;
- the pursuit of energy pricing policies geared to attaining Community energy objectives.

While these guidelines were fixed for the Community as a whole, the Council in its Resolution underlined the importance of increased consistency among national energy policies in their pursuit.

The projected pattern of energy supply and demand in the Community in 1990, as it emerges from the latest available information supplied by Member States, is summarized in Tables III and IV. If the assumptions turn out to be correct:

- the share of oil in gross energy consumption should fall to 43% in 1990 (compatible with the 1990 guideline);
- the share of coal and nuclear in electricity production should rise to 81% (above the guideline);
- the energy coefficient would fall to 0.67 in 1990 (an improvement on the guideline);
- the share of new and renewable energies would rise from 1.6% to 2.3% of total inland energy demand.

The Energy Council of 4th November 1983 stated that by publicly announcing and constantly reaffirming the Community's commitment, the 1990 objectives make clear to consumers, producers, investors and third countries that the Community is seriously concerned with the energy situation and determined to achieve restructuring.

---

[1] OJ C149, 18 June 1980

Among the main areas of interest - other than the solid fuels which will be discussed later on - we will consider the following:

3.2 Pricing

The Commission has underlined the necessity to increase the transparency of energy prices so that consumers and investors can be better informed. As a contribution to transparency the Commission is now publishing a six-monthly Bulletin of Energy Prices. Discussions are underway between the Commission, the industries and Member governments on other measures to achieve acceptable transparency in each sector.

3.3 Research, Development and Technological Demonstration

In December 1982, the Commission presented to the Council a Framework Programme for Community R&D, which gave renewed emphasis to the energy sector. A proposal for a 3rd Energy R&D Programme has been submitted to the Council and will be discussed by the Energy Council planned for 29th May 1984; a proposed programme of activities by the Joint Research Centre which will include a sizeable energy component will follow shortly.

In the demonstration field, progress in setting a stable framework up to 1985 was achieved in March 1984. The Council agreed that for the period 1983-1985, the amount estimated to be necessary for this action is 265 MioECU. The programme covers the following sectors[2]:

- liquefaction and gasification of solid fuels
- geothermal energy
- solar energy
- biomass
- wind and ocean energy
- hydro-electric power
- energy saving (building, industry, energy industry, transport)
- substitution of hydrocarbons by solid fuels
- substitution of hydrocarbons by use of electric power
- substitution of hydrocarbons by heat transmission, distribution and storage.

---

[2] OJ n° L195, 19.07.1983, p.1-13.

In 1983, further to the publication of an invitation to submit projects, 867 applications were presented to the Commission. The total cost of all the projects submitted is 4155 MioECU. The Commission has selected 187 projects with a total cost of 816 MioECU. The amount engaged by the Commission in this action for 1983 is 108 MioECU. The success of the demonstration programme proves that there is a real necessity for such action. Many projects are ready for the demonstration stage, but due to the high risk, financial support of up to 49% - half of it repayable in case of commercial success - encourages the applicant to start the project.

3.4 Nuclear

Despite the fact that some Member States have made no commitment to nuclear power programmes, all Member States acknowledge that:

- nuclear, along with coal, must bear the main burden in helping to diversify Community energy supplies between now and the year 2000;

- nuclear power has economic advantages, giving operators access to competitive sources of energy supply.

These agreements are a useful complement to the 1990 objective of increasing the contribution of nuclear and solid fuels to electricity production.

3.5 Natural gas

Three questions deserve particular attention:

- the implications of increased use of natural gas in power stations. Additional recourse to gas must be only at the expense of oil and must not be allowed to slow down progress in the introduction of additional solid fuels, nuclear or hydro power station capacity;

- natural gas supply and security in the 1990s. The most pressing elements are the need for early negotiations between Norwegian and Community companies on the exploitation of the Troll gas field and greater attention than in the past to the cost and security benefits of improved transnational pipeline systems;

- the long-term rôle of natural gas in helping to reduce dependence on oil. More analysis is needed now on the long-term costs of producing and transporting gas around the world and the implications for the expansion in the longer-term of its economic use.

4. SOLID FUELS POLICY IN THE EUROPEAN COMMUNITY

We have seen that the Council acknowledged that major progress in diversifying the Community's sources of energy supply could be made between now and the year 2000 only by increased use of coal and other solid fuels, as well as nuclear energy.

In the spring of 1983, the Energy Ministers agreed that the discussion should focus on the following four matters:

- obstacles to greater use of solid fuel;

- ways and means of promoting greater use of solid fuel through research, development and demonstration;

- the influence of Community production of solid fuels under satisfactory economic conditions on the security of energy supplies;

- what the Community can do to resolve the social and regional problems closely linked to the future of the coal industry.

These four items, together with the Commission proposals, will be discussed below.

4.1 Greater use of solid fuels

It is a fact that consumption of hard coal in the European Community has not increased significantly since the first oil crisis of 1973/74. It then totalled 310 million tonnes, rising to all of 314 million tonnes by 1980. Since then it has fallen again; Table I shows that in 1983 solid fuels represented only 23.7% of the total energy consumption of the Community. The steel crisis and the low level of economic growth are mainly responsible.

Obviously, the large consumers (power stations and cement works) have substituted oil by solid fuels to a considerable extent. But the competitivity of solid fuels has not been sufficient to enable a substantial increase in solid fuel consumption in other sectors; among the obstacles to the increased use of solid fuels are:

- the uncertainty of solid fuel prices: there is for the moment pressure on heavy fuel oil and this trend will increase if crude oil prices continue to decrease;

- high investment costs in infrastructure (roads, harbours, transport, etc);

- influence on the environment and acid rain.

The Community already has a number of instruments and further measures have proposed. Two Council Directives adopted in 1975 prohibit the construction of new oil or gas power stations. In 1983, the Council addressed two Recommendations to the Governments of the Member States calling upon them to promote investment in the conversion of oil-fired boilers to solid fuel in industry and to support the use of coal for the heating of public buildings and in district heating systems.

4.2 Research, development and demonstration

The improvement of the present processes for solid fuels utilisation and the development of new technologies must be emphasized. If solid fuels are to be substituted for hydrocarbons, their use must be more practicable and more attractive for industrial utilisation. A huge effort has to be made at R, D&D level in order to facilitate the combustion, handling, transportation, treatment and storage of solid fuels and their waste products. New technologies entail risks, however, and so the decision to employ them is deferred, or only taken if there is sufficient financial incentive.

4.2.1. Research and Development

Within the framework of technical coal research under Article 55 of the ECSC Treaty, the medium-term guidelines have defined the priorities which must now be concentrated on production problems:

- mining engineering: the main part of research aims at a better utilization of machines thanks to modern management methods in the field of development work in coal and stone, coal getting, transport, methane studies, ventilation control and mine climate, rock pressure and supports;

- mechanical coal preparation to improve the quality of the final products;

- coking.

In the multiannual non-nuclear research programme under discussion at the moment by the Council, a sub-programme is reserved for new methods of solid fuels utilisation. In particular it deals with techniques for production of heat and electricity, and with environmental questions.

The objectives of this sub-programme are to discover and disseminate knowledge of more cost effective, convenient and environmentally acceptable techniques for the use of solid fuels and to strengthen basic scientific knowledge of solid fuel composition and utilisation. This mainly involves:

- developing new and improving existing combustion techniques (fluidized bed combustion, solid fuel/liquid mixtures, burners);

- reducing the effects on the environment (disposal of solid waste, flue gas cleansing, emission of trace elements and compounds);

- developing advanced power generation systems (combined cycle);

This programme should be adopted shortly.

4.2.2. Demonstration projects

Since 1978, the Commission has suppported demonstration projects in the field of liquefaction and gasification of solid fuels. Between 1978 and 1982 ten projects received support and the total aid was 70 MioECU. In 1983, the application was extended to pilot industrial projects. Nine projects received support in 1983 totalling 27.7 MioECU. In 1984, 17 MioECU should be allocated to this action. The areas covered are fluidized bed, entrained bed and fixed bed gasification together with direct and indirect liquefaction (this action is extensively reported in another lecture given during this course).

Since 1983, demonstration projects relating to the substitution of hydrocarbons by solid fuels have also been supported. These projects deal with new or improved technologies for the handling, transportation, combustion, treatment and storage of coal, lignite and peat and their waste products. A sector considered as being particularly important is the one which deals with fluidized bed combustion, combustion of mixtures of solid fuels with liquids and combined cycles. The Commission has allocated 20 MioECU to 16 projects selected from 68 proposals. In 1984, 13 MioECU should be allocated to this action.

4.3 Production of hard coal in the Community.

The focal point of the solid fuels policy which the Commission has proposed to the Council is the question as to how the production of hard coal in the Community can be made more competitive. Only a small proportion of the existing production capacity can compete with imported coal, and only an industry operating on a sound economic basis can contribute towards the security of energy supplies.

For this reason, the Commission has proposed that aid should be made available from the Community Budget each year for 5 years in order to promote investment designed to increase productivity. A non-repayable subsidy of up to 25% of the investment costs is envisaged.

In addition, investments designed to increase capacity in the peat and brown coal industries in Ireland and Greece should also receive a Community subsidy of this kind in order to enable these two Member States, which are among the least prosperous, to reduce more quickly their dependence on oil and the related costs. Progress on this matter is being made and decisions are expected by July 1984.

The aid proposed by the Commission to cover the above measures is for 300 MioECU per year over the 5-year period.

4.4 Social measures

The Commission has proposed to the Council that additional funds totalling 60 MioECU for each of the next three years should be transferred to the ECSC budget. It is likely that the restructuring of the coal industry, and in particular the closure of the most unprofitable pits, will result in increased spending on social measures. Article 56 of the ECSC Treaty which provides for various types of aid for the workers affected, will continue to play a very important rôle. In addition, the scheme for creating new jobs for ECSC workers through the granting of subsidized loans to industrial firms which has also proved its worth, will be continued.

5. PROSPECTS FOR SYNTHETIC FUELS

Despite the lull in the current oil supply situation, the Community oil supply system remains fragile. It is necessary to repeat that in 1983, crude oil represented 47% of primary energy consumption in the Community. If another oil shock arose suddenly - and there is nothing to say that it is no longer possible - what will the alternatives be ? Coal/water mixtures can be used promptly but are obviously not sufficient. Alas, synthetic fuels technologies will not be in place quickly enough in an emergency situation. This is why the European Commission is aware of the strategic necessity to ensure timely and effective availability of a broad range of commercial synthetic fuels technologies. These processes, once they are mature, can produce a secure and steady supply of synthetic fuels for a long period, relatively independent of any oil crisis. This means that due to huge technical and economic difficulties, decisions must be taken to carry on this progress before any crisis occurs. The exact moment when these technologies could become commercially viable varies from case to case and depends on the development of energy prices. Furthermore, it is probable that certain processes will be applied, at least in the first instance, in third countries, rather than within the Community, due to inherent site advantages and proximity to cheaper sources of raw materials. The large industrialized countries outside the Community have a diminishing interest in these processes. In this situation, the Commission considers that the Community must itself ensure the research, development and demonstration of this technology, which it will certainly have need of in its vulnerable energy situation. Because of their size and long lead times, projects in this sector are particularly costly, there is thus very real advantage in a joint development and demonstration activity at Community level.

For these reasons, the Commission intends to pursue this action diligently, together with improvements designed to take account of the special features of this sector.

A REVIEW OF THE CURRENT STATE OF THE COAL MINING INDUSTRY

P.G. Tregelles
Mining Research and Development Establishment
National Coal Board
UK

SYNOPSIS. Coal mining is an extractive process and the method of doing it depends on how and where the coal lies in the ground. If it is near enough to the surface in a suitable place the rock on top of it can be scraped off, the coal taken away and the rock put back. In other cases deep mines have to be used. It costs half as much again (on average) to deep mine the coal.

Coal differs in characteristics. Some is very rich in calories and chemicals eg bituminous, while some is mostly water eg lignite.

Deep mining is the major source of production in the UK, Belgium and France. There is substantial lignite in Germany.

Deep mining is traditionally represented as a very dangerous and unhealthy occupation as indeed it used to be. This is no longer necessary or true. It is much more dangerous to be, for example, a construction worker, a fisherman, farmworker or steeplechase jockey.

The methods of mechanising deep mining operations were engineered during the 1930/50s. They are now well established and will sustain the industry for the next 20/30 years.

Advance in productivity and of profitability will be a consequence of the application of control technology. That is the use of management information systems and automation. These lead to higher utilisation of the mining machinery.

The succeeding advances will depend on progress in the in situ conversion of coal.

THE OCCURRENCE OF COAL

Coal is a stratified deposit and lies either near the surface from which it can be scraped (opencast), or deep where it has to be extracted by burrowing.

The coal may be in thin seams of less than a metre, or in very thick seams of many tens of metres thickness.

Coal is surprisingly varied in type and characteristics: anthracite coal has very high calories and is low in volatiles.

G. E. Beghi (ed.), Synthetic Fuels, 35–47.

Lignite, on the other hand, is very low in calories. In between there is a wide range of coals, called bituminous, with a wide range of qualities and characteristics. These features determine the value of the coal and the kinds of use to which it can be put.

The use to which coal is put in Europe is given in Table 1.

Table 1

MARKETS	UK		EEC	
	(m) Tonnes	% Consumption	(m) Tonnes	% Consumption
Power Stations	86.2	75	182	53
Coke Ovens	7.6	7	100	29
Industry	8.8	8	26	8
Domestic	7.3	7	14	4
Export	7.1			

THE MANNER OF EXTRACTION

The two ways of extracting coal are opencast and deep mining. The opencast operation, consisting of exposing the seam by removing the overburden, extracting the seam and replacing the overburden, is very capital intensive, using very large earth moving machinery.

It is a much cheaper operation than deep mining. In the UK, for example, the cost of opencast production is only two thirds the cost of deep mined production. In the USA, Figure 1, where there is more scope for larger scale operation the cost of opencast is only half the cost of deep mined coal. In the UK, however, opencast accounts for only 12.5% of total production because most coal is too deep for mining by that method. Table 2 shows the extent to which other coal producers are better situated in this regard.

U.S. COAL MINING

AVERAGE PRODUCTION COSTS: 1981

DEEP-MINED	$35.78 PER SHORT TON
OPENCAST	$20.60 PER SHORT TON

FIGURE 1.

FATAL EXPLOSION TRENDS

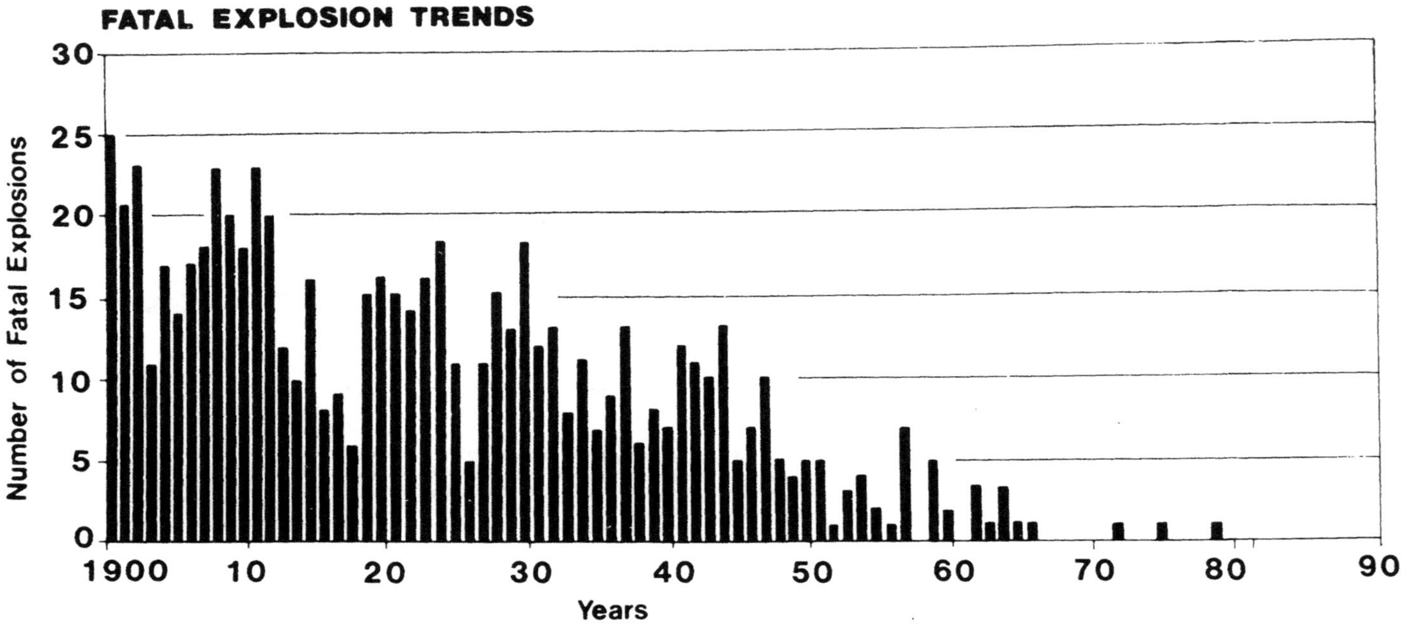

FIGURE 2.

Table 2

	(m) Tonnes		Proportion Opencast/Deep Mined
	Opencast	Deep Mined	
EEC	148	230	39%
USA	496	334	60%
S Africa	32	97	25%
Australia	84	50	63%
Poland	36	163	18%
USSR	273	431	39%

The main method by which seven eighths of the UK output is produced, the underground operation, involves tunnelling to reach the coal partly by vertical shafts but mainly also by horizontal or inclined tunnels. The coal is then taken out from between layers of rock which form a roof and floor. Except for the access tunnels which are supported, the roof is allowed to collapse. The coal is transported through these access tunnels or roadways to the surface. Here the coal is prepared for the market by removing most of the rock particles that have become mixed with it and by sorting it into various sizes.

The access tunnels are not only required to allow men and equipment to get into the mine but also to act as a vehicle for air to ventilate it. This is needed to flush out toxic gases and supply air for life support, to carry heat out of the mine to keep temperatures tolerable, to dilute the explosive methane gas and to dilute and carry away dust.

SAFETY RECORD OF THE UNDERGROUND MINES

Safety in the underground mines has been radically improved during this century:

Figure 2 shows that explosions which caused loss of life on an alarming scale are now eliminated as an uncontrolled risk.

Figure 3 shows that fatalities from all causes are closely approaching the asymptote.

Figure 4 illustrates the relative place of mining in the risk-of-employment league.

Figure 5 shows that pneumoconiosis has been mastered in the recent past.

This progress should not be interpreted as meaning that all environment and safety matters are now beyond further progress. As the mines get deeper, as the speeds of production increase, so the problems of maintaining a safe mine are exacerbated. As machinery increases in

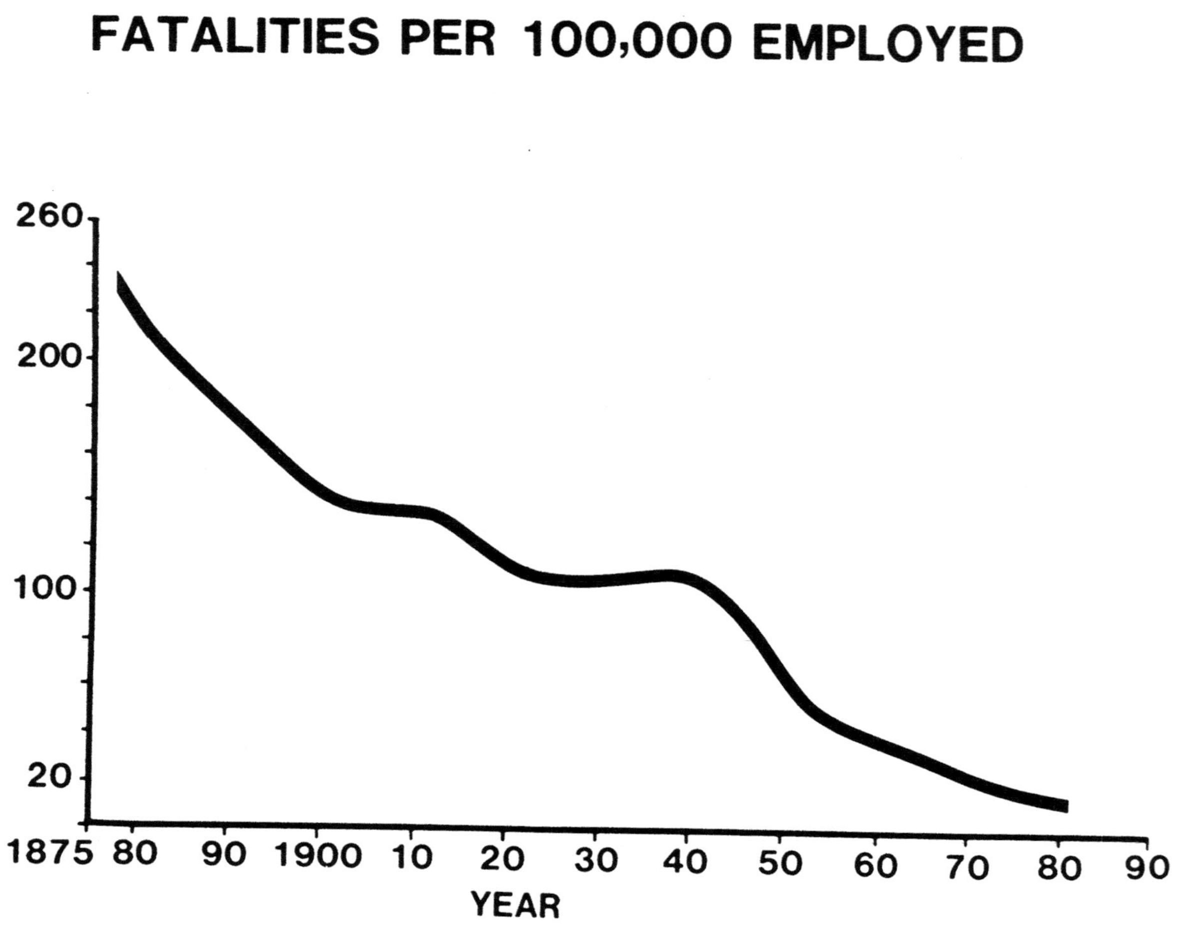
FATALITIES PER 100,000 EMPLOYED
260
200
100
20
1875 80 90 1900 10 20 30 40 50 60 70 80 90
YEAR

FIGURE 3.

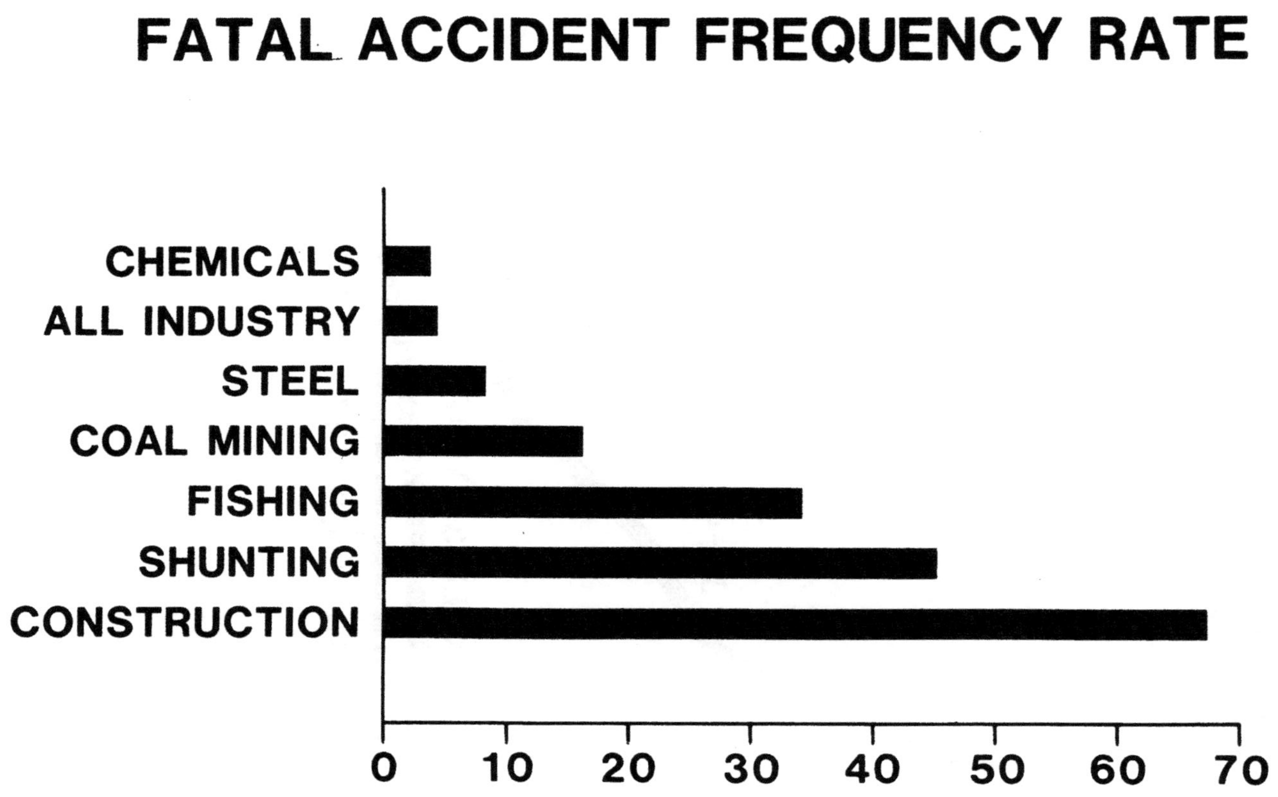
FATAL ACCIDENT FREQUENCY RATE
CHEMICALS
ALL INDUSTRY
STEEL
COAL MINING
FISHING
SHUNTING
CONSTRUCTION
0
10
20
30
40
50
60
70

FIGURE 4.

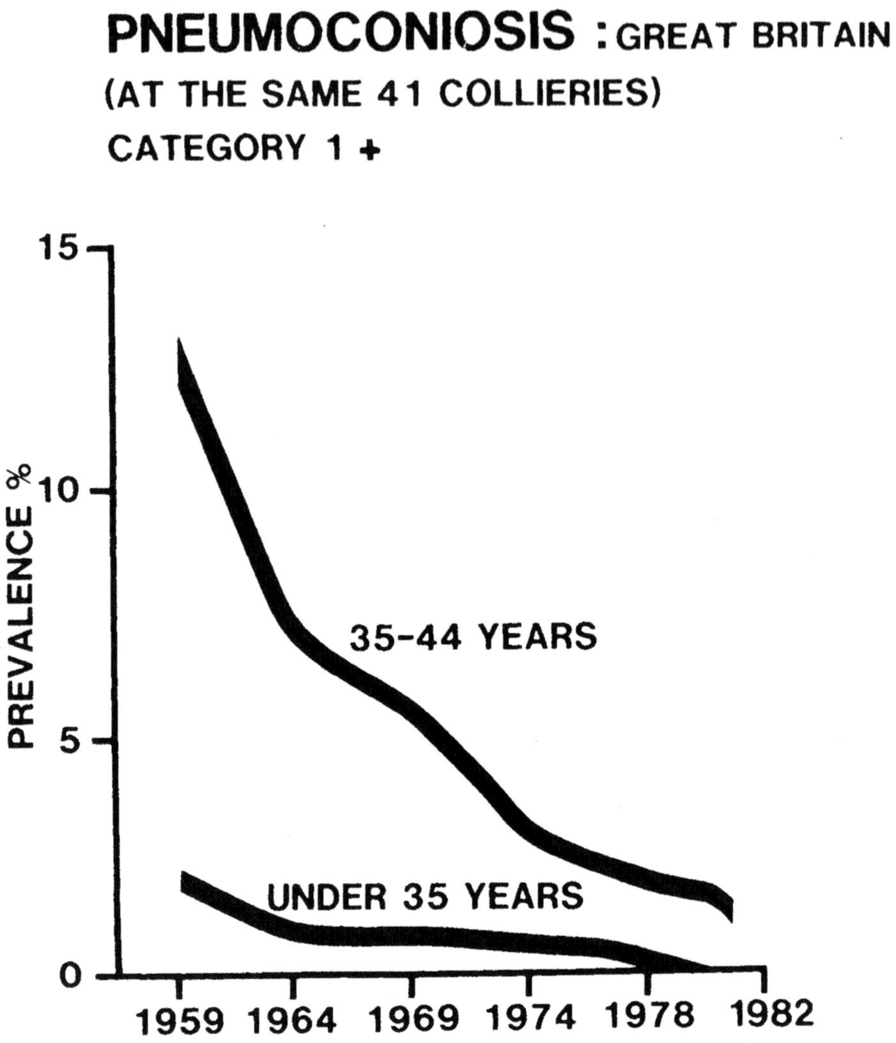
PNEUMOCONIOSIS : GREAT BRITAIN
(AT THE SAME 41 COLLIERIES)
CATEGORY 1 +
15
10
5
0
PREVALENCE %
35-44 YEARS
UNDER 35 YEARS
1959
1964
1969
1974
1978
1982

FIGURE 5.

power there are consequential problems of noise and of heat emission. These trends ensure a continuing need for work which will lead to ways and means of coping safely with the increasing difficulties.

The achievement of the highly rewarding advance in safety is due in part to the advances in mining technology, especially regarding the knowledge of the natural phenomena associated with mining; in part to the availability of measurements to ensure that adequate standards can be planned for and measured in the event, and in part to the high standards of mining discipline inculcated by management and reinforced by Miners Unions and by the Government's regulatory authorities.

THE METHODS OF PRODUCTION

The method of production of deep mined coal has been mechanised during the 1950/60s. It can be said that the methods of rock tunnelling, coal winning and transport and coal preparation are now finally decided and that there is no new mining method in contemplation which will have an impact on the mining world for at least 20 years.

During the last ten years the mechanised mining equipment has been increased in capacity and in robustness. As can be seen by the Table 3, there is now considerable capacity available in all categories. Furthermore the machinery has undergone considerable redesign to make it more reliable despite the increasing demands made on the design engineer by the production management.

Table 3

	Production Capacity tonnes/hr	
Item	1973	1983
Shearers	650	1500
AFC	760	1200
Conveyors	1200	3200
Washboxes	250	500

Further developments will continue to improve the range, capacity and the reliability of the equipment. The progress will however be incremental and directed towards betterment of those elements which are clearly shown to be weak. This information will come from intensive application of life-cycle costing of the equipment which is now extremely expensive.

The range of results from mechanised faces is very considerable. The best machinery in Europe on the best seams are capable of producing at the rate of 1,500 tonnes per hour. Those best faces to sustain a production of 6,000 tonnes a day. However the average for Germany is

1480, and the average for the UK is 760 tonnes per day.

The reason for the low average performance is in general a combination of three causes:

First, the capacity of the coalfaces has in the UK outstripped the capacity of the mines. In the UK the average mine size is 2250 tonnes per day. In Germany it is 8000 tonnes per day. Clearly the small mines are incapable of digesting the high capacity machinery. The solution rests either with reconstruction of existing mines or with the building of new mines.

Second, the mechanised mining methods are extremely intolerant of bad geology. Coal mines in geological disturbed strata cannot sustain high outputs from the coalfaces; the management therefore hedge against bad geology by having spare capacity. Because of the dynamic nature of the mining environment the spare capacity has to be operated so that the equipment moves into new ground before it becomes damaged. Hence there is a low shifting index (utilisation rate) in those circumstances. Those mines which are high cost because of natural conditions have to be closed.

Third, the high rates of extraction bring with them problems of control which exceed the capabilities of men. The need for control aids becomes imperative. It is in this field that R & D can make a valuable contribution.

CONTROL TECHNOLOGY

Control technology is directed towards enabling management and the machinery to operate better as a result of being better informed. It is concerned with the measurement of key parameters, their collection and digestion, and the decisions which flow from that additional information.

In recent years there has been a revolution in the field of control as a consequence of advances in the instrumentation fields and because of the development of the computer, in particular the micro-processor.

The develpment of control technology in the mining industry was selective largely because the measuring devices and the computers themselves became available in a piecemeal fashion. The technology was not driven uniquely by the mining industry to solve mining industry problems, it came from world-wide developments and pressures. Hence there was no preconceived direction for the developments to take, except the tow aims of (i) improved reliability of machines and of systems and (ii) automation of the mechanised processes.

After ten years of development the work has now settled into a perceived pattern. This pattern enables R & D workers to concentrate on developing ever more secure modular packages which managements can exploit. It enables managers to apply the technology selectively and to make judgements as to the profitability of the modules. It enables managements to work out effective codes of practice before full exploitation. It enables suppliers to market a product and software to a well defined activity and to further develop and extend the modules' capabilities.

The perceived pattern of the application of control technology in fact emerged from the experience described earlier in this paper. It can be summarised as follows.

DESIGN AND PLANNING

Designing, planning and organising are the precursors to any action. There is of course much interaction from previous experience which ensures that mistakes are not repeated and that defects in otherwise good machines or systems are eliminated. The new technology enables this phase to be improved by the computer's capacity to store and to iterate alternatives. The work of operational research scientists enables best options to be sought and tested. The seismic searches ensure better foreknowledge of areas to be mined. The test engineers evaluate the predicted reliability of equipment submitted for test. The whole system can be 'worked' in a computer and presented for criticism in graphic display. In operations which carry a heavy burden of safety, eg the ventilation circuit of a mine, the plans have to be lodged so that in the event of a mishap the cause of a failure can be explored against the background of a verified design. As the importance of productivity of capital equipment grows, this safety practice is being extended to production considerations.

MACHINE CONDITION MONITORING

Machines, once manufactured and installed, have to be commissioned and maintained to meet the performance capabilities prescribed by the production engineers. The instrument and the computer make a considerable contribution to this part of the pattern. Instruments are used to ensure that the electrical, hydraulic and mechanical specifications are being met. They are used to ensure that the key control parameters or indices are maintained within acceptable standards during operation. They are used to diagnose troubles as they are predicted by adverse trends in the control indices or when machinery is perceived to be in trouble. The computer's role is to store and digest the data, and to present it in a form appropriate for the operator or the engineer to use. As with the information collected by one's eyes, most is only of passing interest. It is the significant factors that have to be recognised and brought to the brain's attention: the rest can be discarded.

SYSTEMS PERFORMANCE MONITORING

The system in which the machine operates consists of a number of interrelated activities and men. The sensors and the computer can aid the operator and the manager by watching the elements of the operations and, by appropriate digest, giving them better visibility of what is happening and of what has happened. The operator can see clearly where the various component machines and men are in the cycle of work and adjust to ensure that the system runs smoothly and to plan. The manager can have an analysis of down-time from which he can perceive recurrent difficulties which he must eliminate either by change in equipment or by an adjustment of his plan. The design engineer can learn where the basis equipment or system has inherent faults which call for his professional skills. The senior manager can establish some degree of accountability by regarding trends in the effective machine running time.

AUTOMATION

As the equipment and the system improve in reliability (and that is always as a consequence of the learning process, which in its turn is enhanced by improved measurement) a point is reached at which it is practicable to make the machines automatic and to automate the system. The instrument engineer has to make or adapt further sensors and transducers to provide those measurements or judgements previously made by man's eyes and other natural sensors, and the computer scientists have to provide some capability for supplementing the work done by the man's brain. The resulting operation is improved mainly by the control function no longer being dependent on a man. That is not to say that the man is less effective than the automated operation: it is to acknowledge that man has limitations of capacity to absorb and react to continuous on-line information and does not have the consistency of performance, either when comparing one day with another, or comparising him with another operator. There is another advantage, which is traditionally seen as the main one for automation, namely that less men are employed. However, delegates should recognise that the main prize is that of improved control. The man is often still necessary, to supervise the automatic system and adjust the parameters to which it must work.

MANAGEMENT INFORMATION SYSTEMS

Industry does not stand still and hence the manager must always be seeking how best to improve his operations. With the aid of the computer the mass of information which is now collected on all aspects of the mine can be brought together and digested in a Management Information System. The information he can interrogate is no longer only that collected by his book-keepers and bureaucrats but also that collected from his control systems. He therefore has the possibility of supplementing his intuitive skills with additional measured information so as to reduce the areas in which he must make intelligent judgements. The advantage of this is the reduction of the risks which he has to take in assessing the future. It does not reduce the need to make judgements: it reduces the risks in their making.

It will be appreciated that these component packages are in fact very interactive: the planning process is improved by information gained from condition monitoring, and so on. Indeed the whole control process is cyclic. In the longer term they will become wholly applied at all coal mines.

NEW TECHNOLOGY IN THE LONGER TERM

For the present methods of mechanised mining to be replaced the successor will need a development span of 25 years before making a significant impact. There is no prospect of such development coming forward until the next century.

The progress meantime will be by the application of the new technology of control and information, ie sensor and the microprocessor, to increase the effective performance of the present generation of mechanised methods.

The longer term option of underground coal gasification is at the moment the only prospect in view, this has already had many years of sporadic investigation and development. It has so far failed because the problems of initiating and then controlling the underground gasifier have proved insuperable. However the new technology of guided longhole drilling coming from the oil industry and techniques for keeping a coal drill in the seam by use of natural gamma sensors again opening up the possibility of this technique being a serious contender to longwall mining. The work on this is confined to a very small expenditure in terms of R & D in France and in Belgium and co-operation with Germany. The United Kingdom has reached a proposal stage which has yet to receive financial backing.

CONCLUSIONS

The paper shows that the coalmining industry has a wide range of mining methods and opportunities, undoubtedly the lowest cost production comes from open pit work but in the deep mine section low costs are achieved by high production levels with the new mining machinery in good geological conditions.

The mining methods are well fixed and whilst they will continue to mature the great change will come from the application of modern control technology to these mechanised methods.

During the next 20 or 30 years the expansion of the control technology in all industries through the world will be applied to mining and the growth in productivity will stem from this field of development. At the same time a more fundamental advance will probably be made with regard to underground gasification. A great deal has yet to be proved in this field, and it is unlikely that other than field experiments, which will be extremely expensive, will have taken place before the turn of the century.

DEVELOPMENTS LEADING TO SUCCESSFUL UNDERGROUND COAL GASIFICATION IN WESTERN EUROPE

A.C. Bailey
Mining Research and Development Establishment
National Coal Board
UK

ABSTRACT. Underground Coal Gasification has potential both as an alternative to conventional mining and as a means of exploiting unmineable coal. After a brief early history of UCG development, the objectives and status of current experiments in Belgium and France are given. The mechanisms and techniques involved in UCG are described to explain the attraction and potential of the process, and the problems to be overcome. Practical developments and theoretical studies have been conducted by the National Coal Board. A technique for drilling long holes in coal has been developed which provides the means for reliable construction of underground gasifiers. Extensive analytical work is described in which process behaviour and gasifier development are shown to be functions of depth and operating conditions.

1. INTRODUCTION

Although Western Europe is fortunate to have a large coal resource, its mining by conventional deep mining methods is uneconomic in comparison with imported coal and with other alternative energy sources.

Underground Coal Gasification (UCG) is appealing for many reasons. Men would not be required to work underground, there would be no spoil, and large deposits of coal which are currently unmineable or unattractive to mine would become accessible for exploitation. The process also offers potential as an alternative, more economic method of exploitation than conventional mining. The potential for Western Europe is therefore twofold:

(i) as a method of low cost energy production, and

(ii) as a means to greater self sufficiency by the exploitation of otherwise unmineable coal.

Experiments were carried out in several European countries in the 1940's and 1950's - in the UK, Belgium, Morocco, Czechoslovakia and others. Development of the process was also active in the USA during

G. E. Beghi (ed.), Synthetic Fuels, 49–61.

this period. The experiments and trials had the objective of gasifying shallow seams by low pressure air gasification. The gas produced in all the trials had low heating value and all work on UCG in Western Europe had ceased by 1960 with the advent of cheap oil.

Interest in UCG reawaked in 1973 with increasing uncertainties of the stability of oil supplies following the Yom Kippur War in the Middle East. Since 1973 many trials have been conducted in USA, a country particularly dependent on Middle East Oil at that time. Experiments related to underground gasifier construction began in Belgium, and two experimental programmes, a French and a Belgian/German collaboration, are currently in progress supported by EEC funds. No field work is in progress in the UK but extensive developments and assessments have been carried out by the National Coal Board with the aims of achieving successful gasifier construction and process behaviour.

Why then has the commercialisation of UCG eluded workers in many countries for so long and what are the problems to be overcome? The answers to these questions require an understanding of the fundamental mechanisms and processes involved in UCG.

2. ACCESS

Access to the coal is required to supply and deliver reactant and product gases to and from the seam, and channels must be constructed in the coal through which gases can pass between the access points. Two methods of access are available - preliminary mining and surface drilled wells. Preliminary mining involves the construction of shafts and roadways from which holes can be drilled in the seam to form the necessary gas flow paths. This method of access requires men to work underground, it leads to the production of spoil and is unattractive on operational and logistical grounds. Gasifier construction by means of wells drilled from the surface is much more attractive and has been the method of access adopted in all recent field trials.

3. GASIFIER CONSTRUCTION

In well systems of UCG, linkage channels must be created in the seam for gases to flow between the process wells. The inability to construct reliable connecting channels between wells has been a major difficulty in many trials. Various techniques have been attempted including electrolinkage, hydrofraccing, reverse combustion and deviated/in-seam drilling.

Electrolinkage requires high power because of the inherent high resistance of the coal and it is an unreliable and ineffective technique over well spacings required for commercial operation. Hydrofraccing is currently under investigation by the French GEGS team as a preliminary technique for establishing initial linkage prior to subsequent reverse combustion. The nature of fracs and the direction of their propagation are unpredictable in coal and the technique, in isolation, is unlikely to prove sufficently reliable for successful linkage.

In reverse combustion, coal is ignited at the base of a well and combustion is subsequently sustained with air injected at a second well. The air passes through the coal to the combustion zone which burns back, against the direction of air flow, to the injection well. The technique has been demonstrated in sub-bituminous coals and lignite but its use is limited in bituminous coal because of the inherent low permeability. The maximum linkage distance achieved by reverse combustion in deep bituminous coal was over a distance of 18 metres, at the Pricetown trial conducted by METC, W.Va. in 1979[1]. Reverse combustion trials by the French GEGS team (at Bruay, over 65 metres, in 1980/81, and subsequent to hydrofraccing) and by the Belgian/German IDGS team (at Thulin, over 35 metres in 1982) have not resulted in channel formation with sufficient flow capability to proceed to gasification.

Deviated/in-seam drilling has been conducted in three trials in USA. A deviated/in-seam hole was drilled at Pricetown by METC in 1976 as the first of three such holes to develop the 'longwall-generator' concept of UCG[2]. The in-seam section of hole was 150 metres long. After drilling the hole, a policy decision was taken by the US Department of Energy to abandon the concept and to drill vertical holes to test reverse combustion linkage. This decision led to the Pricetown trial referred to previously.

A slant deviated/in-seam hole was drilled at Hoe Creek, Wyoming, by Lawrence Livermore Laboratory in 1978 to link three vertical process wells in a very shallow seam[3]. Considerable problems were encountered in drilling the in-seam hole but gasification was eventually effected. The gasification phase of this experiment was severely hampered by excessive ingress of ground water into the gasifier.

The most successful trial to date has been that conducted at Rawlins, Wyoming in 1981 by Gulf Research and Development Co[4]. The test used in-seam drilling in a 7.5 metre thick sub-bituminous steeply dipping seam to create a 65 metre long drilled gasifier in the seam at a maximum depth of nearly 200 metres. Steam/oxygen gasification converted over 8500 tonnes of coal and produced a gas of average dry heating value 3000 kcal/m^3.

4. REACTANT

The composition and heating value of the product gas depend on the choice of gasifying media, the coal type, and features of gasifier operation. With air, the maximum heating value of the product gas by gasification is approximately 1100 kcal/m^3; steam/oxygen mixtures will produce gas in the range 2700-3500 kcal/m^3, similar to that produced from surface gasifiers. Product gas quality increases as the operating pressure is raised, the pressure which can be used being limited by the depth of the seam and the nature of the overburden. An operating pressure at or just below the hydrostatic pressure in the seam (aproximately 1 bar/10 metres depth) should prevent excessive water ingress to gasifiers and preclude leakage of gas into overlying strata.

The costs of access to seams increases with depth and explains the proliferation of shallow UCG trials in the USA in recent years.

Unfortunately the choice of shallow seams dictates that only low pressures can be used, it promotes the likelihood of excessive water ingress and is accompanied by an insufficiency of overburden pressure to effect rapid and complete roof collapse - a situation resulting in 'by-pass' or 'override' and a low quantity of coal conversion.

5. CURRENT EXPERIMENTS IN WESTERN EUROPE

The development of UCG of deep coal in Western Europe is not progressing because of the lack of a reliable means of linking process wells drilled from the surface. There are good grounds indicating that, in deep coal, a high quality gas could be achieved by operation at high pressure, that water ingress to gasifers could be constrained and, as will be described later, that the compaction of collapsed roof strata should promote high overall percentage conversion.

Practical trials began in both France and Belgium in 1978. In the first pilot trial conducted in France, at Bruay-en-Artois in the Pas-de-Calais coalfield, two wells were drilled, from a 1000 metre deep roadway in an existing mine, to a thin seam 200 metres below. The wells were 65 metres apart and aligned in the direction of major cleat in the coal. Hydrofraccing was used to create a connection between the wells and although reverse combustion was subsequently initiated, it was not possible to maintain the process to effect linkage because the injection of air led to combustion of the coal in the vicinity of the injection well. The trial was terminated in July 1981[5]. In addition to practical experiments to investigate the potential of electrolinkage, now proceeding on a site at L'Echaux near St Etienne in central France, a second linkage trial is in progress at Haute-Deule near Lens, also in the Pas-de-Calais region, but in different coal. The project is similar to the Bruay experiment except that access to the coal, at a depth of nearly 900 metres, is obtained by wells drilled from the surface. After sinking exploratory holes, two process wells have been constructed at a separation of 60 metres and a mini-frac has been conducted to establish the direction of frac propogation. The aim of the project is to link by hydrofraccing and reverse combustion, and to proceed to gasification by end 1984[6].

The Belgian/German collaborative project is at Thulin in Southern Belgium. Four wells were drilled in star formation 35 metres apart to a seam at a depth of approximately 850 metres. The objective of the experiment is to establish linkage by reverse combustion, and to gasify with high pressure steam/oxygen mixtures. The coal at Thulin is non-swelling semi-anthracite and the seams are heavily faulted over the test area. Reverse combustion was attempted during 1982 but was not successful in achieving the formation of a linkage channel. As at Bruay, combustion of the coal in the vicinity of the injection well was encountered and fissures in the coal between the wells became blocked by the condensation of tars produced in the combustion zone. During the early part of 1983 modifications were made to the ignition, injection and production systems to attempt to prevent reoccurrence of the problems experienced in the first linkage attempt. Further attempts to achieve linkage are in progress[7].

6. DEVELOPMENTS IN THE UK

Although no practical trials of UCG have been conducted in the UK since 1959, recent work by the National Coal Board has been carried out with the aims of:

(i) developing in-seam drilling technology to provide a reliable means of linking surface-drilled process wells at large separations in deep coal

(ii) analysing UCG process behaviour and predicting the effects of operating conditions and seam and strata characteristics on gasifier development and performance.

7. IN-SEAM DRILLING

In-seam holes can be drilled by geometric guidance in seams which have uniformity of thickness and dip over extensive areas. Such holes have been drilled in the USA for methane drainage prior to conventional mining[8]. Most UK seams however do not have sufficient uniformity to enable long holes (say 1000 metres) to be drilled by geometric guidance. The technique developed by the National Coal Board to drill long holes in-seam uses the difference of natural gamma radiation in coal, roof and floor to locate the position of the drill bit within the seam. The ratio of natural gamma radiation in roof and floor to that in coal is typically 5:1 in UK seams. The instrument developed by the NCB measures natural gamma radiation, inclination and tool angle and incorporates safety features enabling it to be used for drilling in existing mines for purposes related to conventional mining. The instrument is currently in wire-line, pump-down form and work is in progress to develop an MWD version.

The instrument package has been used to drill several holes over 500 m long in-seam at an outcrop trial site. Holes have been drilled both with rotating string and down-hole motor to develop the steering techniques necessary to stay in the seam. The natural gamma instrument has a range of approximately 0.5 metre in the location of position from roof and floor and surveying every 5 metres is sufficient to maintain holes in-seam.

In combination with deviated drilling technology in use in the oil industry, the ability to drill long holes in seams provides the means to connect widely-spaced process wells for UCG, thereby solving the well linkage problem experienced in all deep UCG trials to date. The capability of deviated drilling is exemplified by the recent Elf-Aquitaine LACQ 91 well, a 650 metre deep, vertical start, 90° deviated well, with 365 metres final 'horizontal' section[9]. Steering and surveying techniques are now sufficiently developed to enable reliable underground gasifiers to be constructed by the combination of deviated and in-seam drilling technologies.

8. ANALYSIS OF UCG PROCESS BEHAVIOUR

Initial analysis of process behaviour was directed to the assessment of product gas composition and process efficiency which would result from the use of a range of gasifying agents[10,11]. Air, steam/oxygen, steam/hydrogen and oxygen/carbon dioxide mixtures were the agents considered and it was concluded that high pressure steam/oxygen had the greatest potential for the production of a medium heating value gas at high process efficiency. The benefits of increasing pressure are better and faster reactions, greater turbulence in the gasifier and the production of a gas of high heating value and more stable composition. Higher mass flow rates can also be achieved for a given diameter of well.

The analysis of product gas composition was initially based on equilibrium conditions and subsequently on the rate of approach to equilibrium, taking into account the residence time of gas in the gasifier. In addition to the chemical kinetics involved in the process, recent modelling work has incorporated gas dynamics effects, to explain and predict the effect of depth on the shape of gasifier growth between pairs of widely spaced process wells[12]. The model is based on roof collapse and the compaction of collapsed roof strata, the variation of permeability in the collapsed strata being used to determine the pattern of gas flow within a gasifier from which its shape during growth can be predicted.

Post-burn coring of shallow gasifiers at the completion of experiments has enabled the coal converted to be shown in plan views. In experiments where linkage was successfully effected, the predominant features of gasifier development are of total seam thickness conversion and of radial growth around injection wells, with little coal converted in the vicinity of product wells. Coring after gasification has also shown that the 'cavities' were filled with caved roof strata. The shape of gasifier growth demands that the pattern of gas flow from injection wells must be predominantly radial and it can be shown, from the flow rates through the wells, that gas velocities within the cavities must be low. The conclusion is that the pattern of gas flow leading to the observed shape of gasifiers results from a radially disposed pressure gradient centred on injection wells which arises from the existence of a low pressure region at the gasifying coal face.

The shape of gasifiers can therefore be explained by the passage of injected gas through caved roof strata to a zone of higher permeability in the vicinity of the coal face. Experience from deep mining, of roof collapse and its subsequent compaction by lithostatic pressure, has been used to analyse gas dynamics within gasifiers. Bulking factors of caved roofs and the corresponding permeabilities have been used to predict gas flow patterns and pressure drops within gasifiers as a function of depth. One of the interesting predictions from the analysis is that the shape of gasifier growth in plan can be represented by a single parameter:

$$J = \frac{K_{II} \cdot T}{K_I \cdot L}$$

where T is the seam thickness
L is the spacing of wells in the seam
K_{II} is the permeability of the zone adjacent to the coal face
K_I is the permeability of the caved zone after compaction by lithostatic pressure.

Gasifiers with the same value of J will develop similar patterns of growth. For a fixed seam thickness and well spacing, the ratio $K_{II}:K_I$ increases with depth because of the higher overburden pressure on the caved roof strata, which, it is predicted, will result in a greater degree of radial development and a higher quantity of coal conversion per gasifier. The predicted patterns of growth at three depths, shallow (< 300 metres), 500 metres and 1000 metres are shown in Figures 1, 2 and 3 for a borehole spacing of 100 x seam thickness. The analysis has interpreted successfully the shape of gasifier growth observed in previous field trials (including the poor results achieved by low pressure air gasification of shallow coals).

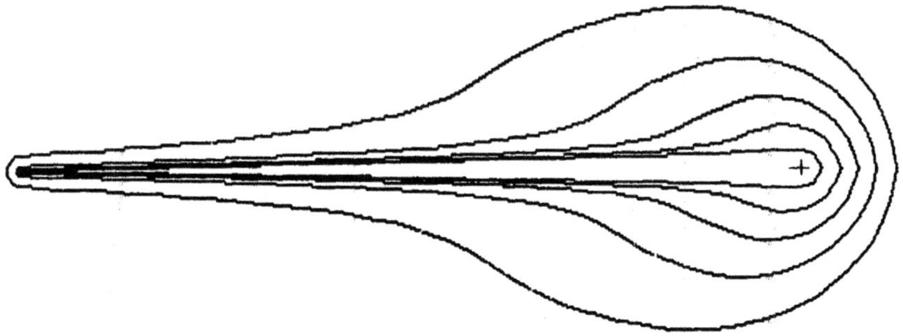

Figure 1. Growth of gasifier at depth of 300 metres or less. (Borehole spacing = 100 x seam thickness)

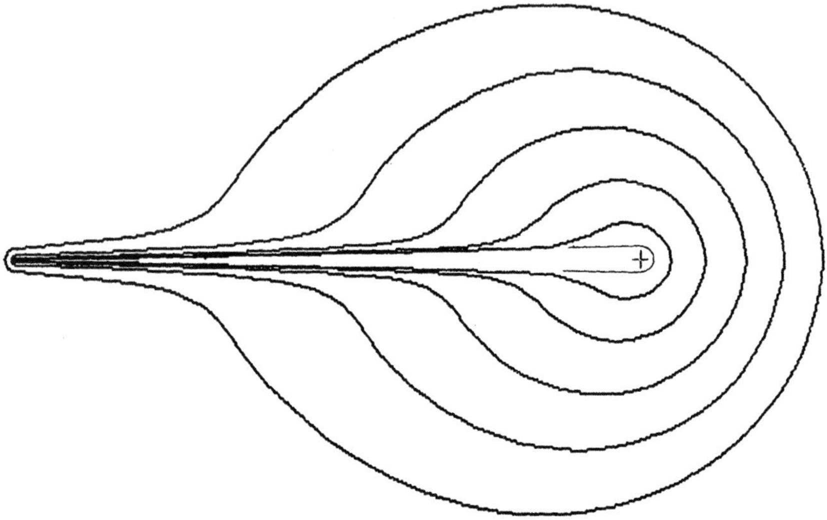

Figure 2. Growth of gasifier at depth of 500 metres
(Borehole spacing = 100 x seam thickness)

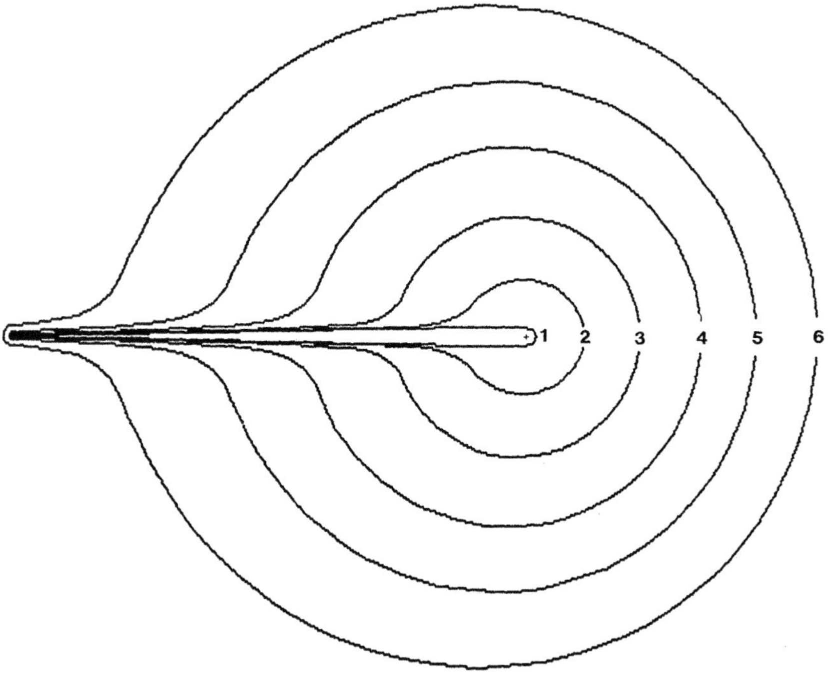

Figure 3. Growth of gasifier at depth of 1000 metres.
(Borehole spacing = 100 x seam thickness)

9. OPERATING CONDITIONS AND PRODUCT GAS

The chemical kinetic and gas dynamics analyses have been used to predict gasifier efficiency and gas composition for a range of operating conditions. Initial trials of linkage and gasification are proposed of seams at depths in the range 500-1000 metres. Operation in this depth range permits the use of established deviated drilling technology for linkage and the use of high pressures. An example of output from the model is shown in Table I.

TABLE I

Operating Conditions		Product Gas Composition (Dry%)	
Coal type	Bituminous	CO	37
Seam thickness	2 metres	H_2	27
Seam depth	650 metres	CH_4	13
Well spacing in seam	200 metres	CO_2	21
$H_2O : O_2$	2:1	Other	2
Injection temperature	350°C	Heating value	326 BTU/scf
Operating pressure	50 bar		~3000 kcal/m^3

It is predicted that gas will enter the product well at a temperature rising to 900-1000°C as the gasifier develops. Considerable attention has been given to the design of wells. Insulated casings are available which are rated up to 350°C and can accommodate the resulting expansion downhole. It is proposed that such casings be used in well construction with short lower sections of high temperature alloy through which the temperature of the product gas will be reduced from ~950°C to 350°C by sparge water. Sparge pipe design and water injection quantities have been estimated and it is predicted that water (steam) composition of the product gas will be in the range 40-50%, a rather wet gas requiring different surface processing than gas from surface gasifiers.

10. COMMERCIAL POTENTIAL AND COSTS

The product gas has potential for enrichment to SNG, for the production of methanol and for steam raising for electricity production.

Costs of raw gas production depend on seam thickness and depth, spacing of process wells in the seam, the ability to produce gas of the predicted composition and most importantly, the layout of gasifiers, the sequence of their operation and the possibilities for areal sweep. A simple pessimistic layout has been adopted for assessment of commercial costs. A pair of gasifiers form a gasifier module; each module comprises two injection wells, a production well and a deviated/in-seam hole linking the three process wells of the module. The layout of the modules is shown in Figure 4.

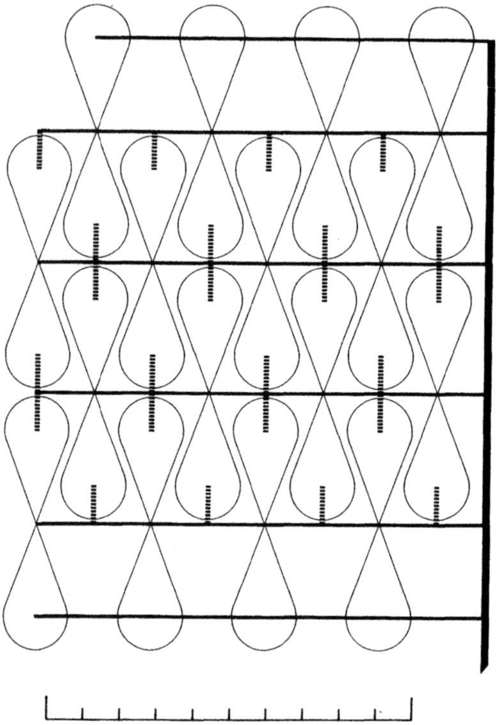

Figure 4. Idealised layout of gasifier modules

Production wells are vertical and injection wells deviated from points on lines joining all wells along which reactant and product pipelines are located. The cost of production of substitute natural gas by UCG has been estimated to be in the range 15-30 p/therm at 1983 prices (dependent on seam thickness, depth, well spacing etc), including all processing costs. The estimated costs are shown in Figure 5. The costs are highly attractive compared to the cost of similar gas produced by surface conversion of mined coal.

The layout is pessimistic because modules are independent and each deviated/in-seam hole links only the three process wells of a single module. In a commercial scheme it is likely that gasification could be relayed between modules thereby reducing the number of process wells required. Furthermore it is possible that all process wells could be deviated with wellheads clustered at drill sites to minimise environmental disturbance. A possible arrangement is shown in Figure 6.

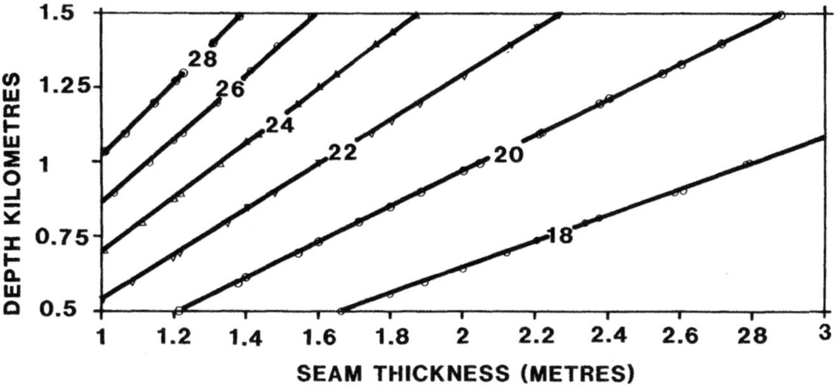

Figure 5. Cost of SNG by UCG (pence/therm) at a well spacing in the seam of 200 metres. The required coal in place is 77 million tonnes

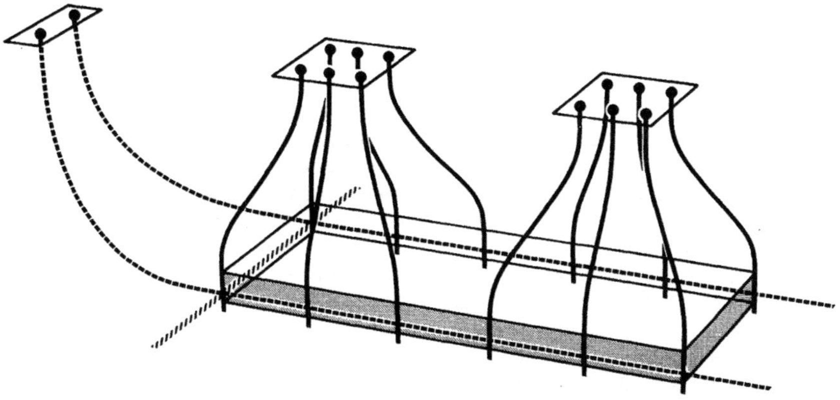

Figure 6. Well configuration in UCG by deviated drilling.

11. CONCLUSIONS

Recent advances in deviated and in-seam drilling techniques provide the means to construct large deep underground gasifiers. High pressure steam/oxygen gasification has been shown to be capable of producing a gas of medium heating value with several potential applications.

The major unknowns are the volume of coal which can be converted per gasifier and the variation of product gas composition with gasifier development ie will the product gas be sufficient in both quantity and quality? Theoretical studies of the gas dynamics and chemical kinetics involved in high pressure operation suggest that satisfactory performance can be achieved. The need now is to repeat the success of the Gulf trial, in near-horizontal bituminous seams at greater depth than at Rawlins. Trials are costly but are a necessary preliminary to commercial development. The potential is enormous; in the short term UCG could be a valuable supplement to the exploitation of coal on-shore; in the long term the technique could be the only means of exploiting the vast deposits of coal under the North Sea - as supplies of oil and natural gas are depleted. Demonstration of successful linkage by deviated/in-seam drilling, and subsequent gasification, would provide the foundation for an alternative and increased utilisation of Western Europe's coal resource.

12. REFERENCES

1. MARTIN, J W and LIBERATORE, A J. A Successful UCG Field Test in a Thin Seam, Swelling Eastern Bituminous Coal. Proc. 6th USA DOE Underground Coal Conversion Symposium, Oklahoma, 1980.
2. SHUCK, L Z and others. Status of the METC In-Situ Gasification of Eastern US Coals Project. Proc 2nd USA ERDA Underground Coal Gasification Symposium, W.Va., 1976.
3. HILL, R W and others. LLL 1979 Field Program. Proc 5th USA DOE Underground Coal Conversion Symposium, Virginia, 1979.
4. DAVIS, B E and others. Underground Gasification for Steeply Dipping Coal Beds. Phase III Report. Results of Rawlins Test No. 2. July 1982. LETC 13108-94 UC-90c.
5. RAFFOUX, J F. Underground Coal Gasification at Great Depth. Proc 6th ICCR Conference, London, 1982.
6. GADELLE, C and others. Major Problems, First Results and Projects of Underground Gasification of Deep Coal in France. Proc 8th USA DOE Underground Coal Conversion Symposium, Colorado, 1982.
7. LEDENT, P. Aims and Present State of the Belgo-German In-Situ Gasification Project, 1983; International Gas Research Conference, London, 1983.
8. OYLER, D C and DIAMOND, W P. Drilling a Horizontal Coalbed Methane Drainage System from a Directional Surface Borehole. Report of Investigations, US Bureau of Mines; 8640.

9. JOURDAN, A P and BARON, G A. Elf Drills 1000 + ft Horizontally. Petroleum Engineer International, September 1981.
10. BAILEY, A C, CHAPPELL, R S and BLADES, M J. Underground Coal Conversion - Feasibility and Potential in the UK, Mining Research and Development Establishment, National Coal Board, UK, 1980.
11. CHAPPELL, R S. High Pressure Gasification of Deep Coal. Meded Rijks Geol. Dienst 33-1/10, September 1980.
12. CHAPPELL, R S and WILKS, I H C. A Theory for the Underground Gasification of Deep Coal by the Linked Vertical Well Method. Mining Research and Development Establishment, National Coal Board. MRDE Report 83/6, 1983.

COAL GASIFICATION PROCESSES

K. Hedden
Engler-Bunte-Institute, University of Karlsruhe
P.O. Box 6380
D-7500 Karlsruhe 1
Federal Republic of Germany

ABSTRACT. Synthesis gases for the production of ammonia, Fischer-Tropsch products, etc. are produced from coal and lignite in large scale gasifiers operated as moving-bed, fluidized-bed, or entrained-flow reactors. In some countries town gas is manufactured from solid fuels by gasification. Lurgi developed the high pressure moving bed gasification process to achieve high heating values. This process is also applied to the manufacture of SNG, substitute natural gas. In new developments gasification under pressure is also applied to fluidized-bed and entrained-flow gasification. It can be expected that in future coal gasification will also play an important role for power production in combined cycle processes. Hydrogasification, catalytic steam gasification, allothermal gasification with high temperature nuclear heat, and gasifiers using a molten iron bath as reactor are under development.

1. INTRODUCTION

Several possibilities for upgrading of coal are shown in Fig. 1: Combustion, Pyrolysis, Gasification, and Hydrogenation (Liquefaction). This paper is dealing with gasification and will treat the following subjects:

- Classification of Processes
- Large Scale Commercial Gasifiers
- Advantages of High Pressure Gasification
- New Developments in West-Germany
- Catalytic Gasification and other Developments
- Costs of Gas Production.

In gasification processes coal or other solid fuels are converted at high temperatures into fuel gases, synthesis gases, or hydrogen. Fuel gases with low, medium, or high calorific values can be obtained by applying different gasifying media like air, oxygen and steam, or hydrogen. The quality of the gas can also be influenced by special process conditions, e.g., high pressure or catalytic gasification to

G. E. Beghi (ed.), Synthetic Fuels, 63–91.

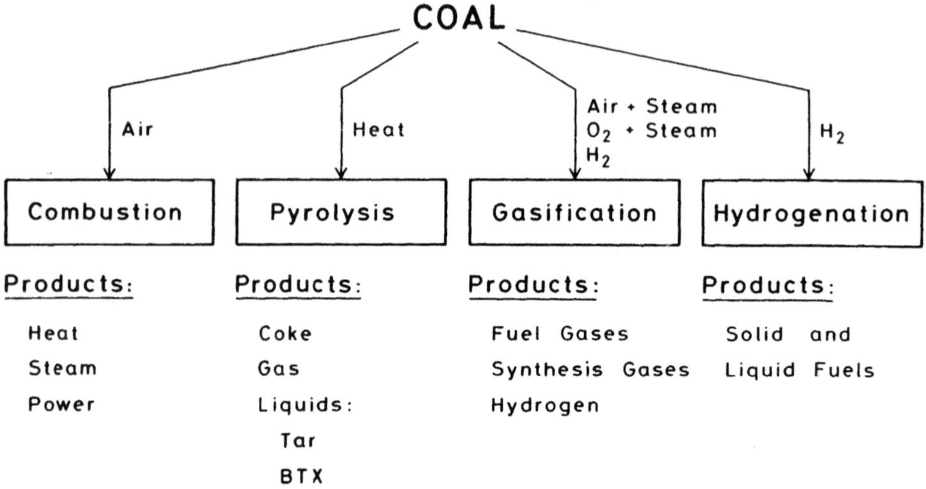

Figure 1. Coal upgrading processes.

obtain high methane contents. Low calorific value producer gas is mainly used for industrial purposes, e.g., for lime burning. Town gas with medium calorific value is produced by pressure gasification of lignite and coal in the German Democratic Republic, in Czechoslowakia, and in China. Substitute Natural Gas - SNG - can be produced by hydrogasification or by gasification with oxygen and steam followed by methanation.

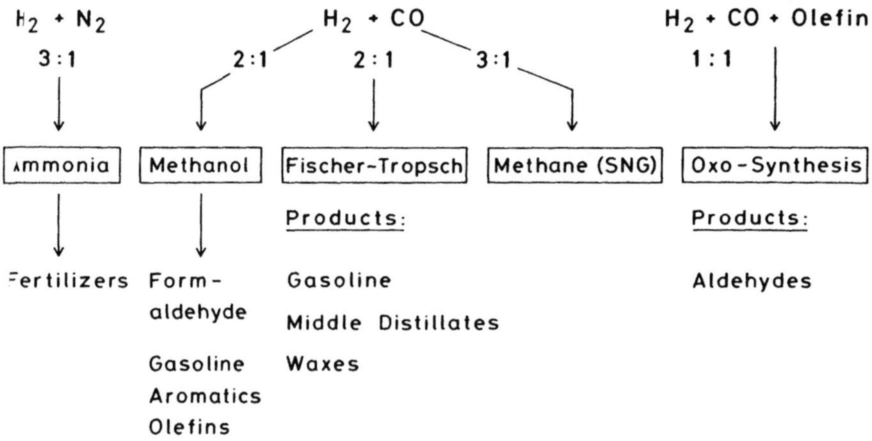

Figure 2. Synthesis gas routes.

The most important products of solid fuel gasification are synthesis gases for the chemical industry. Fig. 2 shows different routes for the use of such synthesis gases. Main products are: Ammonia for the fertilizer industry, motor fuels and waxes by Fischer-Tropsch-Synthesis, Aldehydes by hydroformylation of olefins. Methanol and also SNG should be mentioned as important fuels of the future. In future times also hydrogen produced from solid fuels may become an important raw material, especially for the liquefaction of coal and of other carbon containing materials.

2. CLASSIFICATION OF PROCESSES

In Table I the chemical equations for different gasification reactions are given together with the enthalpies of reactions which are exothermic for combustion, hydrogasification, and homogeneous methanation and which are endothermic for the gasification of carbon with steam and carbon dioxide.

TABLE I Gasification reactions

Reaction	$\Delta_R H$, kJ/mol
Combustion	
$C + O_2 \longrightarrow CO_2$	-393.8
$C + 1/2\ O_2 \longrightarrow CO$	-110.6
$CO + 1/2\ O_2 \longrightarrow CO_2$	-283.2
$H_2 + 1/2\ O_2 \longrightarrow H_2O$	-242.0
Gasification	
$C + H_2O \rightleftarrows H_2 + CO$	+131.4
$C + CO_2 \rightleftarrows 2\ CO$	+172.6
Hydrogasification	
$C + 2\ H_2 \rightleftarrows CH_4$	-74.9
$C_nH_mO_p + (2n - \frac{m}{2} + p) H_2 \longrightarrow nCH_4 + pH_2O$	exothermic
Methanation	
$CO + 3\ H_2 \rightleftarrows CH_4 + H_2O$	-206.3

The equilibrium contents of methane in the exothermic hydrogasification and methanation processes are favoured by high pressure. On the

TABLE II Classification of gasification processes

autothermal - continuous

autothermal - cyclic

allothermal - continuous

by heat exchanger

by heat carrier (solid or gaseous)

contrary the equilibrium composition of hydrogen and carbon monoxide is decreasing with increasing pressure for the endothermic gasification reaction with steam and carbon dioxide. Normally, the rates of reactions increase with increasing pressure.

Gasification processes can be classified in different ways, e.g.,

- Method of supplying the heat for the endothermic gasification reactions (Table II);
- Method of contacting the solid and gaseous reactants in the gasifier (Fig. 3 and 4).

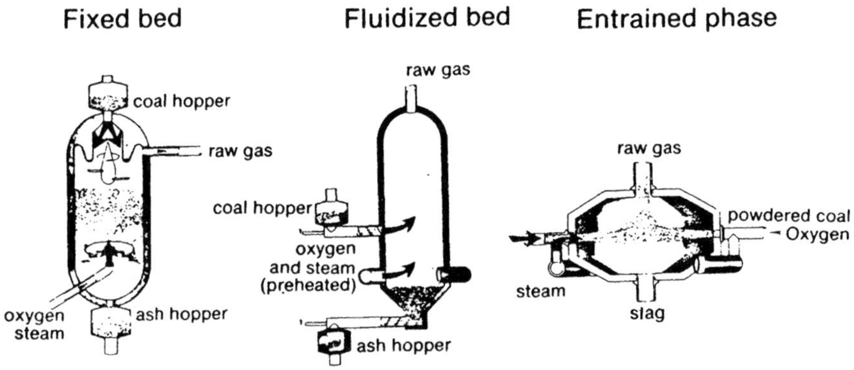

Figure 3. Principles of gasification (9).

For autothermal processes the heat is supplied internally by partial oxidation of the fuel with oxygen (or air). This can be done either continuously or in a cyclic process. During continuous gasification steam is supplied together with air or oxygen, whereas in a cyclic process the fuel bed is heated up in a first step by combustion with air followed by a second step, in which the hot fuel is gasified by steam and cools down. In allothermal continuous processes the heat for the endothermic reactions is carried to the gasifier either by heat-carrying solids or fluids, e.g., superheated steam, or through the walls of the reaction vessel or via heat exchangers located in the fuel bed, respectively.

In Fig. 3 the three main principles of gasification are shown:

- Fixed-bed gasification, normally with countercurrent flow of gas through a slowly moving coal bed. Also cocurrent flow is applied in some cases;
- Fluidized-bed gasification with random distribution of the solid fuel;
- Entrained phase or dilute phase gasification; in this case gasification takes place in a cocurrent flow of gas and suspended fuel dust entrained in the gasifying medium.

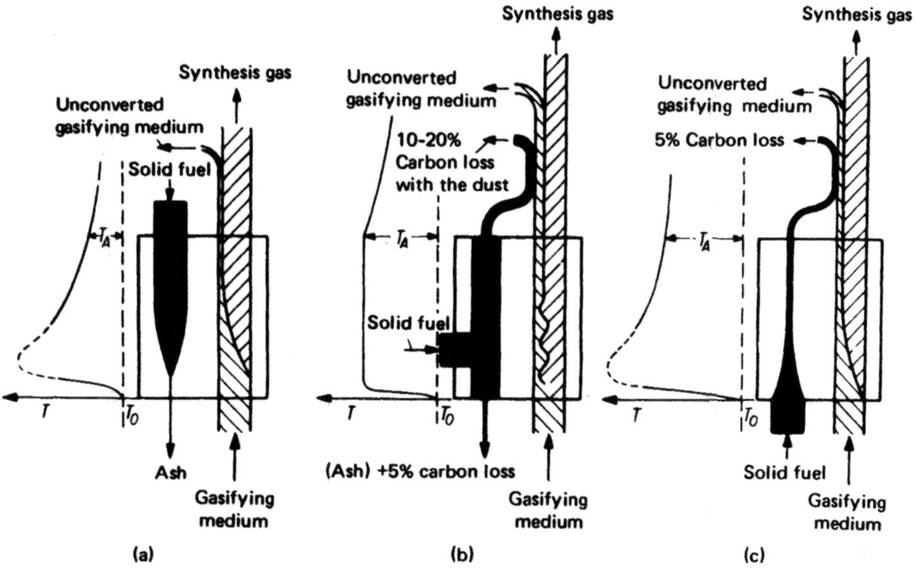

Figure 4. Comparison of the three types of autothermic gas producers: (a) Fixed-bed gasification; (b) fluidized-bed gasification; (c) suspension or entrained dust cloud gasification, where T = gas temperature, T_0 = entrance temperature, and T_A = exit temperature (1).

A more detailed comparison of these three types of autothermal gasifiers is shown in Fig. 4.

(a) In countercurrent fixed-bed gasification with dry ash removal the conditions are ideal as far as heat exchange is concerned: The gasifying medium is heated up by the ash in the bottom of the gasifier and the solid fuel fed to the gasifier from the top is heated up by the raw gas. The temperature is going through a maximum when the oxygen is nearly consumed.
As a result of the good heat exchange the specific oxygen-consumption is low compared with both the other processes. It is of some disadvantage that the tar vapours produced by devolatilization of the coal at temperatures of about 400 °C are leaving the gasifier with the raw gas and have to be removed by condensation. To keep the fuel bed in the condition of a fixed bed and to avoid a higher pressure drop the fuel particles should have diameters above 6 mm.

(b) In fluidized-bed gasifiers the temperature is nearly constant along the fuel bed. Depending on the reactivity of the fuel the reaction temperature is around 900 °C. It is an advantage of the process that the raw gas does not contain tar vapours as these are cracked immediately after devolatilization. The raw gas contains large amounts of dust, therefore, carbon losses are high compared with fixed-bed gasification. As the gasification temperature is high for fuels with low reactivity difficulties may arise when the ash melting point is low.

(c) Entrained-bed gasification needs very high reaction temperatures, as the residence time of the solid fuel is very short (cocurrent flow). Therefore, the raw gas is leaving the gasifier with temperatures of about 1400 °C. The gas is tar free. As a result of the high outlet temperature the oxygen-consumption is high. To avoid higher carbon losses the fuel has to be ground down to less than 0.1 mm of diameter.

3. LARGE SCALE COMMERCIAL GASIFIERS

Germany is the leading country in the design and construction of commercial plants for the gasification of coal and lignite. Lurgi pressure gasification, Winkler fluidized-bed gasification and Koppers-Totzek entrained dust gasification are the processes used worldwide for the production of synthesis gases and of heating gases from solid fuels. Nevertheless, in the Federal Republic of Germany itself no gasification plant is in operation at the time being to produce gas from coal on a commercial scale. The wellknown Lurgi plant at Dorsten operated by Ruhrgas AG to produce town gas was shut down in 1967.

Modern <u>fixed-bed gasifiers</u> work either with a rotating-grate and dry-ash removal or with molten slag removal. In the first case the gasifying medium is fed through the grate, while in the second case tuyeres are in use similar to those of blast furnaces. In both these cases the fuel bed is about 3 meters high.

The Lurgi gasifier has been developed for gasification at elevated pressures up to 30 bar (Fig. 5). Coal or lignite is fed to the gasifier

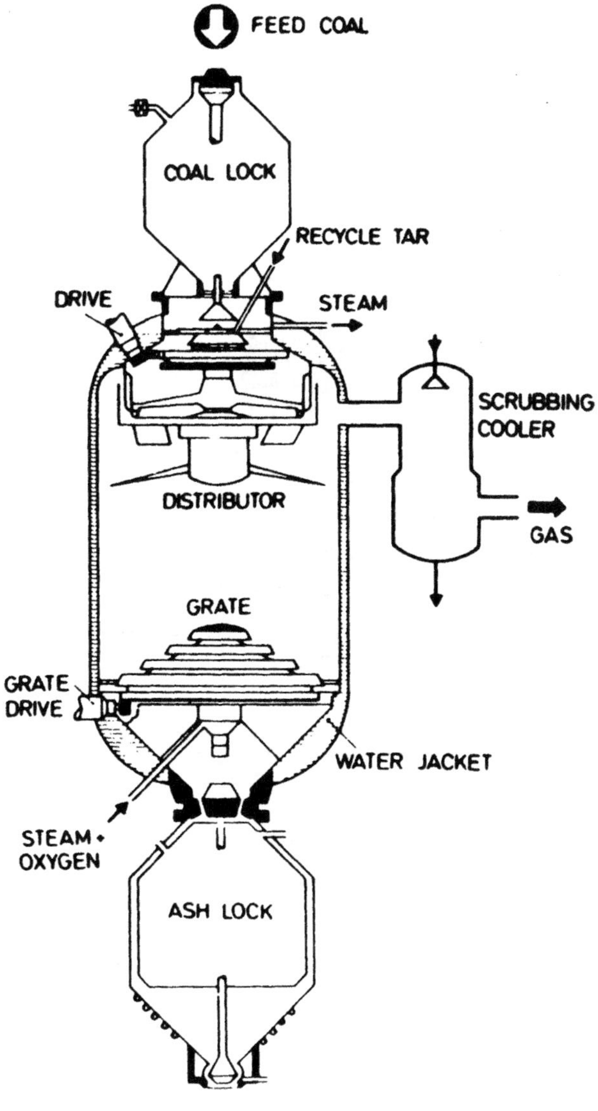

Figure 5. Lurgi moving-bed pressure gasifier with dry ash removal (9,12).

through a lock hopper at the top. The residence time of the fuel is about 1 hour, during which the coal grains move down countercurrent to the gasifying medium. The steam-oxygen ratio is normally high to ensure that ash sintering or fusion is avoided. The dry ash is removed from a rotating grate through the ash lock. To overcome the disadvantage arising from the high-undecomposed steam content of the gas, British Gas developed the British Gas/Lurgi slagging gasifier (Fig. 6).

The estimated performance of high-pressure slagging and non-slagging fixed-bed gasification is given in Table II. As the steam-oxygen ratio is significantly lower under slagging conditions, the formation of dirty gas liquor is also reduced drastically.

The relatively high methane content of the Lurgi gas arises from coal devolatilization and hydrogenation in the upper zone of the fuel

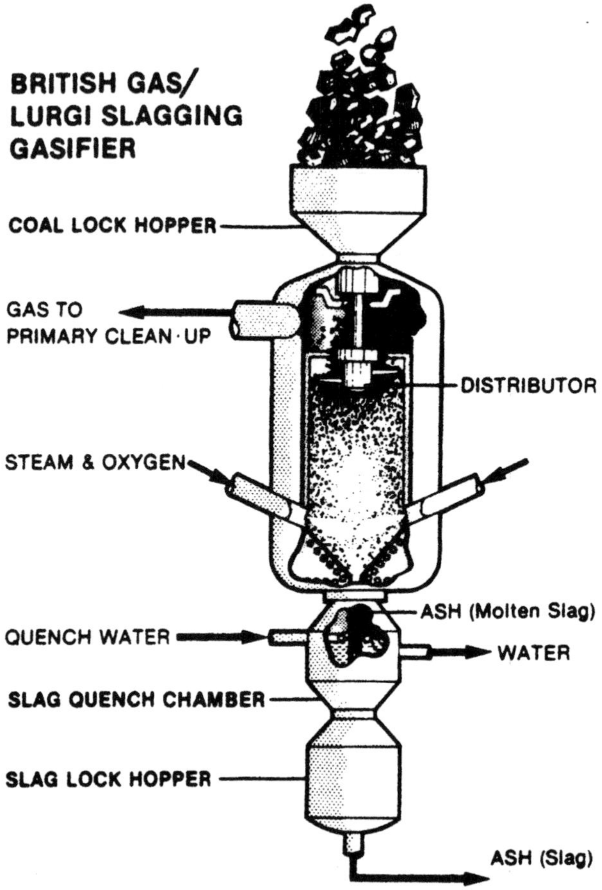

Figure 6. Britisch Gas/Lurgi slagging gasisier (9).

bed. This high methane content is an advantage for the production of heating gases, e.g., town gas or substitute natural gas (SNG). For the

TABLE III Estimated performance for a typical highly volatile weakly caking coal in a slagging pressure gasifier compared with that obtained in a gasifier with dry ash removal (after 3)

	Slagging	Dry ash
Process steam (kg/GJ of gas)	16.3–12.9	77.0
Oxygen (m^3/GJ of gas)	14.5–17.2	12.9
Steam-oxygen ratio (by vol.)	1.5–1.0	8.0
Final reaction temperature (°C)	1,250	750
Steam decomposition (%)	90–76	40
Gas yield (GJ/ton daf[a] coal)	27.6–27.1	28.2
Gas composition (% by vol.)		
Crude		
CO_2	4.7–7.0	28.5
CO	58.8–59.3	18.5
C_2H_4	0.2–0.2	0.2
H_2	26.3–25.6	42.5
CH_4	9.0–6.9	9.6
N_2	1.0–1.0	0.7
Total	100.0 100.0	100.0
Washed to 2% of CO_2		
CO_2	2.0–2.0	2.0
CO	60.4–62.5	25.3
C_2H_4	0.2–0.2	0.2
H_2	27.1–27.0	58.4
CH_4	9.3–7.3	13.1
N_2	1.0–1.0	1.0
Total	100.0 100.0	100.0
Gas calorific value (kJ/m^3)	13,972–13,413	14,903
By-products (GJ/ton daf[a] coal)	4.43–3.48	4.43
Liquor (kg/GJ of gas)	3.9–9	52
Thermal efficiency[b] (%)	89.3–86.2	73.6

[a]daf = dry and ash free.

[b](Heat value of crude gas + heat value of steam generated)/(heat value of fuel feed + heat value of steam consumed).

production of ammonia synthesis gas, however, the methane formation is a disadvantage, as the methane dilutes the synthesis gas as inert gas which has to be removed from the ammonia synthesis loop.

In South Africa more than 80 large Lurgi gasifiers (dry ash removal, 4 meters in diameter, gasification capacity about 35 tons of coal per hour) are in operation to produce synthesis gas for the Fischer-Tropsch-Synthesis. In Great Plains, North Dakota/USA, a plant with 14 such gasifiers will be started up this year to produce SNG from lignite.

Fluidization techniques were first applied by Winkler in fluidized-bed gasification. Winkler-gasifiers (Fig. 7) are mainly used for lignite gasification. The process operates at atmospheric pressure. In some applications secondary oxygen is added above the fluidized bed to increase the temperature and, thereby to increase dust gasification and methane decomposition with the result of achieving a better synthesis gas. In Table IV operating results are given. Fuels low in reactivity and with a low ash-fusion temperature are not suitable for use in fluidized-bed processes, because of lump formation by sintering of ash.

The most important characteristic of entrained phase gasification is its ability to utilize any type of coal, which has to be pulverized. Coking characteristics of the fuel and ash-fusion temperature do not affect the operability of such processes since each particle is discrete and separated from other particles in the flowing gas stream. Because of the very high gasification temperatures, more than 1400 °C, no tar and very little methane is formed, so that this process is ideal for the production of ammonia synthesis gas. A disadvantage, compared with the other types of gasification, is the high oxygen demand.

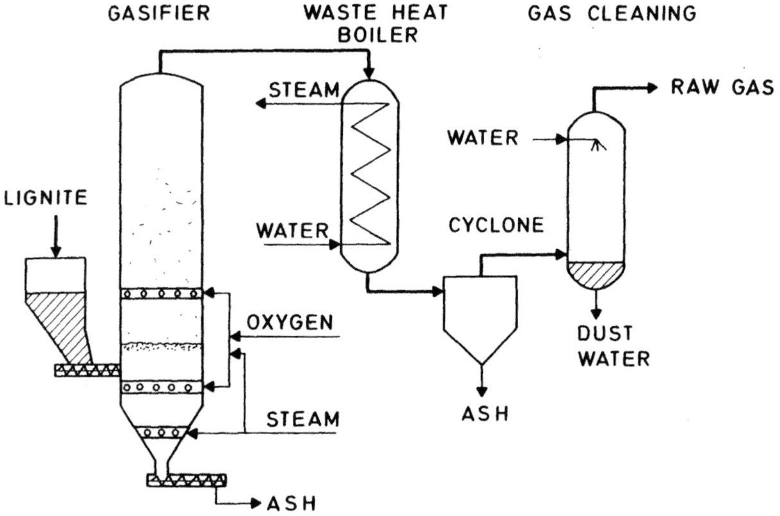

Figure 7. Atmospheric fluidized-bed gasification (Winkler-Process).

TABLE IV Operating results of a Winkler generator (after 3)

Charged coal	
H_2O (wt. %)	8–10
Ash (wt. %)	4–5
Analysis (wt. %)	68.1 C; 5.4 H; 25.2 O; 0.8 N; 0.5 S
Tar content, Fischer (wt. %)	8.7
Grain size (mm)	0–8
Proportion over 1 mm (%)	40
Ash sinter temperature (°C)	900
Ash fusion temperature (°C)	1280
Gas composition at full load, H_2S free (vol. %)	
CO_2	13.8
CO	48.2
H_2	35.3
CH_4	1.8
N_2	0.9
Consumption per 1000 m^3 crude gas	
Brown coal, 8.6% H_2O (kg)	645
Oxygen (m^3)	235
Low-pressure steam (kg)	100
Condensate (kg)	60–110
By-product yield per 1000 m^3 crude gas (kg)	
Dust	40
Steam, 17.6 bar	485
Output, crude gas (m^3/hr)	8,000–17,000
Specific output, normally [m^3/(m^2)(hr)]	2,000
Gasification efficiency[a] (%)	74.9
Thermal efficiency[b] (%)	81.8
Carbon conversion ratio[c]	90.6

[a](Heat value of crude gas)/(heat value of fuel feed).
[b](Heat value of crude gas + heat value of steam generated)/(heat value of fuel feed + heat value of steam consumed).
[c](Mass of carbon in crude gas)/(mass of carbon in fuel feed).

The Koppers-Totzek-Process, performed at atmospheric pressure, is used commercially in many plants to produce ammonia synthesis gas from solid fuels. In Fig. 8 a gasifier with four burners is shown; its capacity is about 25 t/h of coal. A homogeneous mixture of oxygen and pulverized coal is introduced through the burners. Steam is introduced

TABLE V Operating results for the Koppers-Totzek process (after 3)

Type of fuel	Powdered coal, high ash (medium coking)	Powdered coal, low ash (noncoking)	Powdered coke	Fuel oil
Fuel analysis, dry				
C (%)	55.18	74.30	79.9	84.60
H (%)	3.18	4.75	1.0	11.30
O (%)	3.82	13.30	1.5	0.13
N (%)	0.96	1.05	0.3	0.40
S (%)	0.76	0.95	0.8	3.50
Ash (%)	35.10	4.65	15.5	0.07
Gross calorific value (kJ/kg)	22,841	30,215	28,261	43,659
Analysis of Totzek gas (%)				
CO_2 (%)	11.9	11.8	11.8	4.5
CO (%)	55.3	57.6	65.9	48.4
H_2 (%)	31.1	29.0	21.0	45.4
N_2 (%)	1.3	1.2	0.9	0.8
CH_4 (%)	0.1	0.1	0.1	0.1
H_2S (%)	0.3	0.3	0.3	0.8
Total	100.0	100.0	100.0	100.0
Gross calorific value (kJ/m^3	10,313	10,358	10,376	11,312
Gas yield per ton of fuel (m^3)	1,603	2,127	1,995	3,106
Steam generated (tons per ton of fuel)	1.06	1.42	1.40	2.0
Consumption per 1000 m^3 Totzek gas				
Oxygen (m^3)	360	350	370	277
Power, excluding oxygen (kWh)	52.3	46.6	55.8	14.1
Steam, low pressure (kg)	200.2	74.49	205	256.3
Thermal efficiency[a] (%)	71.3	71.7	72.1	79.2
Conversion ratio (%)	96.0	97.7	95.4	95.6

[a](Heat value of crude gas + heat value of steam generated)/(heat value of fuel feed + heat value of steam consumed).

around the burners to protect the burners and refractories from excessive temperatures. The product gas leaves the reaction zone through a central outlet and is cooled down by some quenching and in a waste-heat boiler in which high-pressure steam is produced. Operating results for the gasification of high-ash and low-ash coals, coke, and fuel oil are given in Table V.

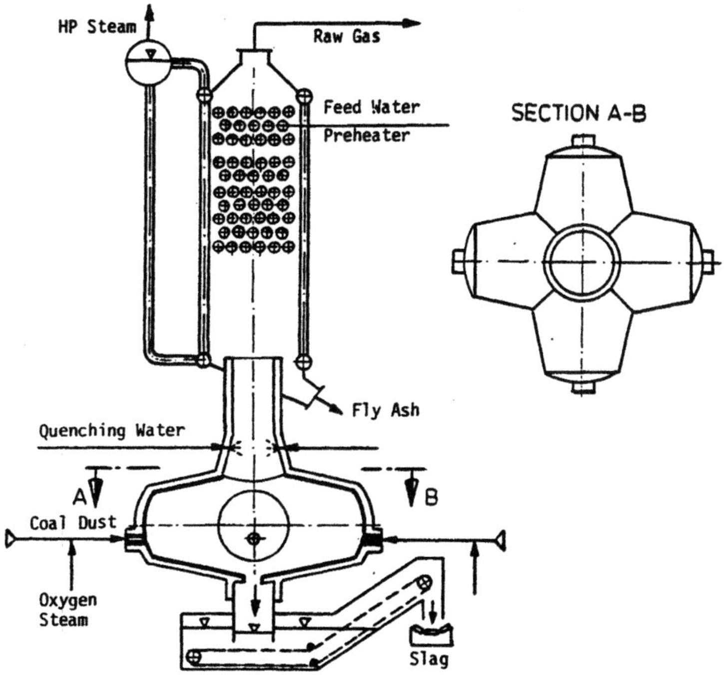

Figure 8. Koppers-Totzek gasifier.

4. ADVANTAGES OF HIGH PRESSURE GASIFICATION

From the above mentioned coal gasification processes only the Lurgi fixed-bed process is operated under pressure. For liquid fuels, e.g., heavy fuel oil, partial oxidation processes under pressure have been developed by Texaco and by Shell, and big commercial plants are in operation at pressures up to 70 bar for the production of synthesis gas and hydrogen. The present aim in the development of coal gasification processes is high pressure operation.

In Table VI the main effects of elevated pressure gasification are listed up:

- The methane concentration increases with increasing pressure, especially at low gasification temperatures (fixed-bed and fluidized-bed gasification). High pressure gasification is, therefore, advantageous if gases with higher calorific value than that of synthesis gas are wanted.

- Overall compression costs are lower when the gasification takes place

TABLE VI Advantages of elevated pressure gasification

Increased methane concentration
(increased calorific value)

Reduction of compression costs
(gas treatment under pressure)
(feed into high-pressure pipelines)
(synthesis under pressure)

Increased gasifier capacity
(reduced flow-velocity, therefore less dust carry over)
(increase of the reaction rates with the pressure of gasifying medium)

Combined cycle power generation
(combination of gas- and steamturbine)

under pressure. In this case only oxygen has to be compressed, the volume of which is about one third of the product gas. As gas treatment, e.g., hydrogen sulfide and carbon dioxide removal by absorption processes often are carried out under pressure one saves the costs for raw gas compression. The main synthesis processes, like ammonia-, methanol-, Fischer-Tropsch-, and Oxo-Synthesis are carried out under pressure and pipeline-gas is also transported under elevated pressures.

- The capacity per unit of reactor volume increases with increasing pressure. If dust carry-over limits the capacity, then higher pressure would lead to lower flow-velocity and, therefore, less carry-over of dust at equal capacity; hence, the capacity can be increased. If the capacity is limited by the rate of gasification reactions than elevated pressures normally allow also higher throughput of coal and gasifying medium, as the chemical reaction velocities increase with increasing pressure.

- Also for combined cycle power stations coal gasification under pressure is wanted, as the gas turbines operate with pressures up to 20 bar.

5. NEW DEVELOPMENTS IN WEST-GERMANY

As far as research and development of coal and lignite gasification is concerned strong efforts have been made especially in West-Germany since 1975. The government of this country has sponsored the development of new technologies in this field with some hundred million marks per year. Several pilot plants have been operated for the production of gases either high in methane with the major objective to produce SNG, or relatively low in methane to be used as synthesis gas in the chemical industry, for the reduction of iron-ore, for combined cycle power stations, etc. Most of the new technologies focus on the application of high pressure gasification technologies. German pilot plant projects with a capacity above 1 ton of coal per hour are listed up in Table VII (heating gases high in methane), and in Table VIII (synthesis gases low in methane).

TABLE VII Pilot plants for the production of gas high in methane

No.	PROJECT	DEVELOPED BY	FUEL THROUGHPUT, t/h	PRESSURE, bar	TYPE OF REACTOR	IN OPERATION SINCE / PLANED FOR
1	Ruhr 100 Dorsten	Ruhrkohle AG Ruhrgas AG Steag AG	4 to 10 (Coal)	max. 100	Lurgi moving bed	1979
2a	Hydro-gasification Wesseling	Rheinische Braunkohlen-werke AG	0.1 to 0.3 (dry lignite)	55 to 95	fluidized bed	1976
2b	Hydro-gasification Wesseling	Rheinische Braunkohlen-werke AG	max. 10 (dry lignite)	max. 120	fluidized bed	1982

For the production of pipeline gas with high heating values (Table VII) there are two different routes of development:

- Gasification with oxygen and steam at pressures up to 100 bar in a Lurgi fixed-bed gasifier;
- Gasification with hydrogen at pressures up to 120 bar in a fluidized-bed gasifier.

In the project Ruhr 100 a Lurgi type high pressure gasifier (see Fig. 9) was tested at Dorsten with good results. As can be seen from Fig. 10 the methane content increased from about 9 to 16 % by increas-

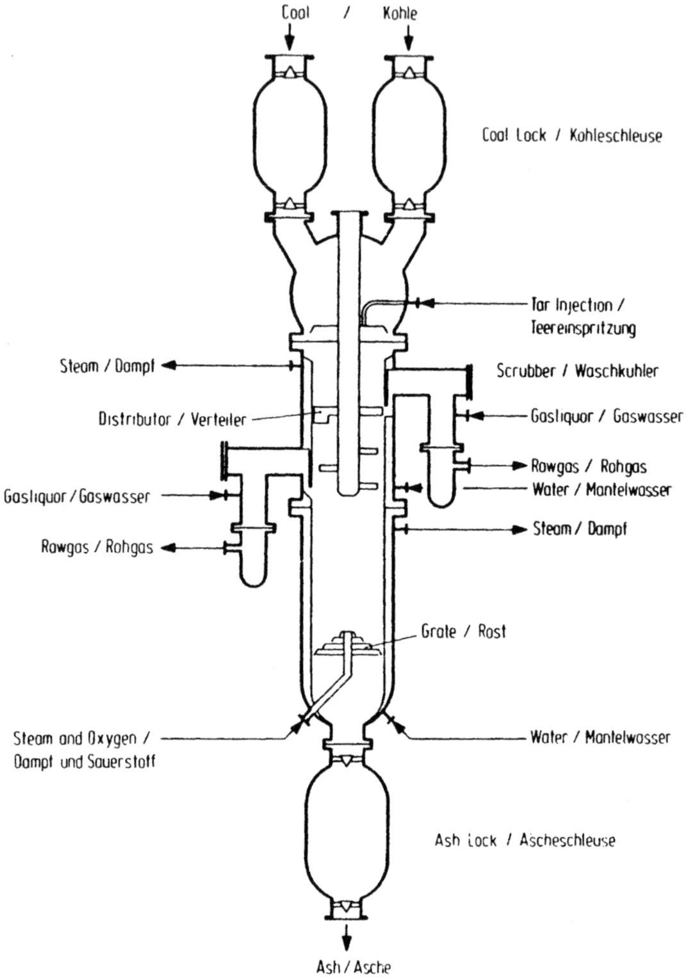

Figure 9. Lurgi pressure gasifier Ruhr 100 (9).

ing the pressure from 25 to 90 bar. As a result of the exothermic formation of methane the specific oxygen consumption decreased from about 19 to 13 m³/GJ with increasing pressure. Beside the 16 % CH_4 at 90 bar the gas contains about 30 % CO_2, 18 % CO, and 35 % H_2. From these figures it can be calculated that the raw gas contains already more than a half of the methane in the final SNG as obtained after methanation and carbon dioxide scrubbing.

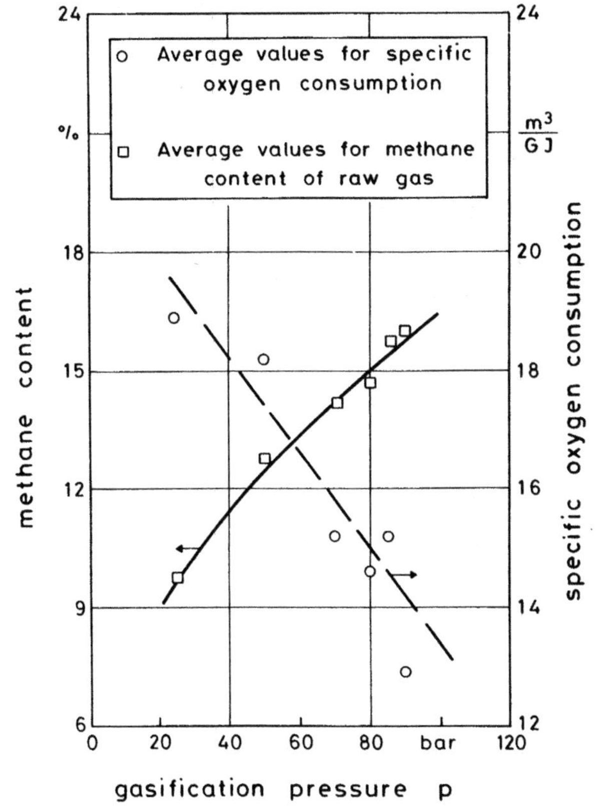

Figure 10. Ruhr 100: Influence of gasification pressure on methane content and specific oxygen consumption (after 13).

Rheinbraun is developing a hydrogasification process to produce SNG. A small pilot plant with a throughput of 0.1 to 0.3 t/h of dried lignite was in operation from 1976 to 1982. A larger plant with a capacity of 10 t/h was put on stream in 1982 and is operating since 1983 with good results. Fig. 11 shows an illustrative flow-sheet of the pilot plant. In the Rheinbraun hydrogasification process dried lignite (12 % moisture) is gasified with hydrogen in an exothermic reaction at a temperature of about 900 °C. At pressures of about 100 bar the raw gas contains about 50 % CH_4, 2 % CO_2, 3 % CO, and 45 % unreacted hydrogen. During a residence time of about 20 minutes 65 % of the lignite carbon is converted to gas. The remaining 35 % leave the gasifier as char. This residual char can be used in the High-Temperature-Winkler-Gasifier (also developed by Rheinbraun, see Table VIII) to produce the hydrogen needed for the hydrogasification step. In Fig. 12 two combinations of the hydrogasification process with the hydrogen production are shown: hydrogen produced by gasification of residual char and pro-

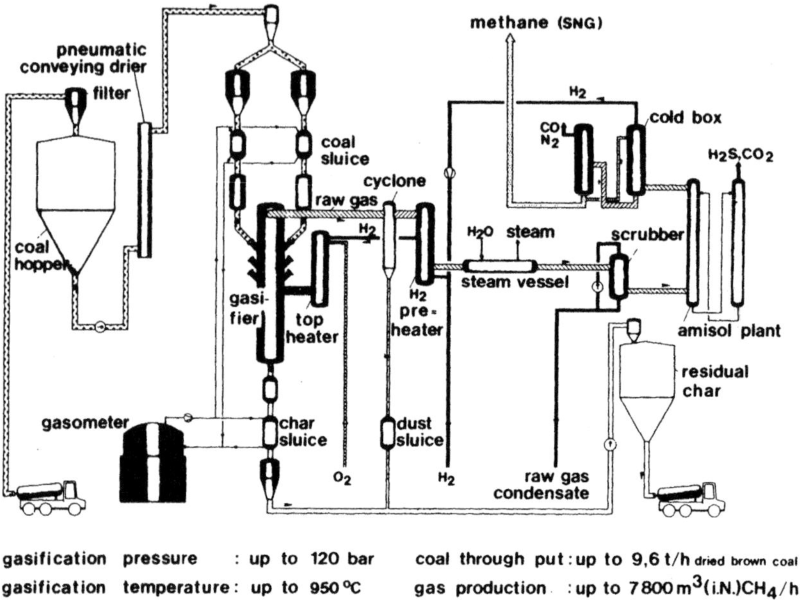

Figure 11. Flow-sheet of the Rheinbraun hydrogasification pilot plant (14).

duction of hydrogen by steam reforming of a part of the produced methane with nuclar heat from a high temperature nuclear reactor.

From the pilot plants to produce synthesis gas (gas low in methane) which are listed up in Table VIII the High-Temperatur-Winkler (HTW) developed by Rheinbraun has already been mentioned. This pilot plant with a throughput up to 1.3 tons of dried lignite per hour is in operation since 1978. To keep the methane content on a relatively low level the pressure was raised only to 10 bar. The results were so promising that now a demonstration unit with a capacity of 37000 m^3/h (STP) of synthesis gas for the production of methanol is under construction and scheduled to go on stream end of 1984.

Two of the three projects for entrained flow gasification (Table VIII, No. 5,6,7) were very successfully operated:

- The S/K Coal Gasification Plant at the Shell Harburg Refinery;

- the Texaco Coal Gasification Plant at Ruhrchemie, Oberhausen-Holten.

The main difference of both these entrained phase gasification processes concerns the supply of pulverized coal to the burners:

In the S/K-process (Fig. 13) developed by Shell and Krupp-Koppers a

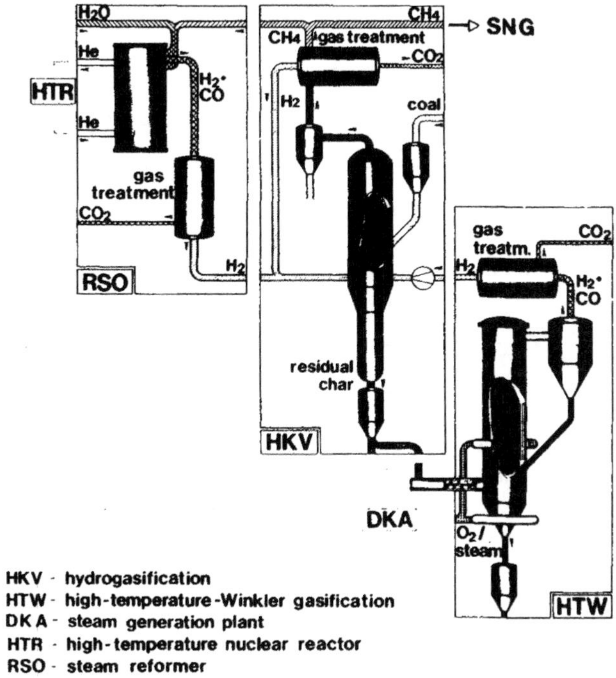

Figure 12. Combination of hydrogasification with hydrogen production either by gasification of residual char or by methane-steam-reforming with high temperature nuclear heat (14).

dry system for the input of coal dust is used, thereby adapting the Koppers-Totzek technique to higher pressures. In the Texaco-Process (Fig. 14) developed by Ruhrkohle AG and Ruhrchemie AG the coal is fed to the burner in a water slurry. A demonstration plant of the Texaco-Process with a capacity of 50000 m^3/h (STP) synthesis gas for the production of Oxo-products is scheduled to go in operation at Ruhrgas AG in 1986.

To avoid tar vapours in the raw gas of fixed-bed gasifiers Kohle-Gas-Nordrhein developed a gasifier (see Fig. 15) with a recycle of the tar containing coal pyrolysis gas being produced in the upper stage of a two stage gasifier. The higher components of the pyrolysis gas are cracked in the hot zone above the rotating grate.

Coal gasification may become very important in combined cycle power generation. Examples are given in Fig. 16 to 18. The flow-sheet in Fig. 16 demonstrates the combination of Shell- or Krupp-Koppers-high pressure entrained flow gasification with power generation. From the gasification of pulverized coal with oxygen results a gas (1400 °C) which is very low in carbon dioxide (2 % CO_2). Therefore, sulfur removal is relatively easy. The CO-rich gas is burnt with compressed air in the gasturbine combustor. After passing the gasturbine

TABLE VIII Pilot plants for the production of gas low in methane

No.	PROJECT	DEVELOPED BY	FUEL THROUGHPUT, t/h	PRESSURE, bar	TYPE OF REACTOR	IN OPERATION SINCE
3	HTW (High Temp. Winkler) Frechen	Rheinische Braunkohlen-werke AG	1 (dried lignite)	max. 10	fluidized bed	1978
4	KGN (Kohle-gas Nordrhein) Hückelhoven	PCV (Flick) Sophia-Jacoba	2 (coal)	1	moving bed, decomposi-tion of tar	1979
5	Shell-Koppers Hamburg	Shell AG	6 (coal)	25 to 35	entrained flow	1979
6	Texaco Oberhausen-Holten	Ruhrkohle AG Ruhrchemie AG	6 (coal)	40	entrained flow slurry feed	1978
7	Saarberg-Otto Völklingen	Saarberg-werke AG Dr. C. Otto	10 (coal)	max. 25	entrained flow slagging bath	1979

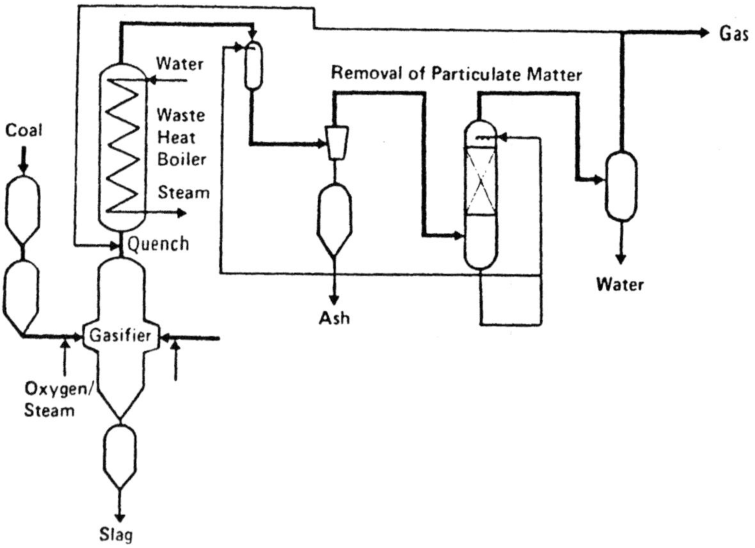

Figure 13. S/K dry coal dust gasification process (9).

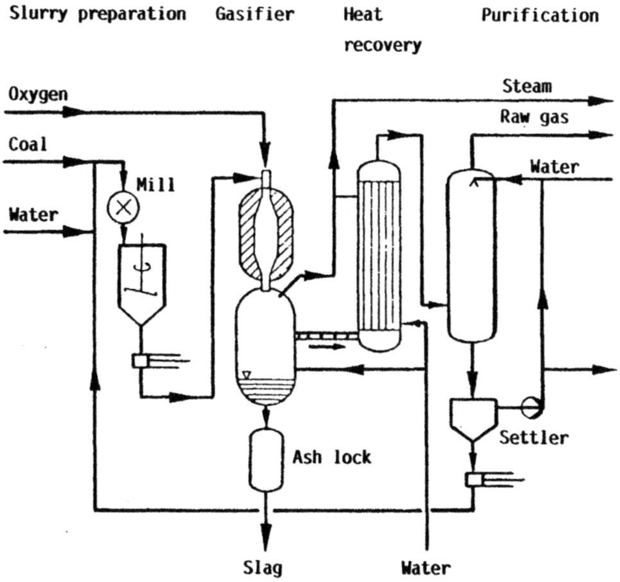

Figure 14. Texaco coal-slurry gasification (9).

the off-heat of the combustion gas is converted into steam in an exhaust boiler. Both this steam and the high pressure steam from the waste heat boiler of the gasifier drive a steam turbine. It is expected that the power generation efficiency for such a plant will increase to over 40 %. For this process-combination oxygen is necessary, because entrained flow gasification with air would leave too much unreacted powdered char in the raw gas.

Steag (Steinkohlen-Electrizitäts AG) developed together with Lurgi the KDV-process (Kohle-Druck-Vergasung) and operated a plant at Lünen with a capacity of 175 MW. In this plant (see Fig. 17) coal was gasified with air in Lurgi-gasifiers at 25 bar. The clean gas was burnt at about 10 bar in a pressurized boiler. The off-gas of this boiler expanded in a gasturbine.

Another process

- VEW-KUV (Vereinigte Elektrizitätswerke Westfalen AG - Kohle-Umwandlungs-Verfahren) -

is under development and a 10 tons of coal per hour plant is under construction (see Fig. 18).

In this case powdered coal is partially oxidized with air in an entrained flow reactor under pressure. The coal is pyrolized and partially gasified, so that a low calorific gas and powdered char are the products of this gasification process. The cleaned gas is burnt with

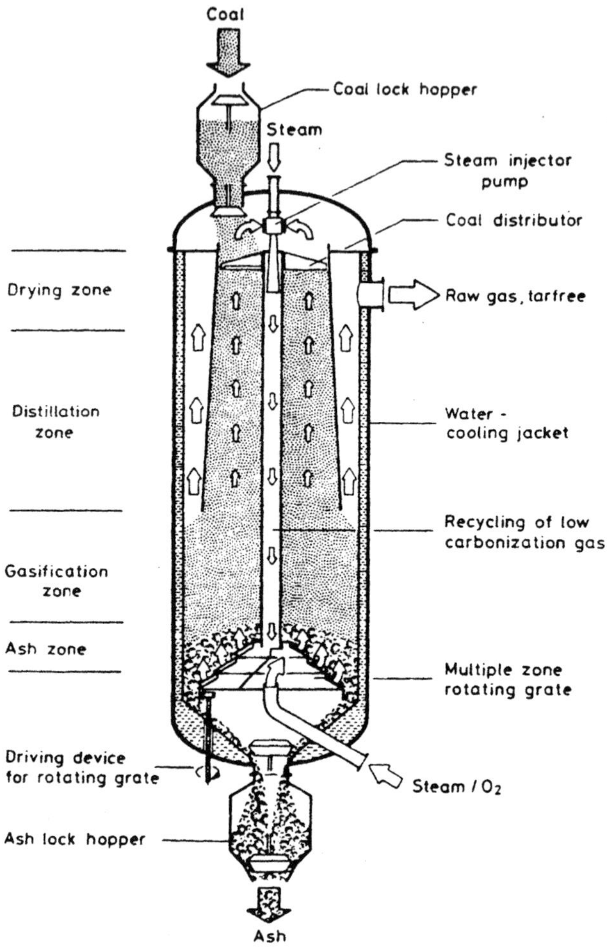

Figure 15. KGN fixed-bed gasification process with tar decomposition (9).

excess pressurized air and used for power generation in a gasturbine, and the oxygen rich hot off-gas is used to burn the remaining char in a conventional boiler to produce steam for the steam turbine. It is expected that this process will lead to an increase in efficiency from 37 to 42 % as compared with a conventional power station.

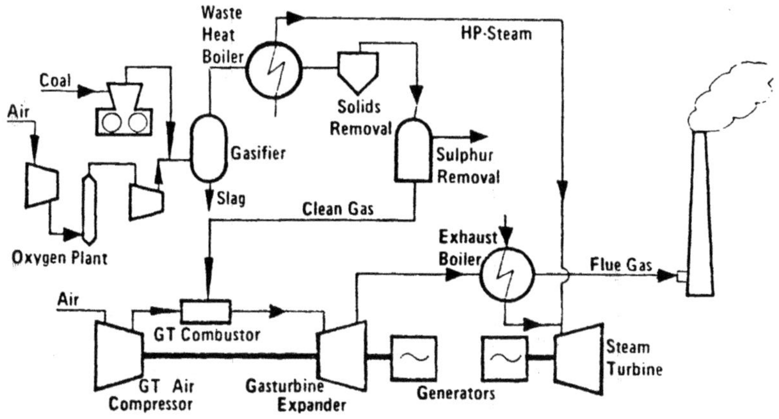

Figure 16. S/K oxygen coal gasification in combined cycle power generation (15,9).

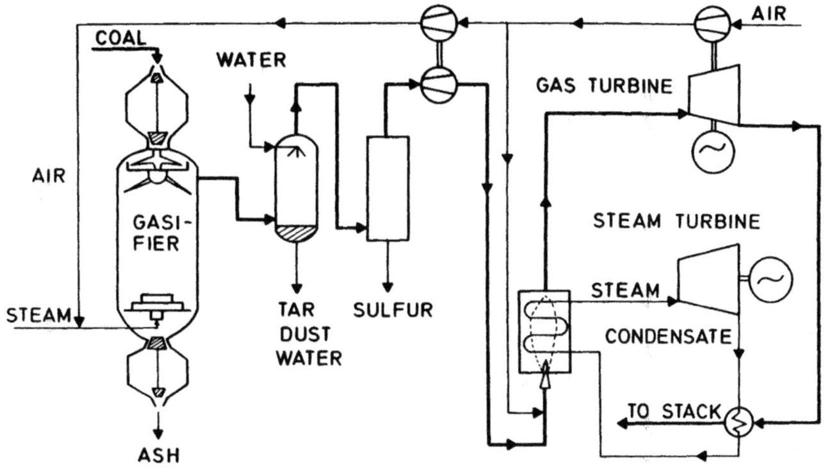

Figure 17. Flow scheme of STEAG-KDV-process.

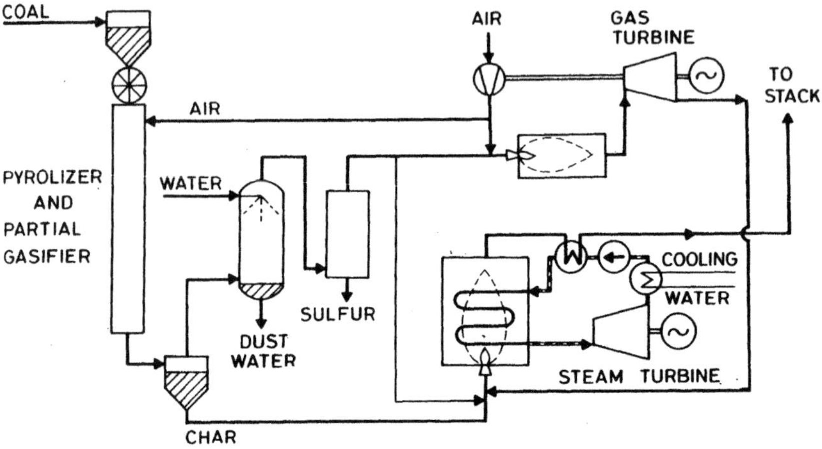

Figure 18. Flow scheme of VEW-KUV-process.

6. CATALYTIC GASIFICATION AND OTHER DEVELOPMENTS

It is wellknown since decades that coal gasification is catalysed by alkaline compounds, especially by potassium oxide. This increase of reaction rates is of high interest, if the reaction has to be carried out at low temperatures. Two applications are briefly described in the following:

- Bergbauforschung : Steam gasification with nuclear heat;
- Exxon : Catalytic steam gasification.

Bergbauforschung GmbH is running a 200 kg coal per hour fluidized bed gasifier since 1976 at temperatures of about 800 °C and pressures of about 40 bar. The heat for the endothermic coal-steam reaction is supplied by a heat exchanger within the fluidized bed. The intention is to use the heat from a high temperature gascooled nuclear reactor (see Fig. 19). The temperature of the helium, which is used as heat carrier, has a temperature of about 900 °C. Therefore, only a small part of the heat of the helium can be utilized in the gasification reaction. This fraction would increase by about a factor two if the reaction temperature could be lowered from 900 °C to 700 °C by catalysis.

Exxon is running a 1 ton coal per day pilot plant to develop a catalytic coal gasification process for the production of methane at low temperature (700 °C) and high pressure (35 bar). A simplified flow-sheet is shown in Fig. 20. The reaction

$$\text{coal} + 2\ H_2O \longrightarrow CH_4 + CO_2$$

is slightly exothermic; low reaction temperatures are necessary to avoid methane decomposition by steam.

A difficulty for the application of catalytic gasification processes is the fact, that the catalyst has to be recovered from the ash because of its high price.

Other significant developments, which are currently active, are coal gasification processes in molten iron baths.

Affiliated companies of Klöckner AG gasified coal with oxygen in

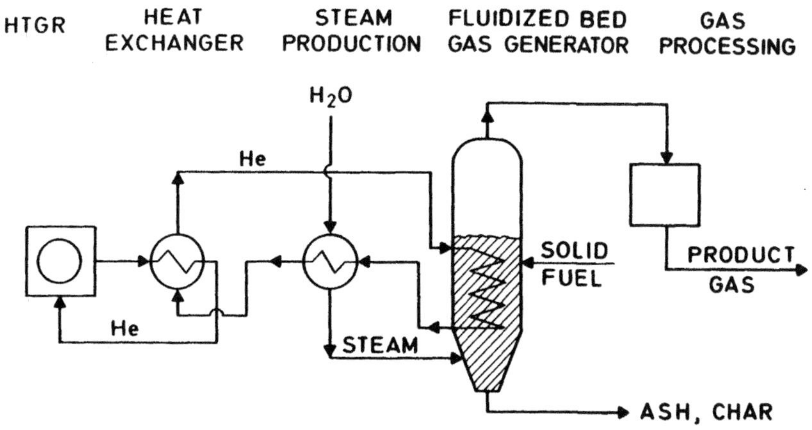

Figure 19. Gasification of coal with steam and nuclear heat developed by Bergbauforschung.

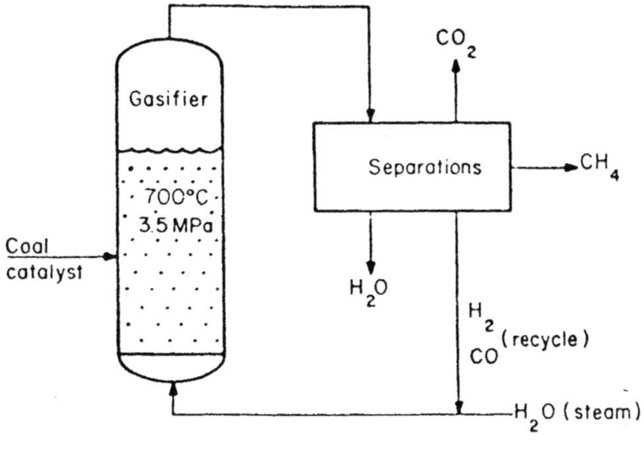

Figure 20. Exxon catalytic coal gasification process.

a 60 ton raw-iron converter at about 1500 °C and atmospheric pressure. The coal throughput was 20 t/h. The slag layer above the iron bath is desulphurizing the gas to about 20 ppm.

Two other companies KHD Humboldt Wedag AG, West Germany, and Sumitoma Metals, Japan, are developing a similar process for the production of low sulphur gas under pressure to be used for iron-ore reduction, as synthesis gas in the chemical industry, or in combined cycle power generation. A pilot plant with a capacity of 10 tons of coal per hour and a pressure of reaction of 3 bar is under construction in Sweden. It is scheduled to go on stream in early 1985.

7. COSTS OF GAS PRODUCTION

Coal gasification is relatively expensive because of its high investment costs, and because of the relatively high energy consumption of the conversion process itself.

This is demonstrated in Fig. 21 and Table IX for the production of hydrogen in plants of 200 tons of hydrogen per day. The curves in

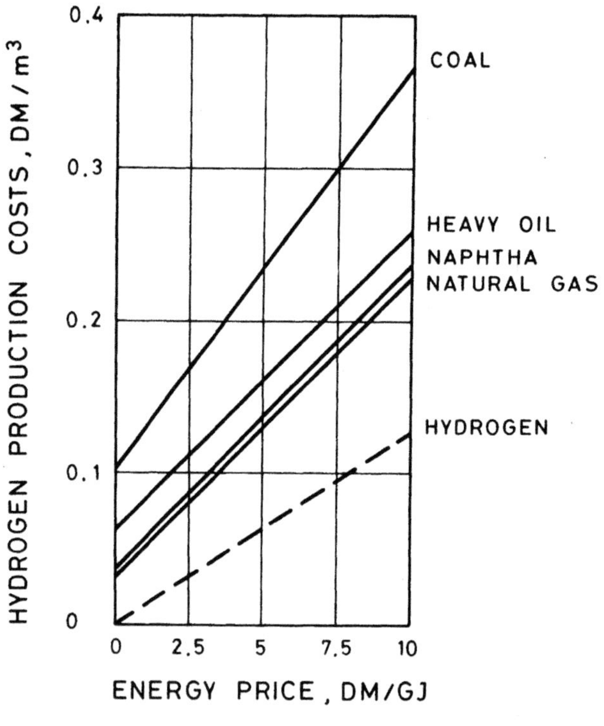

Figure 21. Production costs of hydrogen from different raw materials (capital related cost: 30 % of investment/a).

TABLE IX Investment costs and specific consumption of raw materials for a 200 t/d hydrogen plant

RAW MATERIAL	PROCESS	CAPITAL COSTS (1), 10^6 DM	FEEDSTOCK AND FUEL CONSUMPTION (2) GJ/t H_2
NATURAL GAS	CATALYTIC STEAM	70	224 (6200 m^3 GAS/t H_2)
NAPHTHA	REFORMING	79	230 (4.8 t NAPHTHA/t H_2)
HEAVY OIL	GASIFICATION BY PARTIAL	140	223 (3) (5.1 t OIL/t H_2)
COAL	OXIDATION	236	295 (10.1 t COAL/t H_2)

(1) Includes capital cost of power plant for on-site generation

(2) Includes fuel consumption of power plant

(3) Includes by-product steam credit of 24 GJ/t H_2

Fig. 21 are based on data originally published by Timmins, Hoskyn, and Doelman in a paper entitled "Natural gas versus other raw materials for bulk hydrogen production (13th World Gas Conference, London, 1976). The investment costs listed up in Table IX have in the meantime increased for the various processes by about 70 %.

On the abscissa of Fig. 21 the energy price is plotted in DM/GJ. This price can be related to any feedstock like coal, heavy fuel oil, naphtha, natural gas, or hydrogen itself. The price per m^3 (STP) is indicated on the ordinate in DM/m^3. The dashed line shows the price of hydrogen per volume against its own energy price. This line represents the minimum price of hydrogen which has to be considered in a refinery if hydrogen is produced in the own reformer plant and shall be used, e.g., for hydrocracking or desulphurization.

The next two curves above the dashed line show the costs of hydrogen produced from natural gas or naphtha by catalytic steam reforming. The upper curves represent the production costs for hydrogen from curde oil or heavy fuel oil and from coal or lignite, respectively, by non-catalytic partial oxidation processes.

The minimum value at zero energy price is related to the capital costs, which in this case were assumed to be 30 % of investment per annum. The slope of the curves indicate the specific fuel consumption (compare Table IX). As can be seen from Fig. 21 the energy price of coal has to be about 5 DM/GJ lower than that of natural gas to make coal gasification competitive to methane-steam reforming.

References

1 'Kohlevergasung' in *Ullmanns Encyklopädie der Technischen Chemie*, 3rd ed., vol. 10, pp. 360-450; Urban und Schwarzenberg, München, West-Germany 1958.

2 G.G. von Fredersdorf and M.A. Elliott,'Coal Gasification' in *Chemistry of Coal Utilization*, H.H. Lowry (ed.), pp. 892-1022, John Wiley and Sons, New York, 1963.

3 'Coal Gasification' in *Encyclopedia of Chemical Technology*, R.E. Kirk and D. Othmer (eds.), vol. 10, pp. 353-442, John Wiley and Sons, New York, 1966.

4 'Kohle, Gaserzeugung aus Kohle und Kohlenwasserstoffen' in *Ullmanns Encyklopädie der Technischen Chemie*, 4th ed., vol. 14, pp. 357-474, Verlag Chemie, Weinheim, W. Germany, 1977.

5 H.D. Schilling, B. Bonn and U. Krauß, *Kohlevergasung - Bestehende Verfahren und neue Entwicklungen*, 3rd ed., Verlag Glückauf, Essen, W. Germany, 1981.

6 D. Hebden and H.J.F. Strand, 'Coal Gasification Processes' in *Chemistry of Coal Utilization*, M.A. Elliott (ed.), pp. 1599-1752, John Wiley and Sons, New York, 1981.

7 H.Jüntgen and K.H. van Heek, *Kohlevergasung - Grundlagen und technische Anwendung*, Verlag Karl Thiemig, München, W. Germany, 1981.

8 K. Hedden, 'Coal Gasification' in *Alternative Energy Sources*, J.P. Hartnett (ed.), pp. 117-147, Academic Press, New York, 1976.

9 *Conference Proceedings: Synthetic Fuels - Status and Dircetions*, vol. 2, Electric Power Research Institute, Palo Alto, California and Jülich Nuclear Research Center, W. Germany, 1981.

10 J.A. Lacey and G.F.I. Roberts, 'The application and development of coal gas technology' in *Round Table Discussion 5, Coal Liquefaction and Gasification*, 11th World Petroleum Congress London, John Wiley and Sons, Chiechester, UK, 1983.

11 Round Table Discussion 'La Gazeification du Charbon' 100st Congress of the Association Technique de l'Industrie du Gaz en France, 1983.

12 G.F.I. Roberts, D. Hebden and C.T. Brooks, 'The application of slagging gasification to the production of SNG from coal' presented on the 14th World Gas Conference Toronto, IGU, Paris, 1979; IGU/B2-79.

13 C. Lohmann and J. Langhoff, 'The development project Ruhr 100 - an advanced Lurgi gasifier', presented on 15th World Gas Conference Lausanne, IGU, Paris, 1982; IGU/B4-82.

14 H. Teggers and K.A. Theis, 'The Rheinbraun High-Temperature Winkler and hydrogasification processes', presented on the 15th World Gas Conference Lausanne, IGU, Paris, 1982; IGU/B5-82.

15 H.K. Völkel, G. Eckstein, A. Linke and E.V. Vogt, 'Operation and economics of the Shell-Koppers coal gasifiaction processes', presented on the 14th World Gas Conference Toronto, IGU, Paris, 1979; IGU/B1-79.

16 J.W. Loeding and I.G. Stanfill, 'Application of the U-Gas process for the production of gas of intermediate calorific value', presented on the 14th World Gas Conference Toronto, IGU, Paris, 1979; IGU/B3-79.

17 K. Weinzierl and D. Deggim, 'VEW-Kohleumwandlungsverfahren: 10 t/h-Versuchsanlage', *Brennst.-Wärme-Kraft*, vol. 36, pp. 211-215, 1984.

18 C. Timmins, P. Hoskyn and J. Doelman, 'Natural gas versus other raw materials for bulk hydrogen production', presented on the 13th World Gas Conference London, IGU, London, 1976; IGU/F4-76.

19 K. Hedden and A. Wilhelm, 'Catalytic effects of inorganic substances on reactivity of solid fuels', *Ger. Chem. Eng.*, vol. 3, pp. 142-147, 1980.

20 Papers presented at the International Symposium 'Fundamentals of catalytic coal and carbon gasification', Amsterdam 1982, *Fuel*, vol. 62, pp. 140-264, 1983.

21 R.A. Cardello and F.C.R.M. Smits, 'Catalytic coal gasification development' in *Round Table Discussion 5, Coal Liquefaction and Gasification*, 11th World Petroleum Congress London, John Wiley and Sons, Chichester, UK, 1983.

DIRECT LIQUEFACTION OF COAL TO LIQUID FUELS

Dr. I. Romey
BERGBAU-FORSCHUNG GMBH
Franz-Fischer-Weg 61
4300 Essen 13
Germany

ABSTRACT. For countries that are largely dependent on imported oil investigation work on coal technologies has gained substantial interest again all over the world in the last decade.

For the production of motor fuels in future mineral oil can be substituted partly by coal liquefaction processes which are under development or still available in a commercialized scale.

According to the current state of technology the liquefaction of coal can be accomplished in three different ways

Synthesis
Hydroliquefaction
Solvent Extraction.

The paper discribes and compares the different processes under technical and economic viewpoints and gives an outlook of the further developments in coal liquefaction.

Due to the fact that synthesis differs fundamentally from the direct liquefaction it will be presented in a separate lecture in more detail.

SYNTHESIS. The synthesis of hydrocarbons has been developed in Germany. F. FISCHER and H. TROPSCH provided the foundation for the technological Fischer-Tropsch synthesis during the '20's; the process was subsequently used in commercial operations with coal as a starting material. Prior to 1945 nine installations with a total capacity for producing petrol/gasoline at 750 000 t/a had been constructed. However,the importance of hydrocarbon synthesis was not very great in comparison to that of direct liquefaction of coal.

For the same coal as starting material, the production of motor fuel by direct hydrogenation is more economical than synthesis. Nevertheless, in South Africa there exist vast reserves of a coal which is cheap but has a very high content of ash and is thus unsuitable for hydrogenation. With the use of this coal as raw material, South Africa is endeavouring to become less dependent on imported crude oil.

G. E. Beghi (ed.), Synthetic Fuels, 93–111.

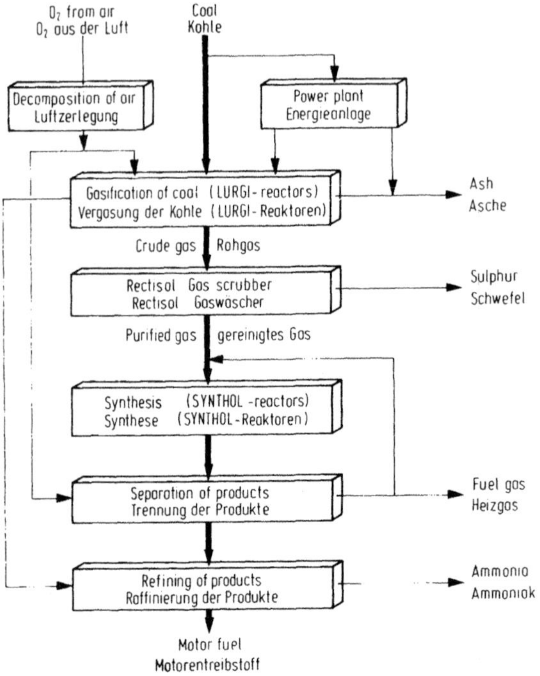

Figure 1: Fischer-Tropsch synthesis (Sasol II plant)

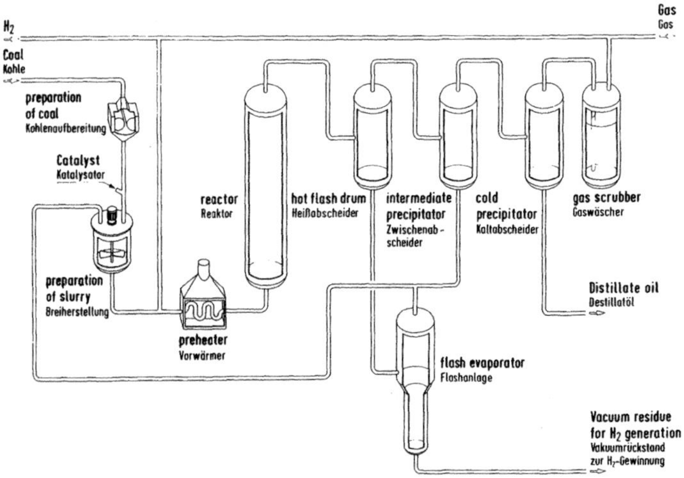

Figure 2: KOHLEÖL process

FISCHER-TROPSCH PROCESS (Figure 1). The method for obtaining liquid products from coal differs fundamentally from extraction and hydrogenation; it proceeds indirectly by way of gasification of the coal. In the first stage synthesis gas is generated from the coal. By means of the Fischer-Tropsch synthesis, carburetor motor fuels and Diesel oil can be obtained. The synthesis step occurs at the surface of a special iron catalyst, according to the general equation:

$$n(CO + 2H_2) \rightarrow n(-CH_2-) + n\ H_2O \qquad -\Delta H$$

Whereas the process of gasification is strongly endothermic, the synthesis reaction is highly exothermic and much heat must be removed.

Quite logically the production of methanol from synthesis gas can be carried out as form of indirect liquefaction of coal too. In addition to the possible immediate use as carburetor motor fuel, the newly developed Mobil M process has increased the interest in methanol. By means of the latter method petrol/gasoline is produced with a good yield at a zeolite catalyst.

The Fischer-Tropsch process is at present the only technique which is commercially applied for liquefying coal.

Despite the greater economic drawback of synthesis in comparison to hydrogenation, South Africa has erected a complex, Sasol I, for synthesis, in 1955; it currently still converts coal at a rate of 10 000 t/d to oil and raw materials for the chemical industry. The decision in favour of synthesis was made in view of the vast reserves of cheap coal with a high content of ash; such coal is hardly suitable for hydrogenation.

A further synthesis plant, Sasol II, is in operation since 1980. The plant converts at full capacity 40 000 st/d coal to petrol/gasoline and Diesel oil (1 500 000 t/a). Operation of a third plant, Sasol III, has started in 1983 so that South Africa obtains motor fuel from coal at a rate of about 3 500 000 t/a and can satisfy about 30 % of its motor fuel requirement from its own coal resources.

HYDROLIQUEFACTION

BERGIUS-PIER PROCESS. The development of direct hydrogenation originated with the work performed by F. BERGIUS from 1911 onward. In 1924 the patents of Bergius were transferred to. I. G. Farben. At the plant of present BASF in Ludwigshafen the decisive progress was achieved under the leadership of M. Pier. As a result of this work the hydrogenation of coal evolved into a large-scale industry within the span of 20 years. This eventually comprised 12 plants for producing motor fuel from coal with a total capacity of nearly 4 000 000 t/a. Independently of I. G., only the Britisch firm ICI succeeded in establishing a plant for hydrogenating coal on a technical scale in Billingham, with a petrol production rate of 100 000 t/a.

Subsequently, the results of the war, the high cost of coal production, as well as the availability of abundant and cheap crude oil, hindered a further development and operation of coal hydrogenation plants for almost 30 years.

In the United States too, a large pilot plant was operated on the basis of German technology in Louisiana, Mo. until 1952, but no further work on coal liquefaction was conducted until the beginning of the 60's.

Four methods of direct liquefaction of coal have been developed to such an extent during recent years that their large-scale technical applicability is now being demonstrated in large pilot plant operations:

- the KOHLEÖL process (German technology)
- the SRC II process,
- the EDS process,
- the H-COAL process.

Besides these there have been several other in part quite different developments. However, they have yet to be thoroughly tested in larger pilot installations before their technical applicability can be apprai-sed.

KOHLEÖL PROCESS (GERMAN TECHNOLOGY). In 1973, after the first major increase in the price of crude oil, the production of oil from coal became a subject of prime importance from the standpoint of energy policy in the Federal Republic of Germany. With both regional and federal government financial support the world-wide status of coal liquefaction was investigated in detail in the Ruhrgebiet and Saarland. Moreover, the design and construction of pilot plants was begun in both districts.

From the inventory of the various processes of hydrogenation it became obvious that the old I. G. liquid phase technique presented a feasible alternative to the American developments. Of course, several decisive modifications of the method had to be undertaken.

Since 1976 this modified liquid phase hydrogenation process is being tested and confirmed at the pilot facilities of the Bergbau-Forschung GmbH in Essen and of the Saarbergwerke AG ind Reden. In 1979 the Rheinische Braunkohlenwerke AG also began operation of a laboratory installation by means of the same process.

On the basis of results obtained from the new pilot facilities and from the former large installations, two large-scale pilot plants have been designed during the interim: Saarberg began operation of a facility for a coal input of 6 t/d near Völklingen in 1981, Ruhr-kohle AG and VEBA Oel AG started operation of a plant for 200 t/d in Bottrop end of 1981.

Developmental work is currently being conducted on preliminary projects for large-scale technical facilities at several companies in the Federal Republic of Germany. It appears possible that one commercial coal hydrogenation plant will already be operating end of the '80s, or begin of the '90s though the construction will be financed preponderantly by the government.

DESCRIPTION OF THE PROCESS (Figure 2). The coal slurry is prepared from recyling oil, ground coal and about 2 % of catalyst (red mud from the aluminium industry). After the addition of hydrogen and

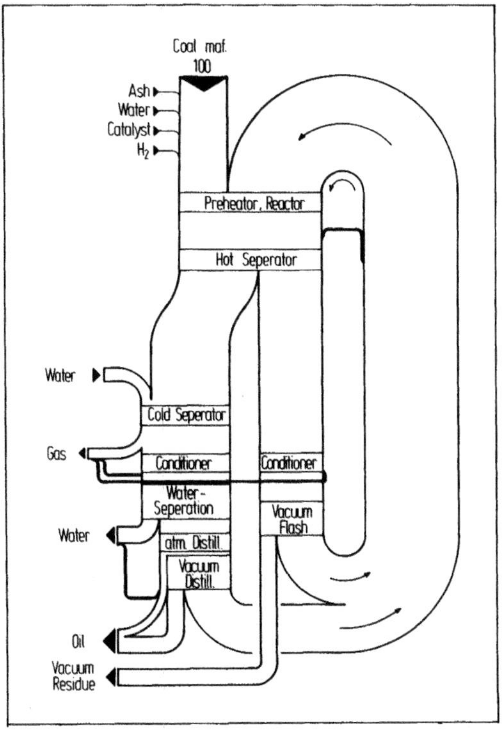

Figure 3: Mass-Flow for Slurry-Phase Hydrogenation
(New German Technology)

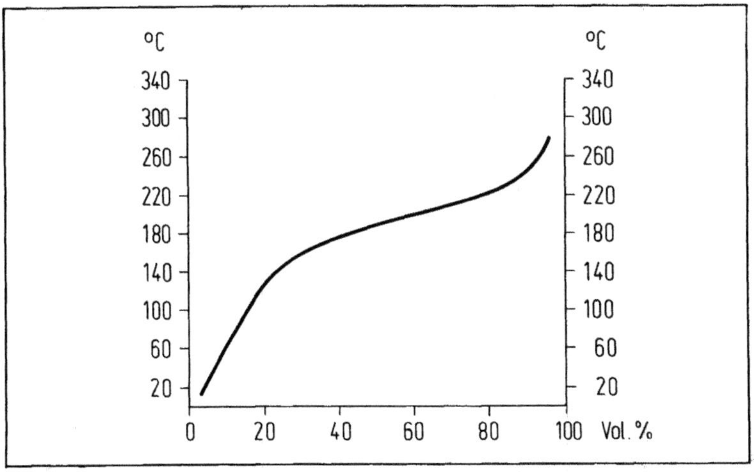

Figure 4: Boiling Diagram of a Typical Product Oil

circulating gas, the slurry is heated and subsequently passes through reaction tubes at a temperature of 460 to 480°C and at a pressure up to 300 bar. The products of the reaction, still under high pressure, are then condensed in precipitators at various temperatures.

The solids, together with the high-boiling and nondistillable oil, are removed from the hot flash drum. By means of flash evaporation in vacuum, oil is distilled as long as the residue remains pumpable in the molten state. This pitch-like residue, with about 50 % solid matter, is gasified by partial oxydation under a pressure of 80 bar. After conversion the gas satisfies the hydrogen requirement of the hydrogenation facility.

The distillate from flash evaporation is combined with the oil from the intermediate precipitator and recycled for use as priming oil for preparing the slurry. In the cold precipitator medium and light oil is collected. This oil is the product oil from the process. It can be further refined to yield fuel oil, motor fuels and raw materials for the chemical industry.

The KOHLEÖL process differs from the basic I. G. liquid-phase method especially in the following respects:

- The formerly necessary elimination of the solid matter by centrifugation and low-temperature distillation of the residue is replaced by a distillation technique.

- This measure provides a distillate free of asphalt as a recycling oil for preparing the slurry in the new process. The asphalt substances formerly imposed a considerable additional load on the reactors. Consequently the operating pressure can be decreased from 700 to 300 bar; at the same time the specific throughput of coal can be augmented.

- The gasification of the liquid hydrogenation residue under pressure permits operation which is self-sufficient and economically favourable with respect to the generation of hydrogen. Coal was formerly gasified in Winkler generators without the application of pressure.

- The only ultimate residue from the new process is the easily disposable, inorganic slag from the gasification. The residue from the older liquid-phase process was a useless distillation waste coke with a high content of both ash and sulphur; its safe disposal was a great environmental problem.

RESULTS Fig. 3 shows a mass-balance of a hydroliquefaction run (new German technology) based on German bituminous hard coal. Typical reaction conditions are a pressure of 300 bar and a temperature of 475°C. Under these conditions it is possible to receive about 55 % distillate oil (boiling range < 350°C) as well as 21 % (C_1-C_4)-hydrocarbon gases. Fig. 4 - 6 show data with the chemical composition of the products received in the process.

The raw oil from hydroliquefaction can be used in the so-called gas-phase for further upgrading to markable products like motor fuels,

		saturates	olefines
methane	p. ct. w/w maf. coal	6.8	
ethane	"	5.7	
ethene	"		0.1
propane	"	5.6	
propene	"		0.3
n-butane	"	2.8	
i-butane	"	0.7	
butenes	"		0.2
total		21.6	0.6
		22.2	

Figure 5: Hydrocarbon Gases

softening point	°C	180
volatile matter	p. ct., w/w	30
ash content	"	26
pyridine insolubles	"	52
asphaltoles	"	9
asphaltenes	"	21
heavy oil	"	18

Figure 6: Flash Distillation Residue Analysis

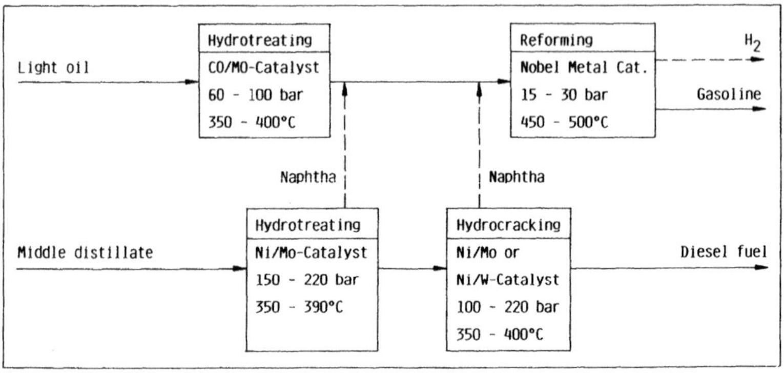

Figure 7: Upgrading of coal oils to gasoline and diesel fuel (source: Ruhrkohle Oel und Gas GmbH, VEBA Oel AG)

			coal oil	Arabian light
Density (15°C)		g cm^{-3}	0.95	0.85
H/C Ratio			1.26	1.77
S		wt %	0.10	1.70
Light oil IBP	- 185°C	wt %	20	20
Middle distillate	185 - 325°C	wt %	80	27
VGO	325 - 500°C	wt %	-	28
Residue	> 500°C	wt %	-	25

Figure 8: Properties of syncrude from coal and Arabian light (source: Ruhrkohle Oel und Gas GmbH, VEBA Oel AG)

		gasoline from coal oil	gasoline from mineral oil DIN 51 600
Density (15°C)	g cm^{-3}	0.84	0.73 - 0.78
RON (unleaded)		101.2	97.4
MON (unleaded)		88.7	87.2

		Diesel fuel from coal oil	Diesel fuel from mineral oil DIN 51 601
Density (15°C)	g cm^{-3}	0.828	max. 0.885
Viscosity (20°C)	mm^2s^{-1}	2.4	2 - 8
Flashpoint	°C	73	> 55
Sulfur	wt %	1 ppm	max. 0.3
Cetane No		47	> 45

Figure 9: Properties of motor fuels from coal oil and mineral oil (source: Ruhrkohle Oel und Gas GmbH, VEBA Oel AG)

	Ruhr Zeche Westfalen (Bituminous coals)	Ruhr Zeche Wulfen (Bituminous coals)	Ruhr Zeche Prosper (Bituminous coals)	Ruhr Zeche Westerholt (Bituminous coals)	Australia New South Wales	U.S.A. West Virginia Pittsburgh Seam	U.S.A. Kentucky No. 9	Kolumbia	Australia Queensland lignite	Rheinland lignite
Volutiles										
% maf	23,4	35,6	37,4	38,2	39,9	40,2	44,0	40,6	49,6	55
Macerals										
Vitrinit, Vol.-%		66	73	64	62	64	82	66	77	
Exinit, Vol-%		15	14	13	7	20	4	8	9	
Inertinit, Vol.-%		18	12	19	28	15	10	24	9	
Analysis										
C, % maf	89,8	85,6	85,0	85,7	85,8	84,3	82,0	84,5	79,7	68,6
H, % maf	5,0	5,4	5,4	5,5	5,4	5,5	5,5	5,8	6,1	5,0
O, % maf	2,3	6,5	7,0	5,9	6,1	7,5	7,6	7,4	12,6	24,5
S, % mf	1,04	0,82	0,97	1,14	0,60	0,90	2,65	0,64	0,39	0,31
N, % mf	1,72	1,66	1,52	1,45	1,8	1,40	1,32	1,66	0,79	1,56
Ash										
% mf	1,8	4,6	4,0	5,1	10,2	8,4	8,1	4,0	9,7	3,2

Table 1: Coals for hydroliquefaction tested at Bergbau-Forschung

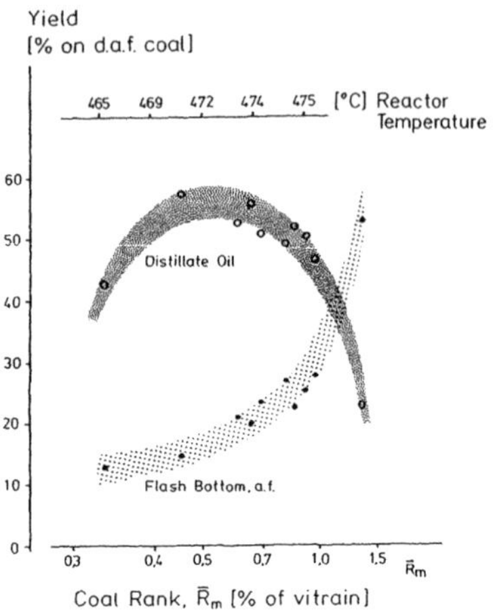

Figure 10: Effect of coal rank on hydrogenation results
(Basis: 4% minerals in dry coal, Pressure, 300 bar)

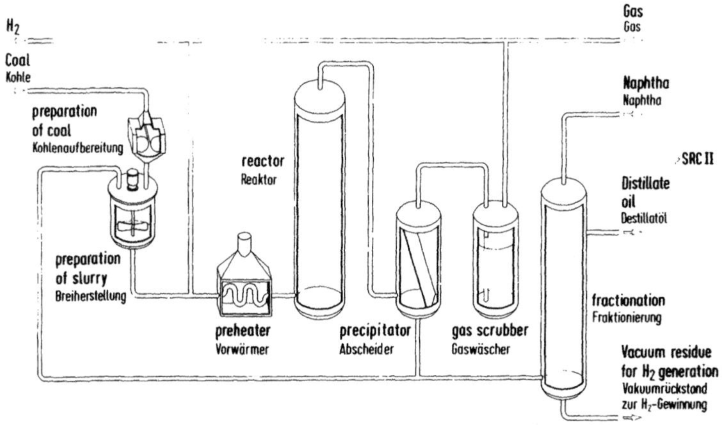

Figure 11: SRC II process

heating oil and chemicals. Investigation work carried out at Ruhrkohle Oel und Gas and Veba Oel have shown that it is possible indeed to upgrade coal oils to petrol/gasoline, diesel fuel and heating oil with the required specifications. Fig. 7 shows in a block diagramm the conditions for upgrading coal oils to gasoline and diesel fuel, in Fig. 8 and 9 properties of coal oil and mineral oil as well as results of upgrading to motor fuels are shown.

To increase the product oil yield in the Kohleöl process as well as to demonstrate the technology for a wide range of coals world-wide further paramters have been investigated at Bergbau-Forschung in Essen. As shown in Tab. 1 hard coals as bituminous, subbituminous as well as lignite can be used as starting material for hydroliquefaction and also the product oil yield is in most cases more than 50 % for the tested coals (Fig. 10).

SRC II PROCESS. The SRC II process (Figure 11) was developed from the SRC I method by Gulf, with participation of Ruhrkohle AG and Steag.

The experimental verification of the data was first conducted in laboratory installations of the Gulf subsidiary Pamco in Merriam, Kansas and at Gulf in Harmarville, Pa. Futhermore, the 50 st/d SRC I facility of Pamco near Tacoma, Wash. was converted for operation according to the SRC II technique.

A 6000 t/d demonstration plant in Morgentown W. Virginia under participation of West Germany and Japan was scheduled to improve the technology in a semitechnical scale. Due to the fact that there was not sufficient funding of the project the further work has been chancelled in 1981.

EDS - PROCESS. The Exxon donor solvent process includes features of both extraction and direct hydrogenation.

The process is under development by Exxon since about 1970. With the help of two laboratory facilities, fundamental data and specifications were obtained for the construction of a large pilot plant. This installation, which is designed for a coal input of 250 st/d, operated in Baytown, Texas between 1979 - 1983.

PROCESS DESCRIPTION (Figure 12). The circulating oil is hydrogenated on a fixed-bed-catalyst in a separate reactor. The use of the donor solvent allows the application of comparatively mild conditions in the reactor for coal hydrogenation; consequently, the operation proceeds at about 450°C and 100 to 140 bar - so far without any additional catalyst.

The products of the reaction are separated by distillation. The residue, which contains the solids, is then transferred in the liquid state to the flexicoker developed by Exxon. In this fluidized-bed coker, fuel gas, fuel oil and low temperature coke are produced; the latter is in turn used for generating hydrogen. Recently, the recycling of distillation residue to the reactor for coal hydrogenation is being carried out; the pressure has thereby been increased to 170 bar. At the expense of the specific throughput of coal, the previously moderate yield of straight-run oil is thereby enhanced. There is a corresponding drop in the amount of low temperature carbonization products of little value.

H-COAL PROCESS. The H-Coal process of Hydrocarbon Research, Inc. (HRI), has originated from the H-Oil process, by means of which pretoleum residues are desulphurized in commercial installations. The H-Coal process has been tested experimentally by HRI in Trenton, NJ in facilities for 20 kg/d and 3 st/d. In Catlettsburg, Ky a large-scale pilot plant with a coal input of 200 st/d (syncrude) or 600 st/d (fuel oil) was in operation between 1979 - 1982. The department of Energy, the State of Kentucky, as well as Ashland, Conoco, EPRI, Mobil and Standard Oil (Indiana), are participating in the project. As a result of German federal financial support, the Ruhrkohle AG also has access to the experience gained from the operation.

PROCESS DESCRIPTION (Figure 13). A special feature of the H-Coal process is the ebullated bed reactor, in which a bed of catalyst in agitated and maintained in a state of turbulence by the motion of the circulated reactants. Thus clogging is prevented and a good material and heat exchange is provided. Hence the catalyst remains in the reactor chamber, while the products of the reaction are drawn from the top of the reactor and subsequently refined. Cobalt molybdate on an alumina carrier, in extruded form, thereby serves as catalyst. For the purpose of maintaining the activity, spent catalyst can be continually removed and replaced by fresh catalyst.

Typical conditions include values of a pressure around 200 bar and a temperature around 450°C. At a low specific throughput of coal the main product consists of distillates (syncrude process); at a high throughput an oil containing residues is the main product (fuel oil process). In the fuel oil process the separation of the solids is accomplished not by distillation, but rather by precipitation of the asphalt substances (antisolvent deashing). The residue from distillation or precipitation is gasified for the purpose of generating hydrogen.

SOLVENT EXTRACTION

POTT-BROCHE-PROCESS. In the Pott-Broche process, the typical method of extraction, the hydrogen is provided exclusively by the donor solvent.

A large - scale pilot plant operated by RUHRÖL in Bottrop-Welheim between 1938 and 1944 produced extract at a rate of 80-90 t/d. A mixture of tetralin and cresol as well as middle oil from a pitch hydrogenation plant was used as solvent. Referred to coal free of water and ash the yield of extract was about 75%. The material had a solftening range around 220°C and an ash content of 0,15 - 0,2 %. Originally further processing of the extract by means of liquid-phase hydrogenation to gasoline had been intended but was not successful. Therefore the extract was utilized for preparation of high grade electrode coke.

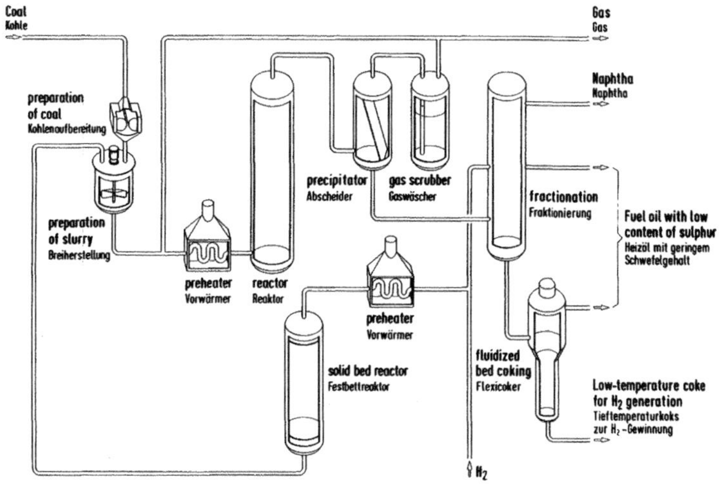

Figure 12: EDS process

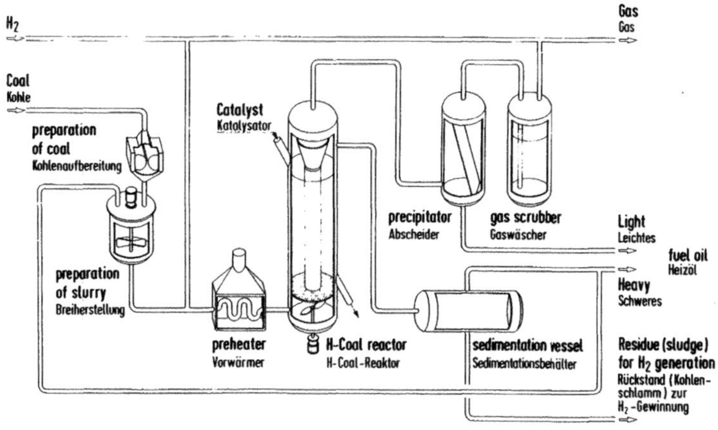

Figure 13: H-Coal process

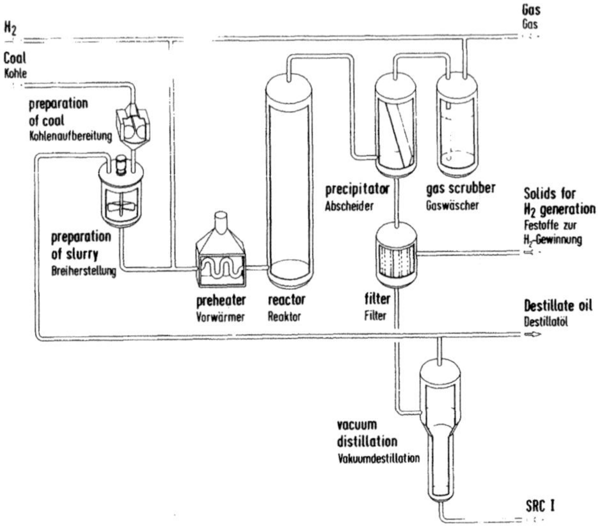

Figure 14: SRC I process

PROCESS:	EDS (Exxon)	H-COAL (HRI)	SRC II (Gulf)	KOHLEÖL (RAG/BF)
Feed Coal	US Subbitum.	US Subbitum.	US Subbitum.	German Bitum.
Volatiles, maf basis	45 %	46,5 %	43 %	38 %
Operating Pressure, bar	150	190	150	300
Products, % on maf coal		(Syncrude Mode)		
Hydrocarbons C_1-C_3	7,3	10,7	16,3	18,3
Distillate Oil	38,8	49,7	42,8	54,2
Residue, af	41,8	29,5	34,4	23,9
Water & Inorganic Gases	16,4	15,0	11,3	9,7
Total (100+H_2 reacted)	104,3	104,9	104,8	106,1
FBP of Distillate, ca. °C	530	520	450	350

Figure 15: Comparison of coal hydrogenation processes

In contrast, the I.G. Uhde process, which evolved at about the same time, utilized hydrogen under high pressure and under hydrogenation conditions in order to produce the extract.

The technology is similar to the recent developments, the SRC I processes in the United States and Japan and the two stage liquefaction in the United Kingdom.

SRC I PROCESS. The development of the process started in 1960 by the Spencer Chemical Corp., which was later bought by the Gulf Oil Corp. Since 1974 the Pittsburgh and Midway Coal Mining Co., likewise a subsidiary of Gulf during the interim has operated a pilot plant for a coal input of 50 st/d in Tacoma, Wash. (The installation has being used later for the development of the SRC II process).

Since 1976 Catalytic Inc. a subsidiary of Air Products, together with other companies in Wilsonville, Ala., is operating a 6 st/d pilot plant for producing extract by the SRC I methed. In 1980 the Department of Energy (DOE), Air Products and Wheelaborator Frye signed a contract for constructing an SRC I plant for a coal input of 6000 st/d. It was planned to build the installation near Newman, Ky in 1984/85, but newer information say that the project has been chancelled in the meantime.

SRC I, the main product of the process, is to be employed as a solid or molten power plant fuel free of ash and with a low content of sulphur. Corresponding combustion tests on a larger scale have been conducted. As with all extracts from coal, the conversion to special coke for metallurgical purposes is possible too.

Recently a further treatment by means of hydrocracking is under development. The intended product is a distillate at $T < 450°C$, to provide feedstock for making fuel oil and motor fuel.

In the SRC I process (Figure 14), which operates in the usual way with recycling oil for preparing the coal slurry, the hydrogen is applied at 100 to 150 bar and about 450°C. The entire liquid product is separated from the solid matter by filtration, or by extractive technique. The circulating oil is subsequently separated by vacuum distillation. The residue of distallation - SRC I - is granulated. Its softening point lies around 200°C. Provided certain types of coal with a high ash content are used as starting material, the operation can be conducted without any additional catalyst.

COMPARISON OF THE PROCESSES

The yield efficiencies and product distributions of the hydrogenation processes described are summarized in Figure 15. As a consequence of the more severe conditions, the generation of gas and the consumption of hydrogen are greater for the KOHLEÖL process than for the other processes. Both the yield and quality (boiling range!) of the product oil, however, are substantaily better with the KOHLEÖL process.

The coal which occurse in Germany is more difficult to hydrogenate than that in the United States; this is already evident from the comparative content of volatile matter, However, the German coal

Process	Operating Companies	Location State	Row coal input t/d	Commencement of operation	Process discription	Products
EDS	Exxon, U.S. Department of Energy, EPRI, Phillips, Arco, Ruhrkohle AG, Japcr. Coal Ltd.	Baytown Texas USA	220	1980	Coal hydrogenation with donor solvent Solvent regeneration with solid bed catalyst	Distillate oil Refining to - fuel oil - motor fuel
H-COAL	HRI, EPRI, Ashland, Conoco, Amoco, Mobil, State of Kentucky, U.S. Department of Energy	Catlettsburg Kentucky USA	550 (fuel oil) 180 (syncrude)	1979/80	Coal hydrogenation with Co Mo catalyst in ebullated bed reactor	Heavy fuel oil (fuel oil) or distillates (syncrude) Refining to - fuel oil - motor fuel
KOHLEÖL	Ruhrkohle AG, VEBA ÖL AG, State of Nordrhein - Westfalen	Bottrop Ruhrgebiet W. Germany	200	1981	Coal hydrogenation with suspended iron catalyst	Distillate oil Refining to - chemicals - fuel oil - motor fuel
SRC I	International Coal Refining Comp., U.S. Department of Energy, State of Kentucky	Newman Kentucky USA	5500	1984/85	Coal extraction Hydrogenation of cleaned extracts	- solid extract - fuel for power plants - carbon materials Distillate oil refining to - fuel oil - motor fuel
FISCHER-TROPSCH SYNTHESIS	SASOL	Sasolburg South Africa	10000 (Sasol I) 36000 (Sasol II) 36000 (Sasol III)	1955 1980 1983	Coal gasification (Lurgi) and Synthesis (Fischer-Tropsch), synthol fluidized dust reactors for Sasol II/III	- products for chemical industry - petrol / gasoline - tar products

Figure 16: Coal liquefaction plants in operation under construction or in preparation for construction for coal input more than 100 t/d (1980 - 1985)

Country \ Process	Hydrogenation Solvent extraction	Fischer-Tropsch Synthesis	Methanol Synthesis Mobil-Process	Pyrolysis
Australia	■	■		
Canada	■	■	■	■
Germany F.R.	■		■	
Japan	■			
S.Afrika		■		
U.K.	■			
USA	■	■	■	

Figure 17: Preferred coal liquefaction processes under development

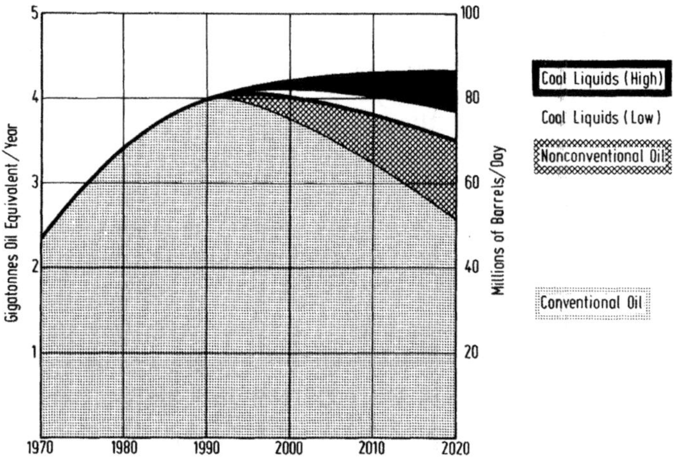

Figure 18: Prospective production of liquid hydrocarbons from coal compared with conventional and unconventional oil production (Study World Energy Conference)

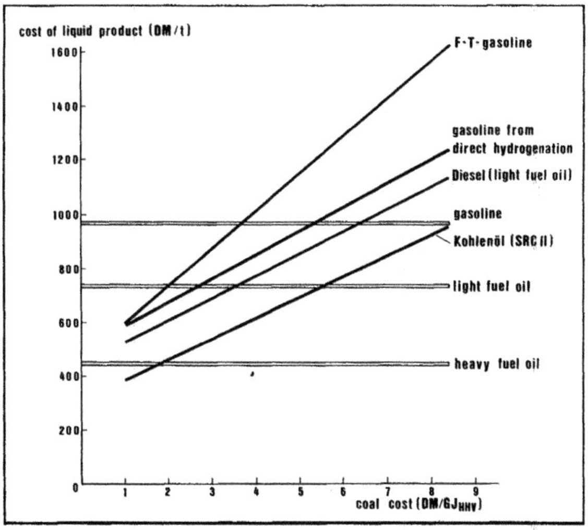

Figure 19: Production cost of liquid fuels from coal

with its higher degree of coalification, also generates considerably less water and inorganic gases. The higher presure prevailing during the KOHLEÖL process also allows the liquefaction of comparatively inert types of coal with a good yield. With the use of a reactive Australian or American coal the pressure could be decreased considerably. It is quite probable that the operating pressure of the American processes will be augmented in the future, in order to achieve a greater yield and throughput.

Figure 16 gives an overview about coal liquefaction plants under construction or in preparation for construction for a coal input more than 100 t/d and for the time scale 1980 - 1985.

CONSTRAINS ON COAL LIQUEFACTION

A study of the World Energy conference (Coal Liquefaction Task Force) showed that the timing of the start of commercial scale coal liquefaction depends on the following factors

- demand for synthetic fuels
- availability of suitable technology
- availability of coal
- resources for establishing a liquefaction industry
- environmental acceptibility of coal liqeufaction technology
- economics of coal liquefaction

In some countries, strategic considerations of security and the use of own resources will be important, but in general coal liquefaction will only become widerpread as a response to limitations on the supply of crude oil and the resultant rise in prices.

Countries with large coal production may also manufacture synthetic liquids for export, particularly where coal is of low calorific value so that transport costs of the coal itself are high.

The greatest common interest for synthetic oil is in the transport fuel market. Only Japan foresees also the production of coal liquids for firing boilers. (Figure 17)

The perceived time of commercial start-up varies from country to country. Only South Africa still has liquefaction plants in a commercial scale available today. The earliest time of introduction else where appears to be in the late 1980's in Australia, Canada the United States and Germany.

Based on informations from national reports the relative contribution that coal liquefaction could conceivably make to the world liquid hydrocarbon supply is shown in figure 18.

ECONOMIC CONSTRAINS

The general result of national studies is that coal liquefaction is not economic at present oil prices, except where coal is very cheap. It is also unlikely that an improvement is the economics of liquefaction will come about as a result of technological advance.

At the present price level for Western European hard coal and also for imported hard coal production costs of liquid hydrocarbons versus coal feedstock are shown in Figure 19. The liquid products have to be compared with present market prices. Based on the market prices in W. Germany gasoline made from coal costs 0,7 - 0,8 DM/l more than gasoline from mineral oil. The raw material costs for gasoline form coal are not longer essentially above those for gasoline from mineral oil, but operational or conversion costs are significant higher for gasoline from coal.

CONCLUSIONS

From the technical viewpoint there are no insurmountable barriers which prevent coal liquefaction making a major contribution to future energy supplies.

However, a boom in the development of coal conversion technology or a substitution of the resulting products for crude oil cannot be expected under present conditions; there are to many impediments:

- The production of coal, which declined severely under economic pressure during the past decade, can be augmented only slowly.
- Not only the prices of oil and gas have risen sharply since 1973; coal too has become much more expensive.
- The technology of coal liquefaction has to be improved in a commercial scale.

It would appear, that the earliest date coal liquefaction is likely to become commercial outside South Africa is the 1990's but that by 2000 - 2020 coal liquefaction could be a well established technology, making an important contribution to liquid fuel supplies.

Fischer-Tropsch synthesis for fuel production from coal

Dr. D. Frohning
Ruhrchemie Aktiengesellschaft
P.O. Box 13 01 60
4200 Oberhausen 11
Germany

ABSTRACT. Syngas can be converted directly into hydrocarbons under appropriate reaction conditions in the presence of iron containing catalysts. The reaction proceeds via a series of consecutive steps, not all of which are completely elucidated, and leads to a broad product distribution. The selectivity of the reaction can be influenced to a limited degree by adjusting the process conditions and the catalysts.

The overall reaction is highly exothermic and requires careful control. Thus only two processes have proven technical reliability so far: the gas-phase fixed-bed (ARGE) and the entrained-bed (SYNTHOL) process, both of which are operated by Sasol in South Africa. While the ARGE-process has been designed to produce mainly long-chain hydrocarbons, including diesel oil, the SYNTHOL-process yields predominantly gasoline. At present there are one ARGE and three SYNTHOL plants in operation with a total production capacity of ca. 4,3 mio t/a hydrocarbons from ca. 25 mio t/a of low quality coal. The mixture of primary products needs separation and upgrading which is achieved in refinery-like down-stream units.

The potential of the Fischer-Tropsch synthesis is directly related to the costs of synthesis gas. The product range can be adapted to almost any demand and there are no principal restrictions in plant size.

1. Introduction

Production of synthetic motor fuels is only worth considering if it is a valid alternative to the processing of mineral oil. Commercial realisation of the synthesis is thus influenced by technical, economic and political factors, which are closely connected with the price and availability of mineral oil. The two recent crises in the price and supply of mineral oil have not only given weight to the political aspects of motor fuel production from coal but also encouraged the resumption of development work which had previously been suspended. Since supply and demand of mineral oil are relatively balanced at present (Fig. 1) and the technical potential of the synthesis processes has been brought up to the current level of technology, the economic aspects are the decisive factor in a choice between the two alternatives. As is well known, this is also the greatest argument against the construction of coal conversion plants – quite apart from any other influencing factors which are still to be discussed – as from the present-day point of view they cannot be operated economically or if so, then only to a limited extent. In any case political considerations have so far not been great enough to overcome economic reservations.

G. E. Beghi (ed.), Synthetic Fuels, 113–134.

Fig. 1

World Oil Supply 1980 – 2000 (Mio bbl/d)

	1982	1985	1990	2000
Demand				
– OECD	32,4	35–36	34–37	33–43
– OPEC	2,9	4	5– 6	8– 9
– Other developing countries	7,2	9–10	11–13	17–22
Total	42,5	48–50	50–56	58–74
Supply				
– OPEC	19,5	23–26	27–29	24–28
– Non–OPEC	22,0	24–25	23–25	25–27
– Eastern Bloc countries				
– Net exports (imports)	1,0	1–(1)	0–(2)	0–(2)
Surplus demand	0	0	0– 4	9– 21

Sources: IEA (1982); OECD (1982)

An undoubtedly positive effect of the mineral oil supply crises was the resulting incentive to revise existing techniques and develop new ones. Today, there are three main techniques for the synthesis of motor fuels, which must either be seen as interrelated or in direct competition with each other: coal liquefaction, the Fischer-Tropsch synthesis and the Mobil process. From the chemistry point of view the direct hydrogenation of coal is the hydrogenative degradation of carbon polymers, which are non-uniform in their structure. It depends, therefore, to a certain extent on the quality of the raw material, whereas the Fischer-Tropsch synthesis and the Mobil process are based on the uniform, defined raw material, synthesis gas. The link in the chain, coal gasification for the production of hydrogen or synthesis gas, has been so vastly improved that its availability can now be taken for granted. A comparison of the various methods for the production of motor fuels can therefore be limited to a comparison of the actual syntheses themselves.

The catalytic conversion of synthesis gas to hydrocarbons has been carried out on a technical scale since 1936. However, the exact process employed has depended on the aims and the location. In the following chemical, technical and some economic aspects of the Fischer-Tropsch synthesis are dealt with.

2. Chemistry of the FT-synthesis

There are two border cases as far as the conversion of carbon monoxide with hydrogen is concerned where the final products, methane and methanol, are formed with near 100 % selectivity. All other kinds of synthesis of hydrocarbons or oxygen-containing hydrocarbons by carbon monoxide hydrogenation have a selectivity which is in most cases considerably lower, i.e. a mixture of homologous compounds is formed over a more or less broad spectrum. The random limiting of this spectrum to a small band, comparable with a distillation cut from a refinery, is the aim of selectivity control of the hydrocarbon synthesis. Should success come, the raw material yield and product refining would be considerably improved and simplified, thus decisively improving economic viability. However, such extensive selectivity control has not been attained so far which can be theoretically justified by the lack of knowledge of the mechanism itself.

The most elegant method of selectivity control, a directed intervention in the mechanism, has not succeeded as the so-called real mechanism of the FT synthesis has yet to be discovered. To be more precise: none of the theories put forward so far can give a complete and unequivocal explanation of the course of the reaction on the catalyst surface. This statement should not detract from the fact that knowledge of the interactions between catalyst and reactants has been considerably deepened over the past few years, above all through the use of sophisticated spectroscopic methods.

All theories on the mechanism have three consecutive steps in common: the formation of a primary complex, chain growth through linking reactions and chain termination with subsequent desorption of the product. The three proposals mainly under discussion are shown in fig. 2.

The real difficulty in giving a conclusive interpretation to the mechanism of the FT-synthesis is the growth stage which involves the formation of at least one carbon-carbon-bond. As the momentary formation of a molecule with a large number of carbon atoms is sufficiently improbable and unprecedented in heterogeneous catalysis, it was logically assumed that there is a sequence of individual steps, the first member of which is the formation of an activated carbon-containing species. The formation and structure of this species and its possibilities for further reaction were always the basis of any theories.

For the primary complex three structures were discussed as an intermediate product capable of growth: an "activated" carbon atom (F. Fischer), an oxygen-containing, enolic intermediate product (R.B. Anderson) and a surface carbonyl (H. Pichler). Fischer's

Fig. 2

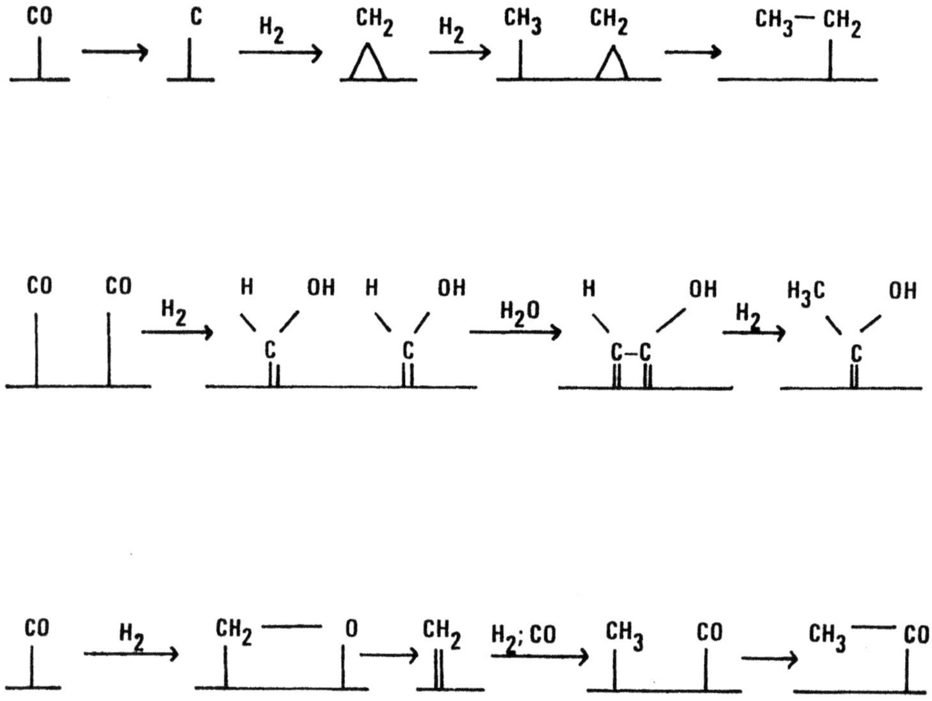

Schematic representation of three different chain growth mechanisms

Fig. 3

$$CO_{(g)} \rightleftharpoons CO_{(ads)} \longrightarrow C_{(ads)} + O_{(ads)} \quad (1)$$

$$H_{2\,(g)} \rightleftharpoons H_{2\,(ads)} \longrightarrow 2\,H_{(ads)} \quad (2)$$

$$O_{(ads)} + CO_{(ads)} \longrightarrow CO_{2\,(ads)} \quad (3)$$

$$O_{(ads)} + H_{(ads)} \longrightarrow OH_{(ads)} + H_{(ads)} \longrightarrow H_2O_{(ads)} \quad (4)$$

$$C_{(ads)} + n\,H_{(ads)} \longrightarrow CH_{n\,(ads)} \quad (5)$$

Single steps of the hydrogenation of carbon monoxide

Fig. 4

$$m_p = (\ln^2 \alpha) \cdot P \cdot \alpha^p \qquad (1)$$

$$\alpha = \frac{r_p}{r_p + r_t} \qquad (2)$$

$$\log \frac{m_p}{p} = \log (\ln^2 \alpha) + P \cdot \log \alpha \qquad (3)$$

Product distribution by chain growth

interpretation of the course of reaction is that dissociative chemisorption of carbon monoxide initially leads to activated carbon which is converted by partial hydrogenation to methylene radicals which can grow to form longer hydrocarbon units by mutual bonding. As some doubts were voiced about Fischer's explanation, other concepts were subsequently developed by analogy with known organic chemical reactions such as condensation, dehydration and hydrogenation, from which a network of primary and subsequent steps could be derived.

With the revived interest in the FT-reaction after around 1975 modern measuring and test procedures were increasingly implemented which produced a large number of results. At first difficulty was experienced in correlating them and only in the last few years has a clear picture been formed. Surprisingly Fischer's theories of 1926 have been largely confirmed (Fig. 3).

In the primary step of the reaction activated carbon is formed by dissociative chemisorption of carbon monoxide (Eq. (1)) and this is converted into an intermediate product by hydrogenation (Eq. (5)). The intermediate product is capable of combining with other similar products so that chain growth can take place which leads to hydrocarbons (Fig. 2).

The initial starting materials for mechanistic investigations were "simple" catalysts which contained for example Ni or Ru as active species. The term "simple" catalysts is to be understood as the opposite of iron or cobalt catalysts since hydrogenation of carbon monoxide on the latter is much more complex owing to the large number of possible and also observed reactions (Boudouard reaction, water gas conversion, formation of oxygen-containing compounds).

When the findings from so-called simple catalysts are transferred to iron catalysts the following conclusions can be made: the occurrence of carbon in activated, carbidic or graphitic form on iron surfaces does not take place in distinct phases but is always characterised by smooth transitions in the above-mentioned sequence. Rate and completeness of the transitions are greatly influenced by the type and composition of the catalysts. Molecules of carbon monoxide are already chemisorbed on iron at room temperature and begin to decompose with increasing speed to carbon and oxygen, both in activated form, at about 100°C.

Elementary "activated" carbon is the primary product of the reaction of carbon monoxide with the catalyst and the chain growth can best be explained by the insertion of carbenoid-bound methylene groups into a metal-carbon-bond. Thus in the FT synthesis the products are generated by a step-wise insertion of primary products into a growing chain. Such a procedure is comparable with a polymerisation reaction and the criteria which govern the composition of the final products can in this case also be applied to the FT reaction. The composition of the products obtained, which depends on the catalyst and reaction conditions, can be described by the chain growth rate α (Fig. 4). Each value for α corresponds to a max. producible amount of the products desired as well as the corresponding amount of by-products. The interdependence of α and the product composition is shown in fig. 4. For conventional syntheses α is between about 0.7 and 0.9, from which the hydrocarbon mixtures occurring can be derived (Fig. 5).

So far there have been only limited possibilities to influence the product composition obtained by intervention in the chain growth on the catalyst. Therefore every variant of the FT synthesis produces per force a mixture of desired products and by-products which have to be separated in the further course of the process.

3. Commerical realisation of the FT synthesis

Three aspects are of primary importance in the realisation of the FT synthesis:

Fig. 5

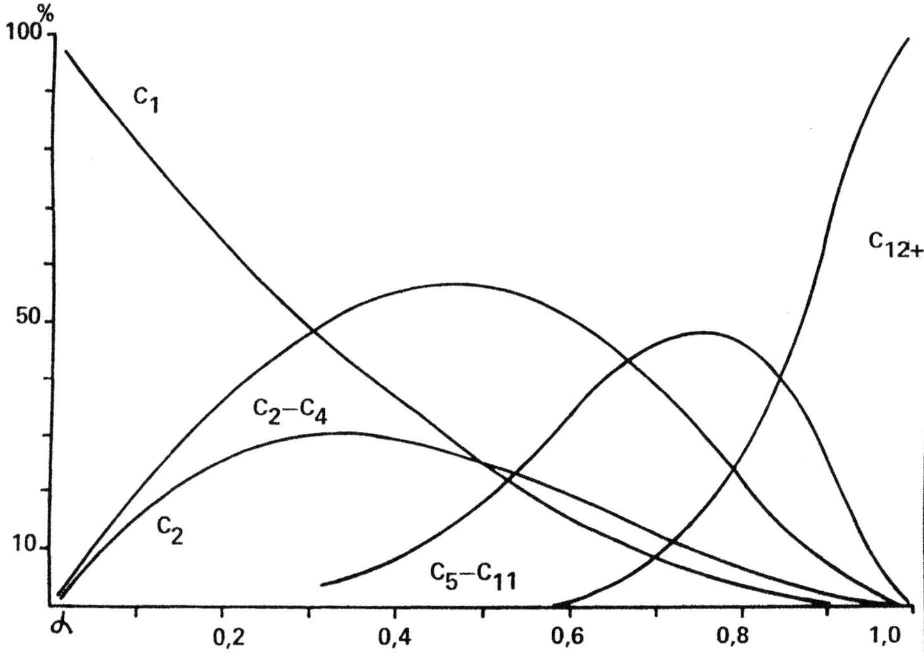

Carbon number distribution vs. α

Fig. 6

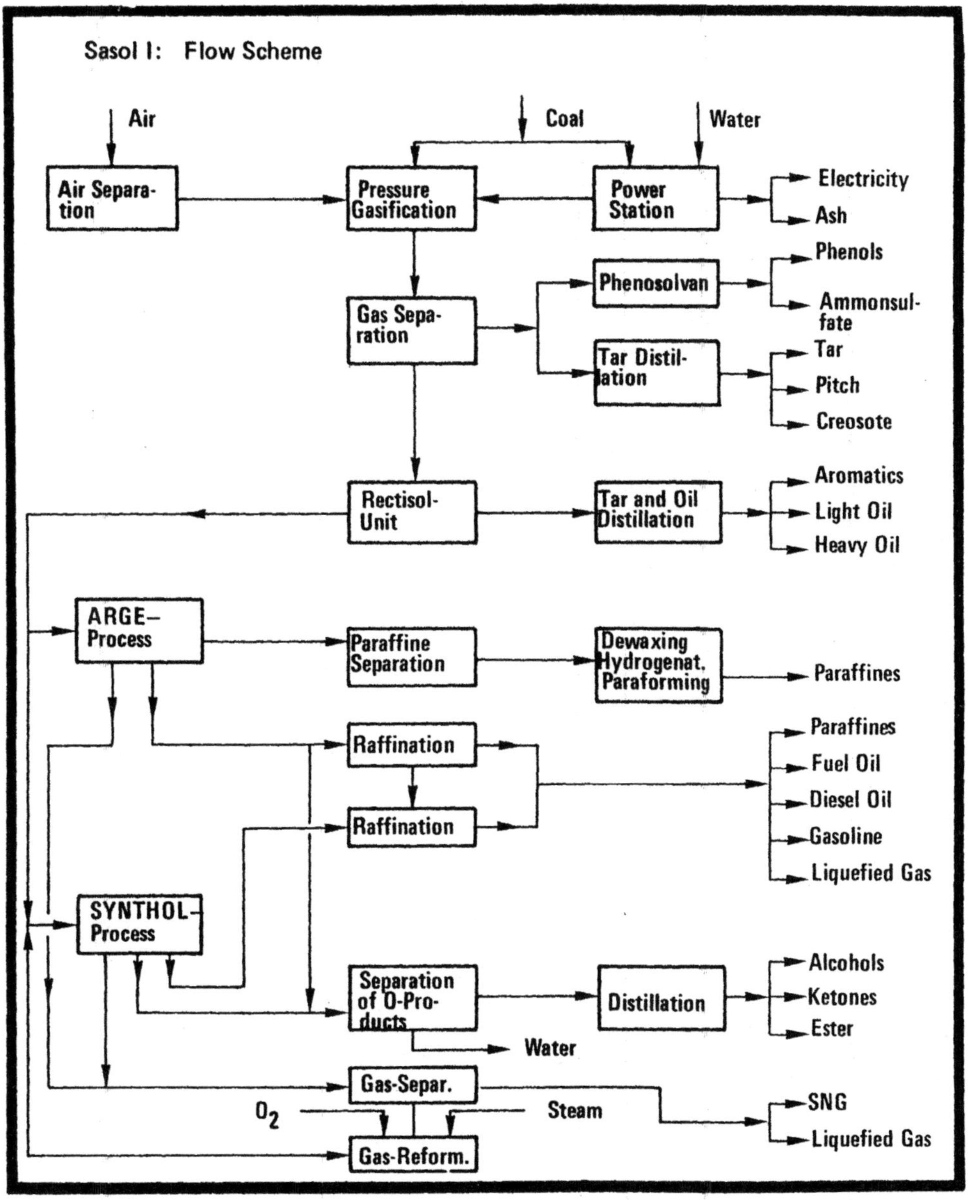
Sasol I: Flow Scheme
Air
Coal
Water
Air Separation
Pressure Gasification
Power Station
Electricity
Ash
Phenols
Phenosolvan
Ammonsulfate
Gas Separation
Tar Distillation
Tar
Pitch
Creosote
Aromatics
Rectisol-Unit
Tar and Oil Distillation
Light Oil
Heavy Oil
ARGE–Process
Paraffine Separation
Dewaxing Hydrogenat. Paraforming
Paraffines
Raffination
Paraffines
Fuel Oil
Diesel Oil
Raffination
Gasoline
Liquefied Gas
SYNTHOL-Process
Separation of O-Products
Distillation
Alcohols
Ketones
Ester
Water
Gas-Separ.
O_2
Steam
SNG
Liquefied Gas
Gas-Reform.

- the reaction is catalysed heterogeneously
- the amount of heat to be dissipated is relatively high (about 2500 kJ per m^3 of reacted synthesis gas)
- temperature peaks are detrimental to the catalysts and influence the product composition

To satisfy the requirements arising from the aforementioned points, a number of process principles have been examined with varying success. Technical problems have impeded the success of the fluidised-bed process while the liquid-phase process using the bubble-column principle has not been carried out on a commerical scale owing to unfavourable economic circumstances. Present-day technology is therefore limited to two process types: the gas-phase fixed bed (ARGE) process and the entrained-bed (SYNTHOL) process. Both are operated by Sasol Ltd. (Johannesburg, Republic of South Africa) and have reached a stage of full technical maturity.

A synthesis plant consists of the synthesis gas production and purification, the actual synthesis and the product work-up (Fig. 6 and 7). When coal is gasified using the Lurgi process, which is employed in South Africa, naphtha and tar occur as by-products which can be processed to road fuels. The raw gas is freed from sulfur and carbon dioxide and apart from carbon monoxide and hydrogen also contains methane. By-products would no longer occur if the gasification techniques of the 2nd generation were employed.

Hydrogen and carbon monoxide are reacted to form hydrocarbons on iron catalysts. In the ARGE reactor the catalyst is arranged as a fixed bed in 12 m long tubes. The heat of reaction is released to a water cooling jacket. The operating temperature lies between 210 and 240°C. The products range from short-chained hydrocarbons to hard paraffins with 30 or more carbon atoms. In the SYNTHOL reactor a suspension of finely divided catalyst in synthesis gas is circulated and passed over cooling elements. Owing to the higher temperature of 300 – 350°C short-chained hydrocarbons with a relatively high alkene content predominate so that a greater amount of gasoline is formed in comparison to the fixed-bed process.

In both systems the heat of reaction is used to produce steam. The gas from the synthesis is cooled to remove the hydrocarbons. The residual gas is worked-up in a reformer and returned to the synthesis. In the "refinery stage" gasoline and diesel are manufactured using conventional processes.

In the Sasol I plant (Fig. 6) production is distributed between the fixed and entrained-bed syntheses in a ratio of about 1 : 1.7. Gas supply and product recovery of both processes are coupled together in a more or less integrated total complex. To understand the sense of combining two different processes it is helpful to bear in mind the objectives at the time when Sasol I was being planned. In brief they were: the production of gasoline and diesel in amounts proportional to the demand and the recovery of as many value by-products from synthesis and gasification as possible. It is understandable that the recovery and refining to market quality of the products from two processes is extremely costly. In the final analysis these costs are due to the previously mentioned problem of limited selectivity of the FT synthesis.

Thus the original concept (Fig. 7) was abandonned when plans were being made for the construction of further synthesis plants prompted by the oil supply crisis of 1973/1974: Sasol II and III (Fig. 8) are principally designed for the production of road fuels and the unavoidable by-products are only recovered if the costs permit them to be sold at realistic market prices. The refining methods for the primary products are much simpler than those used in Sasol I. Gas production takes place in 36 enlarged Lurgi pressure gasifiers, each with a raw gas production capacity of 46.000 Nm^3/h. Hydrocarbon synthesis is conducted using the entrained-bed process, which was given preference over the fixed-

Fig. 7

	Process	
	ARGE (Ruhrchemie/ Lurgi)	Synthol (M.W. Kellogg)
Process data		
Number of Reactors	5	3
Catalyst Volume per Reactor	40 m^3	135 t
Fresh Gas Input per Reactor (Nm^3/h)	ca. 20.000	ca. 75.000
Daily Production C_{3+} per Reactor (t/d)	55	250
Temperature (°C)	220 – 250	320 – 340
Pressure (bar)	24	22
H_2 : CO (Reactor–Entrance)	1,5 – 1,7	5 – 6
Servicefactor	0,96	0,75
Production data		
Coal Input (Mio t/a)	2,7	
Total Products (t/a)	96.000	170.000
Final Products (t/a)		
Gasoline ca.	158.000	
Diesel ca.	46.000	
Others ca.	62.000	

Sasol I: Process and Production Data

Fig. 8

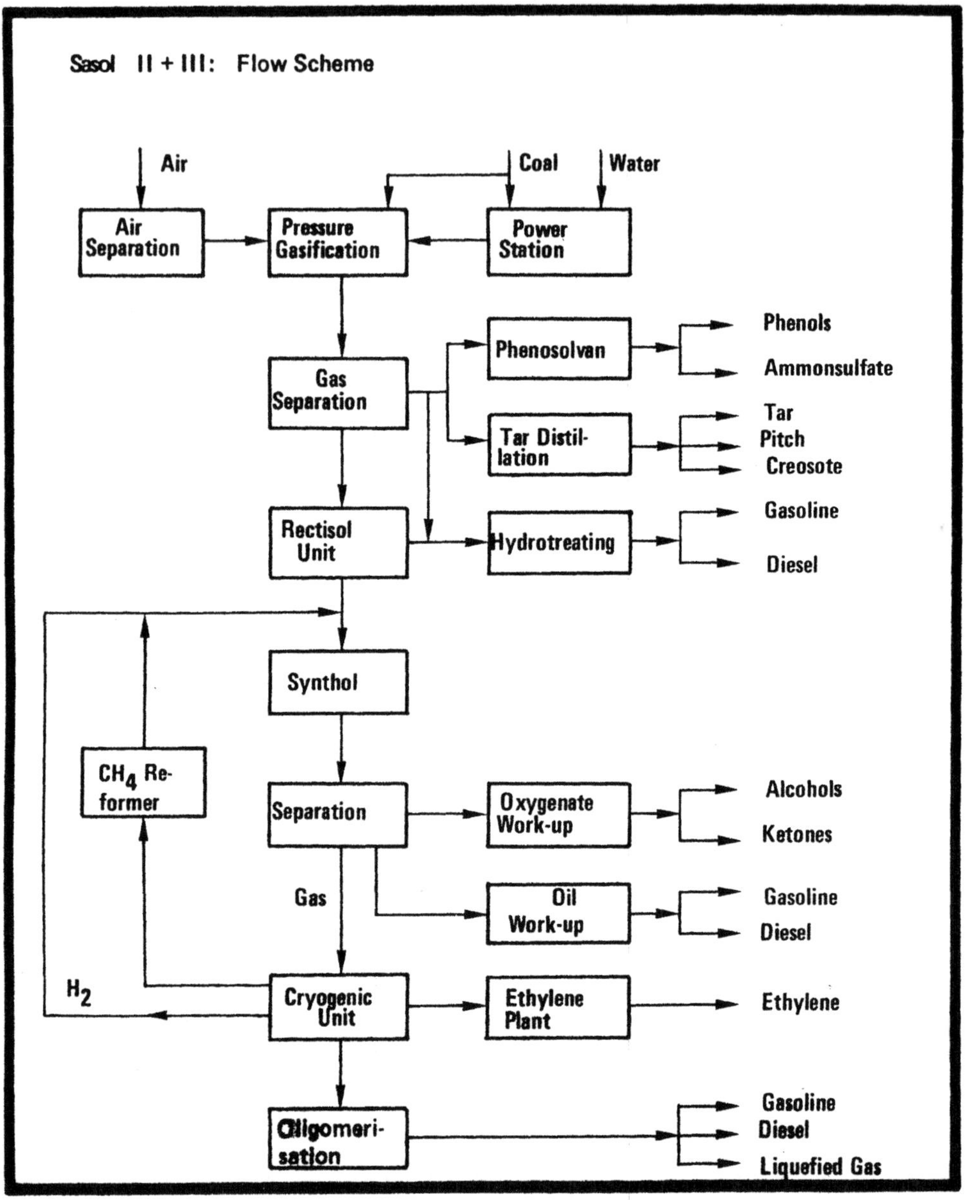
Sasol II + III: Flow Scheme
Air
Coal
Water
Air Separation
Pressure Gasification
Power Station
Phenols
Phenosolvan
Ammonsulfate
Gas Separation
Tar
Tar Distil-lation
Pitch
Creosote
Gasoline
Rectisol Unit
Hydrotreating
Diesel
Synthol
CH4 Re-former
Separation
Oxygenate Work-up
Alcohols
Ketones
Gas
Oil Work-up
Gasoline
Diesel
H2
Cryogenic Unit
Ethylene Plant
Ethylene
Gasoline
Diesel
Oligomeri-sation
Liquefied Gas

bed process for two reasons: the products are richer in alkenes, a result of the higher reaction temperature, and the specific reactor performance is higher. Possible problems with the scaling-up of the entrained-bed reactors were taken into account and have obviously been mastered. The process and production data for Sasol II are shown in fig. 9.

Sasol III is practically identical to Sasol II and was commissioned only a few months ago. Both plants together produce about 3.8 million tonnes of hydrocarbons per annum with a coal input of about 27 million t/a. The coal used is of poor quality and has an ash content of up to 40 per cent by weight.

When the two synthesis processes are compared, the following feature becomes apparent: with the entrained-bed process it is possible to manufacture gasoline hydrocarbons with a relatively high space-time yield. The fixed-bed process is relatively limited as far as the space-time yield is concerned, but is much more flexible with regard to product composition. This is of special significance if diesel is to be the main product, since it can only be recovered in limited quantities using the Synthol process. A concept prepared by Lurgi is based on the following product composition (Fig. 10):

Diesel	673.000 t
Gasoline	145.540 t
LPG	69.000 t
C_1, C_2, fuel gas	112.460 t
Primary product C_{1+}	1.000.000 t

To produce the above-mentioned quantities, about 0.65 million Nm^3/h gas is required, 94 % of which is converted is three stages in 25 synthesis reactors with 3500 m^3 of catalyst. The process route is based on tested processes which have been combined to achieve the objective of producing diesel oil with a high selectivity. No development work is necessary and coal of any quality can be used.

4. Economic aspects of the FT synthesis

Apart from the direct high-pressure hydrogenation of coal using the Bergius or a modified process, two other ways of producing synthetic road fuels can be considered: the Fischer-Tropsch synthesis and the Mobil process, which uses methanol as an intermediate product. Both processes are independent of the coal quality since this factor has little significance in the modern gasification processes. As methanol only occurs in an intermediate stage in the Mobil process and is then completely converted for purposes of comparison, the combination of methanol synthesis and the Mobil process can be considered as a unit and assessed as such.

Some data on the road fuels which can be manufactured by the hydrogenation of hard coal or by the FT-synthesis are listed in fig. 11. Owing to their chemical composition gasoline and diesel from the FT synthesis are very light. The cetane number of the diesel is very good whilst the octane number of the gasoline does not meet the requirements of premium gasoline. All FT products are completely free of sulfur. On the other hand gasoline from coal has sufficient pre-ignition resistance (knock rate) owing to its high aromatics content whilst diesel from coal barely fulfills the required quality standards. In view of the product quality it makes sense, therefore, to mix products from the FT synthesis with those from coal hydrogenation and in doing so improve both.

The idea has been incorporated in the plans for an Australian coal up-grading project, where hydrogenation and synthesis products are to be produced in a ratio of about 55 : 45.

The gasoline produced using the Mobil process is of premium quality without any after-treatment. However, diesel can only be produced by indirect means with this process

Fig. 9

Process data

Number of Reactors		7
Catalyst Volume per Reaktor	(t)	360
Fresh Gas Input per Reaktor	(Nm^3/h)	ca. 260.000
Daily Production C_{2+} per Reaktor	(t/d)	660

Production data

Coal Input	(Mio t/a)	13,5
Total Products	(t/a)	2,1 Mio
Final Products	(t/a)	
Ammonia		100.000
Sulfur		90.000
Ethylen		160.000
Chemicals		50.000
Tar Products		200.000
Motor Fuels		1.500.000

Sasol II: Process and Production Data

Fig. 10

Schematic diagram of a Fischer–Tropsch motor fuel synthesis

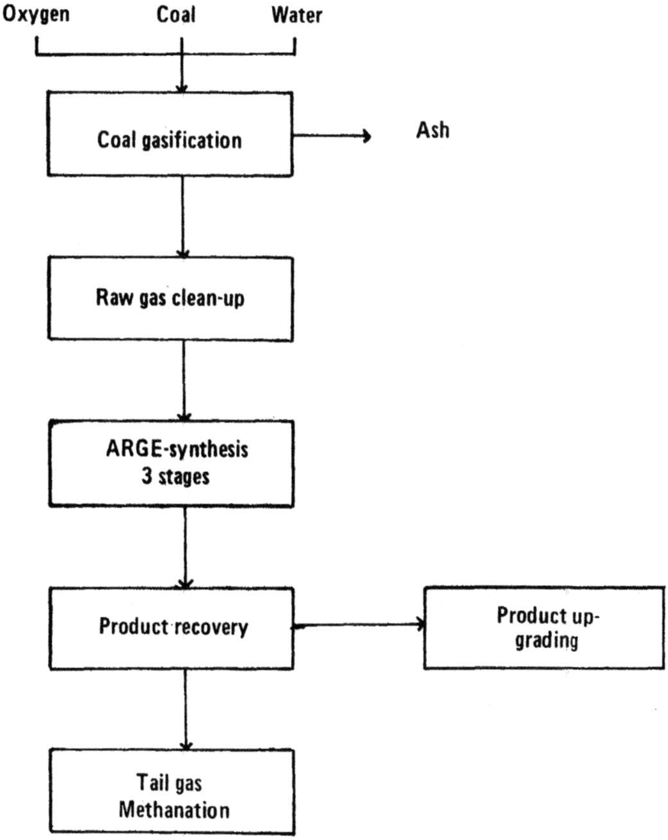

Diesel	673,000 t
Gasoline	145,540 t
LPG	69,000 t
C_1, C_2, fuel gas	112,460 t
Total	1,000,000 t

K.-H. Eisenlohr, H. Gaenslen (1982)

Fig. 11

Characteristic data of synthetic motor fuels

		ARGE-synthesis	Coal hydrogenation (Bituminous Coal)
Gasoline			
Spec. density	g/ml	0.71 – 0.72	0.82 – 0.85
Aromatics	% b.w.	10 – 20	70 – 75
RON (0.15 ml TEL/l)		95	98 – 100
Diesel oil			
Spec. density	g/ml	0.79 – 0.81	0.855
Aromatics	% b.w.	$<$ 10	8
Naphthenes	% b.w.	$<$ 5	56
Cetane-Nr.		70	44 – 46

Fig. 12

Thermal efficiencies

	Efficiency %	GJ coal per GJ product
Fischer-Tropsch synthesis		
– Production of motor fuels	40	2.50
– Production of motor fuels and pipeline gas	58	1.72
Methanol synthesis and Mobil process		
– Gasoline and LPG	45	2.22
– Gasoline, LPG and pipeline gas	61	1.64
Coal hydrogenation		
– Production of motor fuels	56	1.79

Basis: Bituminous Coal; LHV
K.-H. Eisenlohr, H. Gaensslen (1981)

Fig. 13

Production costs (1980; 1 US-$ = 1.80 DM)

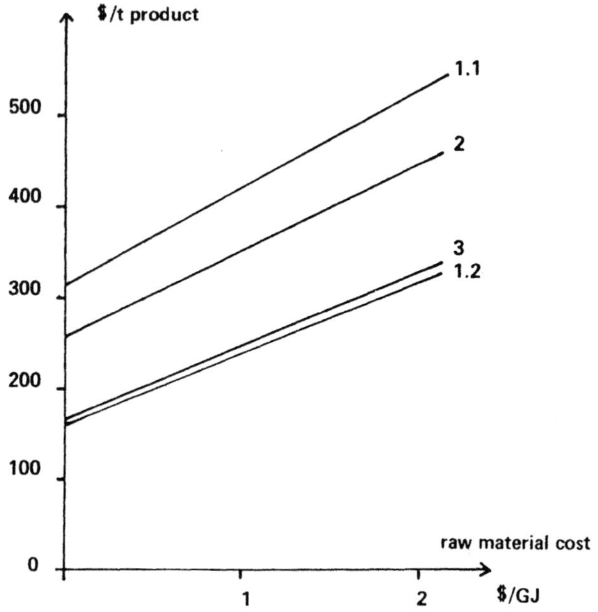

1. Fischer-Tropsch synthesis
   - 1.1 motor fuels
   - 1.2 motor fuels and pipeline gas
2. Methanol synthesis and Mobil process to gasoline
3. Coal hydrogenation to motor fuels

K.-H. Eisenlohr, H. Gaensslen (1981)

Fig. 14

Coal costs for synthetic motor fuels
(1980; 1 US– $ = 1.80 DM)

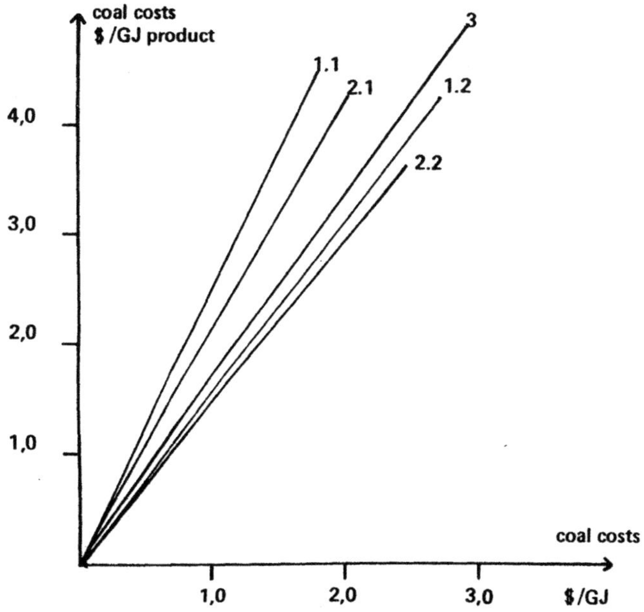

1. Fischer-Tropsch synthesis
   1.1 motor fuels
   1.2 motor fuels and pipeline gas

2. Methanol synthesis and Mobil process
   2.1 gasoline
   2.2 gasoline and pipeline gas

3. Coal hydrogenation to motor fuels

K.-H. Eisenlohr, H. Gaensslen (1981)

Fig. 15

	%
Synthesis gas	70
Catalysts, miscellaneous	3
Personnel costs	4
Energy	12
Capital expenditures	11
	100

Distribution of the manufacturing costs of FT–products

and the method has yet to be tested. Both the fluidised-bed and the fixed-bed method can be used for the Mobil process. From 1000 kg methanol, 391 kg of gasoline (including alkylate), 23 kg of liquid gas and 20 kg of fuel gas are formed — the latter being used in the process. In addition 566 kg of water are formed.

The thermal efficiency is defined as the net calorific value of the products divided by the net calorific value of the coal used both as a raw material and for plant operational requirements. The reciprocal of the efficiency gives the specific heat requirement for the manufacture of the products in question, from which the coal input requirement can be determined. The data on some processes have been compiled in fig. 12.

For the production of gasoline and LPG the thermal efficiency ranges from 40 % for the FT synthesis to 56 % for coal hydrogenation. When pipeline gas is simultaneously produced, i.e. methane and light hydrocarbons are not recycled back into the process, the thermal efficiency of the FT synthesis is increased from 40 to 58 % and that of the Mobil process from 45 to 61 %. Thus the same order of magnitude is attained as for the production of hydrogen (57 %) or for the synthesis of methane (61 %) if both use coal as a feed stock.

Fig 13 gives an outline of the production costs ($/t product) as a function of the raw material (coal) costs. Fig. 14 outlines the influence of the coal costs on the production of synthetic motor fuels by different processes. The figures are based on the assumption that Lurgi pressure gasification is used for the synthesis-gas and hydrogen production. Owing to the relatively high methane content which results, the production costs, e.g. of the FT synthesis, are appreciably reduced if methane can be produced as well as gasoline. This also applies to the methanol synthesis in combination with the Mobil process, whereas high-pressure coal hydrogenation only produces very small amounts of methane. If a modern gasification process were employed, the problems arising from methane formation would be eliminated.

As far as the Federal Republic of Germany is concerned the break-down of production costs for road fuels from the FT synthesis would be more or less the same as that listed in fig. 15. The basis of these calculations is a capacity of 1.5 million t/a of primary products C_{1+}, specific investment costs of DM 1400/t and a coal price of DM 230/t. By far the highest percentage of the costs, 70 %, is for the raw material, synthesis gas. Therefore its manufacture and optimum use are in the forefront of all efforts to reduce costs.

5. Summary and outlook

The catalytic conversion of synthesis gas to hydrocarbons using the Fischer-Tropsch synthesis is an almost 60-year-old process, which has been improved in the last few years by increasing the plant size. In principle the FT synthesis can be used to produce gasoline of adequate quality and diesel of extremely good quality. The nature of the processes and the catalyst which are characteristic of the FT synthesis means that a relatively wide range of products always occurs which must undergo further treatment using processes which are as such well-known and unproblematic.

As far as the product quality is concerned, the FT-fixed-bed synthesis is characterised by its particularly high-grade diesel oil which is excellently suited for mixing with lower-grade products such as those obtained from coal hydrogenation. The optimisation of the fixed-bed process with the aim of producing diesel as the main product can be achieved by the simple combination of time-tested techniques. In a three-stage method of operation a selectivity of 74 % (based on C_{3+}) diesel is achieved for a 94 % conversion of synthesis gas.

The thermal efficiency of the FT synthesis is influenced by the gasification process selected and the product composition required. The highest efficiencies are attained when only small amounts of methane are produced during gasification and synthesis and these can be taken off as SNG.

In Western Europe with the present coal prices the FT synthesis is not an economic alternative to mineral oil. For countries where coal is available at lower prices the FT synthesis is already an economic proposition.

METHANOL PRODUCTION BASED ON SYNTHESIS GAS FROM COAL AND BIOMASS

Dr. Rainer Reimert
Lurgi GmbH / Process Development Department
P.O. Box 11 12 31
D-6000 Frankfurt am Main 11
Federal Republic of Germany

ABSTRACT. Processes for both wood and coal gasification are described as far as the composition of the different raw gases is concerned. This is followed by a discussion on the downstream gas conditioning processes and the methanol synthesis focussing on the features of the LURGI Low Pressure process. Based upon total process schemes methanol production costs are considered. Methanol from coal is competitive in the case of cheap coals only. Wood as feedstock results in costs close to those calculated for cheap coal. From a technical point of view methanol from coal plants could be built right away, whereas wood treatment and gasification units of the necessary size are not yet available.

1. INTRODUCTION

During the last few decades, methanol has become a top-ranking mass chemical and production capacities may be expected to be further extended in the future. In regard to quantity, world methanol production takes third place behind ammonia and ethylene as shown in Fig. 1, /4/5/. This growth in production capacity is, of course, accompanied and initiated by an increasing consumption or a corresponding demand expectation. In addition to conventional applications as a chemical raw material to manufacture formaldehyde, solvents and dimethylterephthalate - to mention only the most important fields -, new applications were also found in the motor fuel and nutrients sector (proteins).

Table I lists current and possible future uses of methanol as a fuel. Modern technology as regards the application for motor fuel purposes can be described in an exemplary fashion as follows:

- o The addition of methanol to gasoline (abt. 3 %) is already standard practice and its further development leads to ever-increasing percentages of methanol supported by homogenizing agents.
- o The application of pure methanol as a motor fuel has been commercially demonstrated.
- o Methyl-tertiary-butyl ether (MTBE) and tertiary-amyl-methyl ether (TAME) are used instead of tetra-ethyl lead as octane boosters.

G. E. Beghi (ed.), Synthetic Fuels, 135–163.

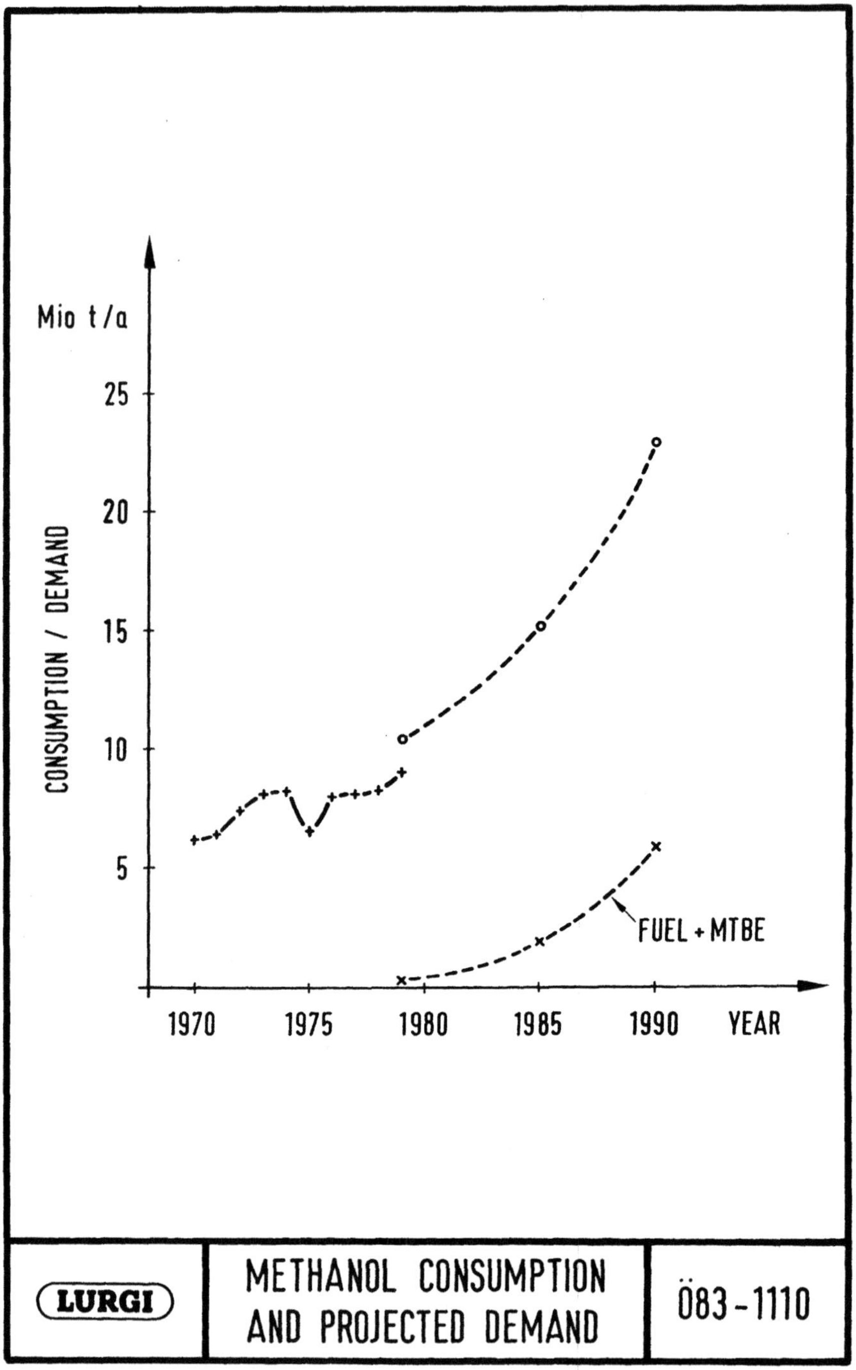
Mio t/a
CONSUMPTION / DEMAND
25
20
15
10
5
1970
1975
1980
1985
1990
YEAR
FUEL + MTBE
LURGI
METHANOL CONSUMPTION
AND PROJECTED DEMAND
Ö83-1110

Figure 1

Tab. I METHANOL APPLICATION IN THE FUEL MARKET

- ADDITION TO GASOLINE
- SUBSTITUTE FOR GASOLINE
- M T B E - AND T A M E - PRODUCTION
- ADDITION TO DIESEL OIL
- SUBSTITUTE FOR DIESEL OIL
- CONVERSION INTO GASOLINE

- FUEL FOR PEAK LOAD GAS TURBINE
- FUEL FOR FUEL CELLS
- CONVERSION INTO SNG, TOWN GAS

- LIQUID FOR COAL SLURRIES

Tab. II

C O A L	B I O M A S S
- ABUNDANT RESOURCES AVAILABLE	- RENEWABLE OR WASTE ENERGY
- RESOURCES IN INDUSTRIALIZED COUNTRIES AVAILABLE	- AVAILABLE IN THIRD WORLD COUNTRIES
- DIRECT USE OF COAL INCONVENIENT	- EVEN GROWABLE IN SEMI-ARID REGIONS
	- CONTAINING LOW SULFUR AND ASH

POINTS OF INTEREST FOR THE USE OF COAL AND WOOD AS FEEDSTOCK FOR METHANOL

- o The admixture of methanol to diesel fuel and its substitution still require further development work.
- o Demonstration plants using the Mobil Process to produce gasoline from methanol will be started up shortly.

Dry wood distillation was the dominant methanol production process until the 1930's. Knowledge of catalytic methanol synthesis from CO and H_2 goes back to the beginning of this century. The development of catalysts can be characterized as follows.

1913 : Iron catalyst
1920/23 : Zinc oxide activated with chromic acid; high-pressure synthesis
1970 : Copper-zinc catalyst, various promoters; low-pressure synthesis.

The wide distribution of catalytic methanol production was permitted by the following conditions:

- Availability of low price primary energy sources (natural gas, heavy residues)
- Availability of efficient, reliable technology for synthesis gas production (steam reforming, partial oxidation)
- Availability of an efficient, reliable synthesis process (low-pressure methanol synthesis) applicable on a large scale.

Modern plants are built to the low-pressure process using the above copper-zinc catalyst and the presentation is based on this process. Here, coal and biomasses ought to be considered as raw materials. These raw materials can be of interest for reasons of energy policies as well as economics (Table II).

2. THE LOW-PRESSURE METHANOL SYNTHESIS PROCESS

2.1 Synthesis Reaction and the Catalyst

Methanol synthesis is an equilibrium reaction consisting of the gross reactions listed in Table III. Due to its exothermal character and to the volume contraction, low temperature and high pressure favour the equilibrium state in regard to the desired product. As opposed to this, low reaction temperatures demand high catalyst activity, and a high reaction pressure makes the synthesis plant more expensive, reduces the energy efficiency of the process and promotes the formation of byproducts. Thus, the reaction conditions are selected in response to the reaction kinetics - determined by the catalyst effectiveness -, to thermodynamic equilibrium and to economic boundary conditions.

Low-pressure methanol synthesis, which takes place under the conditions stated in Table IV with a highly active copper-zinc catalyst, is the result of this optimization process.

This process benefits from the basic copper catalyst and from the fact that the sulphur contents in the synthesis gas can be reduced, for example by the Rectisol process, to sufficiently low levels as demanded by the copper catalyst.

In addition to the main reactions listed in Table III, there are diverse side reactions, some of which are given in Table V. The selec-

Tab. III METHANOL SYNTHESIS REACTIONS

MAIN FUSION REACTIONS

$$CO + 2H_2 \rightleftarrows CH_3OH$$

$$\Delta H_{25\,^{o}C} = -90{,}84 \text{ KJ/MOL}$$

$$CO_2 + 3H_2 \rightleftarrows CH_3OH + H_2O$$

$$\Delta H_{25\,^{o}C} = -49{,}57 \text{ KJ/MOL}$$

SIMULTANEOUS REACTION

$$CO + H_2O \rightleftarrows CO_2 + H_2$$

$$\Delta H_{25\,^{o}C} = -41{,}27 \text{ KJ/MOL}$$

Tab. IV CONDITIONS FOR THE LOW PRESSURE METHANOL SYNTHESIS

- REACTION PRESSURE	50 - 100 BAR
- REACTION TEMPERATURE	$< 280\ ^{o}C$
- REACTOR INLET TEMPERATURE	$230 - 240\ ^{o}C$
- FEED GAS COMPOSITION	
$\frac{H_2 - CO_2}{CO + CO_2}$	2.0 ... 2.2*
- SULFUR CONTENT	< 0.1 PPM
- SPACE VELOCITY	$10\,000\ \frac{Nm^3}{M^3\ H}$
- SPECIFIC PRODUCTION RATE	$500 - 1\,000\ \frac{KG}{M^3\ H}$

* Lower figures for Fuel-Methanol

Tab. V PARALLEL AND CONSECUTIVE REACTIONS OF METHANOL SYNTHESIS

$$CO + 3\,H_2 \rightleftharpoons CH_4 + H_2O$$

$$CO_2 + 4\,H_2 \rightleftharpoons CH_4 + 2H_2O$$

$$nCO + (2n-1)\,H_2 \rightleftharpoons C_nH_{2n+2} + nH_2O$$

$$nCO + 2n\,H_2 \rightleftharpoons C_nH_{2n+1}\,OH + (n-1)\,H_2O$$

$$2CH_3\,OH \rightleftharpoons H_3\,C - O - CH_3 + H_2O$$

$$2CO \rightleftharpoons CO_2 + C$$

tion of the catalyst and process conditions also takes into account that the formation of methane, alkenes, higher alcohols and carbon as well as the decomposition of methanol into dimethylether should be prevented. Further details in regard to this complex can be taken from the pertinent literature /1/2/.

2.2 The Synthesis Process

At the heart of every chemical process is the reactor. It enables the mode of reaction determined by previous deliberations on the physical chemistry of the reaction to be realized. When the type of catalyst, the reaction pressure and the most favourable feed gas composition have been established, the main task of the reactor remains to ensure as constant a temperature as possible. This is important in view of the fact that too low temperatures lead to low conversion rates whereas excessive temperatures cause recrystallization and de-activation of the copper catalyst, a process starting above 270 °C.

Two reactor concepts have established themselves technically:

- The tubular reactor cooled by boiling water, the LURGI Process;
- The quench-cooled fixed-bed reactor, the ICI Process.

Fig. 2 shows the temperature profiles along the longitudinal axis of the reactor for the two cooling principles /2/. In the quench reactor, the temperature profile is serrated, reflecting the rise in temperature caused by the exothermal reaction and the cooling by the quench gas. The numbers and locations of the quench gas nozzles therefore have to be dimensioned in such a way that the allowed maximum temperature is not overrun under any operating conditions.

In the case of tubular reactors the temperature level is determined by the pressure of the boiling water surrounding the tubes. In accordance with the known heat transfer conditions, an over-temperature of abt. 11 K establishes itself in the tubes filled with catalyst. Thus, the temperature limits are easily maintained at all times. Steam production (40 - 50 bar) is abt. 1.4 t per t methanol. After some years of operation, the catalyst has de-activated so far that a pronounced temperature maximum can no longer be recognized. The temperature level is increased by a few degrees by raising the pressure of the boiling water to compensate for the decreasing catalyst activity. The LURGI catalyst has a lifetime of 4 - 5 years.

The LURGI Process scheme is shown in Fig. 3, describing the synthesis loop and the methanol production stage as examples /3/. The synthesis gas is compressed to synthesis pressure (normally between 50 and 100 atm.) in a turbo-compressor (1) and then mixed with circulating gas. Should the synthesis gas originate from gasification processes (such as partial oxidation) which operate at pressures above 50 bar, synthesis gas compression is not required.

The mixture of circulating gas and fresh gas is heated in a heat exchanger (2) and then enters the tubular reactor (3).

The reaction mixture leaving the reactor is cooled in a heat exchanger and in cooler (4) with air or water. The condensed crude methanol is separated from the unreacted gas in separator (5), the unreacted gas is returned to the reactor inlet via a recycle compressor

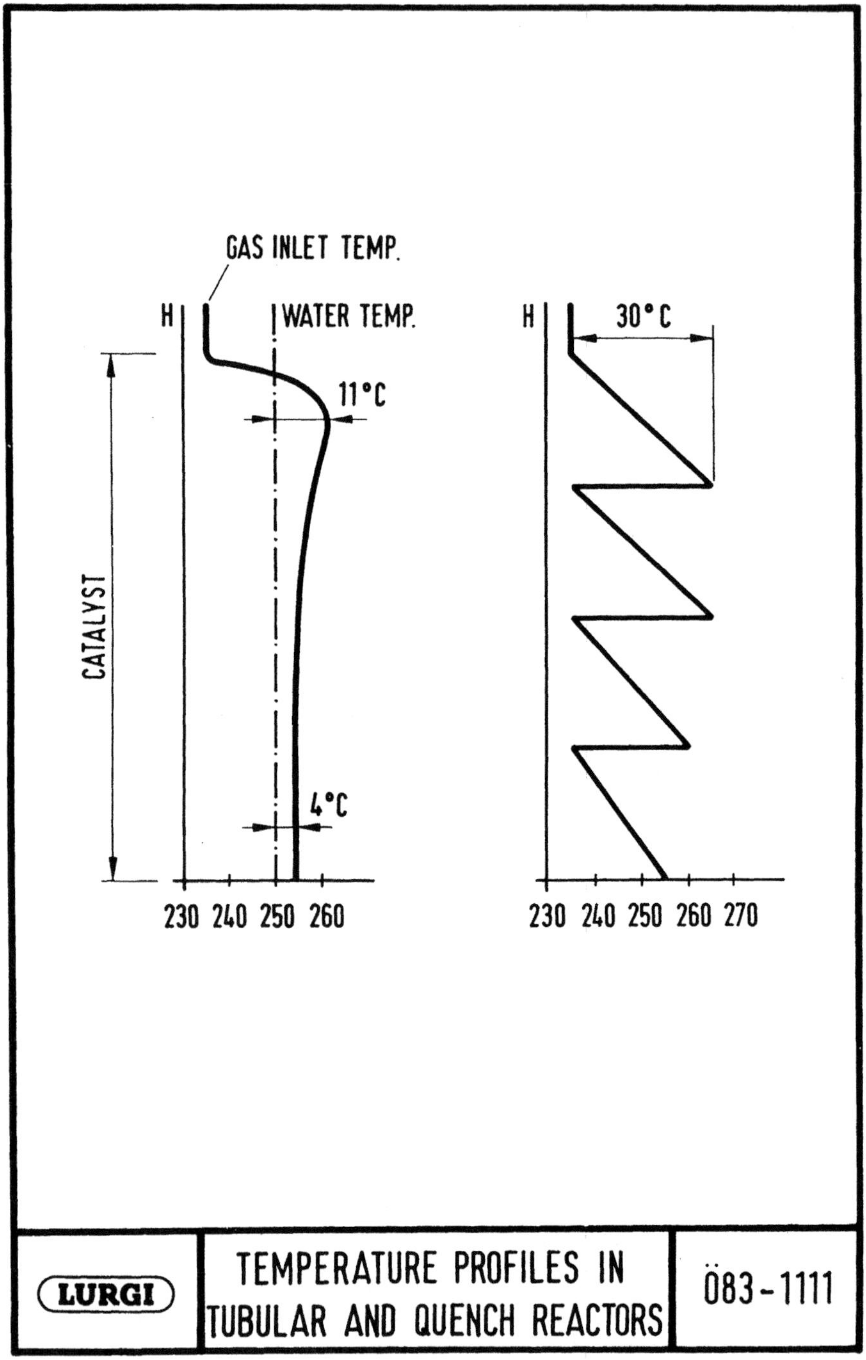
GAS INLET TEMP.
H
WATER TEMP.
11°C
CATALYST
4°C
230 240 250 260
H
30°C
230 240 250 260 270
LURGI
TEMPERATURE PROFILES IN
TUBULAR AND QUENCH REACTORS
Ö83-1111

Figure 2

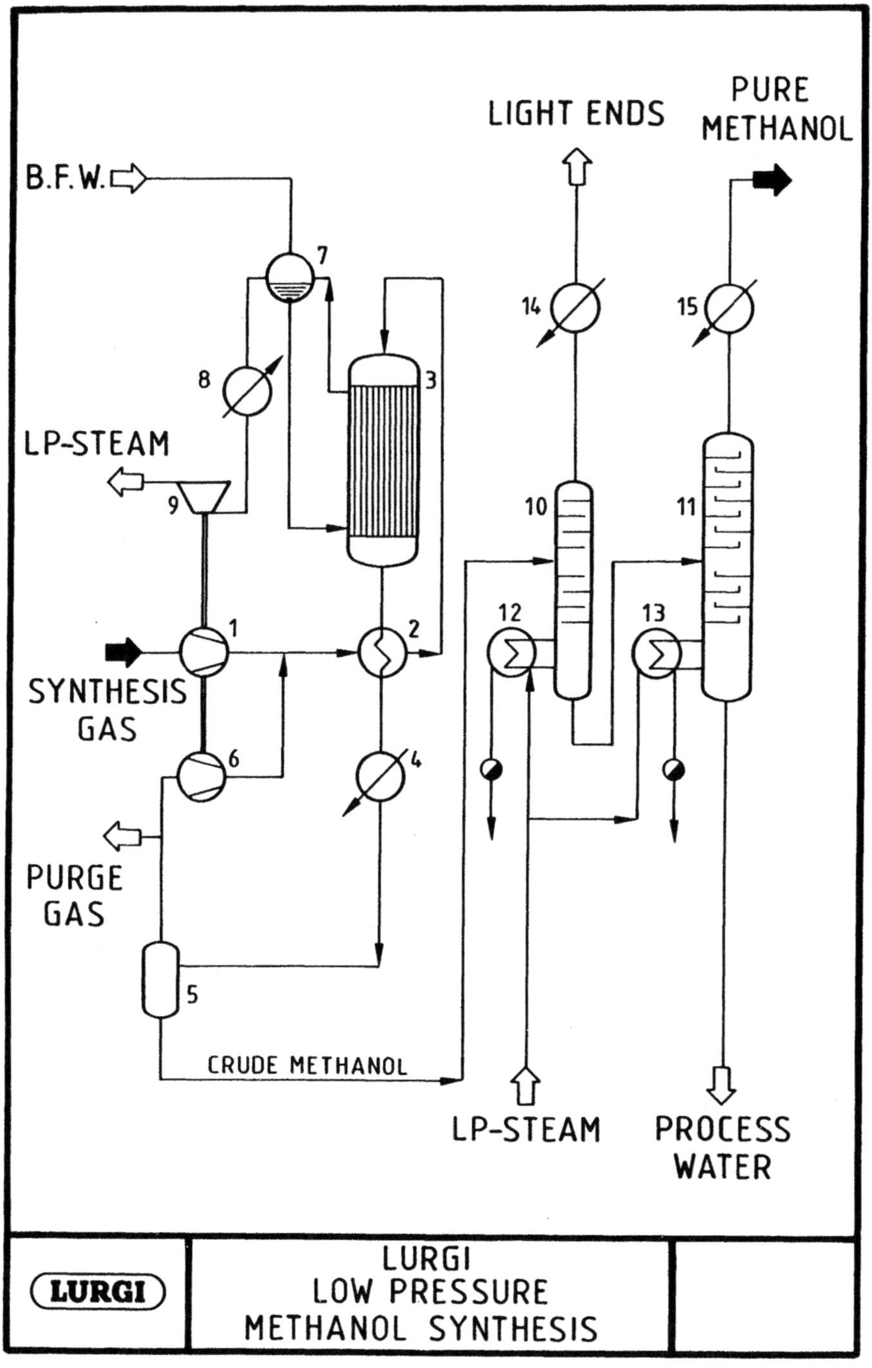
LIGHT ENDS
PURE METHANOL
B.F.W.
7
14
15
8
3
LP-STEAM
9
10
11
12
13
1
2
SYNTHESIS GAS
6
4
PURGE GAS
5
CRUDE METHANOL
LP-STEAM
PROCESS WATER
LURGI
LURGI
LOW PRESSURE
METHANOL SYNTHESIS

Figure 3

(6). A small amount of purge gas is exported from the loop to avoid inert gas accumulation.

The heat from the methanol reaction is utilized for the regeneration of HP steam. The reactor tubes are cooled with boiling water via a natural circulation system and the resulting steam is collected in a steam drum (7). The steam is superheated in superheater (8), which is either associated with the methanol synthesis unit or integrated in the syngas production unit. The superheated steam is used to drive a turbine for the syngas and recycle gas compressors (9). The turbine exhaust steam is utilized for methanol distillation.

The flashed crude methanol from the separator is processed to pure methanol in a subsequent distillation unit. The light-ends-column (10) strips dimethyl ether, methyl formiate and other low-boiling impurities, while further column (11) removes water and higher boiling impurities.

Reboilers (12) and (13) are heated with low-pressure steam from the turbine exhaust (9). The light ends are recovered in condenser (14) and the pure methanol in condenser (15).

For plants having capacities of more than 500 t/d methanol production, LURGI's energy-saving distillation system is being applied. In this case, the pure methanol column is divided into two sections. The first section is operated at elevated pressure and some 50 % of the methanol are distilled overhead. In the second section, which operates at ambient pressure, the balance of the methanol is separated from water and high boilers. While the first section is heated with LP steam, the heat for the second section is provided by condensing the methanol vapours from the first section.

3. SOURCES OF SYNTHESIS GAS

As the gasification processes themselves are dealt with in a special paper, this chapter concentrates on

- o characteristics of the composition of the different raw gases in so far as they are important for methanol synthesis
- o differences in the raw gases derived from coal and biomass gasification.

3.1 Coal Gasification

Although many process developments for coal gasification have been launched in recent years, the number of commercially available processes is small. In regard to the process principle, the following reaction systems are applied:

- Moving Bed
- Fluidized Bed
- Entrained Bed.

Further developments to these existing processes were largely aiming at:

- o the application of pressure (Shell-Koppers, Texaco, HTW, U-Gas)

- o an increase in the reaction temperature (HTW, U-Gas, British Gas/LURGI-Slagger)
- o an increase in the process pressure (Ruhr 100).

A comparison on the various processes is to be found in frequently published papers. In selecting a process, one has to consider the prevailing peripheral conditions such as coal quality, desired products and size of the plant. Needless to say, the overriding aim is to minimize production costs. These costs are mainly dictated by:

- Capital costs being a function of the investment costs and, indirectly, of the efficiency, as a plant with higher efficiency can be smaller for the same output.
- Required utilities and raw materials, being a direct function of the overall plant efficiency and of the secondary processes applied.

In addition, maintenance and repair costs are another factor; they are a function of the technical and development status of a process.

Gas analyses pertaining to the various gasification processes are shown in Tables VI - VIII. The selected processes represent the three aforementioned process principles. Besides, these processes are commercially available or expected to go for commercial demonstration /6 - 12/ as the next step.

In addition to the gas analyses, the tables give the stoichiometric ratios both for the raw gas and for the gas which would be obtained after the - hypothetical - removal of CO_2 down to 3 %. Here, it is to be seen that in the case of the Entrained Bed Process, as well as of the British Gas/LURGI Slagging gasifier, the stoichiometric ratio is clearly less than 1, i. e. a large part of CO has still to be converted. This is due to the application of a higher reaction temperature in conjunction with low steam or water flows leading to a high CO/CO_2 ratio. As opposed to this, the figure reached in the Fluidized Bed Process and, especially, in the LURGI Moving Bed Process is very close to 2, the ratio aimed for in the case of methanol synthesis. Only part-stream conversion is then still necessary to reach the stoichiometric ratio of 2.

This favourable feature results from the fact that - in the case of the Moving Bed reactor, for example, - the conversion reaction occurs to a significant degree simultaneously in the gasification reactor itself.

Another point which should be considered is the content of non-reactive components. These might be inert or poisonous compounds as well as compounds which are not inerts but non-reactive such as CH_4. Here, one has to distinguish between the gases entering the process as impurities via the oxygen and those generated from the coal during gasification.

The former are dependent upon the performance of the air separation unit, where in all cases an oxygen purity of normally more than 98 % and, with little extra complexity, more than 99,5 % can be reached. The non-reactive compounds generated during gasification are largely the sulphur compounds H_2S and COS which have to be removed in any case to protect the catalyst. Besides, methane and nitrogen or

Tab. VI GAS ANALYSIS FOR ENTRAINED BED GASIFICATION

	K-T	SHELL-K	TEXACO
CO	55.9	63.9	45.8
H_2	29.6	28.8	34.2
CO_2	11.9	2.5	17.9
CH_4	0.1	0.1	0.3
REST	2.5	4.7	1.8
STOICHIOMETRIC RATIO	0.26	0.40	0.26
	0.46	-	0.66

Tab. VII GAS ANALYSIS FOR FLUID BED GASIFICATION

	WINKLER*	HTW*	U-GAS
CO	41.2		22.8
H_2	35.9		42.9
CO_2	17.4		29.8
CH_4	3.8		3.7
REST	1.7		0.8
STOICHIOMETRIC RATIO	0.32		0.25
	0.76		1.63

* LIGNITE

Tab. VIII GAS ANALYSIS FOR MOVING BED GASIFICATION

	LURGI	LURGI *	BGC/LURGI SLAGGER	RUHR-100 70 bar
CO	19.0	18.1	58.0	16.7
H_2	41.9	39.8	28.8	35.8
CO_2	28.2	30.5	3.2	30.5
CH_4	9.7	10.2	6.2	14.5
REST	1.2	1.4	3.8	2.5
STOICHIOMETRIC RATIO	0.29	0.19	0.42	0.11
	1.87	1.86	0.42	1.78

* LIGNITE

ammonia are released as well. The ammonia is removed by contact with water. Thus, mainly N_2, A_r and CH_4 remain as inert components in the gas.

A purge gas stream maintains the inert gas content below a certain level. Appropriate to the plant configuration, this purge gas stream is upgraded to hydrogen and recycled to the synthesis stage. Part-streams can be used as clean, sulphur-free fuel gas.

3.2 Biomass Gasification

The term biomass describes all vegetative useful and waste materials insofar as they are available for gasification in a practical manner. An incomplete review is given in Table IX.

In regard to the number of new process developments, the development of biomass gasification is now at the stage at which coal gasification found itself a decade ago, except that in the case of biomass gasification there are not yet any commercially applicable processes as they are required for economic methanol production. A review of on-going process developments is to be found in /13/ and /20/, for instance.

The process principles being applied are mainly

- o Moving bed
- o Fluidized bed.

The entrained bed is used, above all, for flash or hydropyrolysis, yielding mainly gaseous and liquid hydrocarbons. These processes will not be considered here.

In the case of biomass gasification, the moving bed reactor is operated not only in a countercurrent mode (updraft) as usually in coal gasification, but in a cocurrent mode (downdraft) as well. In the downdraft gasifier, the moisture and volatile matter of the biomass are routed through the oxidation and reduction zone of the gasifier bottom part. Exposure to the high temperatures in the oxidation zone and contact with the hot char effectively crack the heavy components into C_3^--compounds. Gasification reactors operating according to this principle had been widely used to produce fuel gas for diesel engines. This type of reactors is associated with the name of Mr. Imbert, and his successors now build apparatus up to a capacity of 3 MW_{th} /14/. Further scale-up is regarded critical due to geometrical considerations /16/ and, therefore, this process does not offer itself for large-scale methanol-from-wood plants.

Fluidized bed reactors include

- o autothermal and allothermal processes
- o stationary and circulating beds
- o catalytic and non-catalytic gasification.

Most of the recent American and European developments for synthesis gas production from biomass make use of the fluidized bed principle /15/. The four processes eligible for financial subsidy by the C.E.C. under its Methanol-From-Wood programme are exclusively of the fluidized bed type.

Tab. X and XI contain gas analyses for moving and fluid bed gasification /17/19/20/21/. With respect to CO and H_2 content, the Down-

Tab. IX SOURCES OF BIOMASS FOR GASIFICATION

WOOD	PLANT DEBRIS
LOGS	RICE HULLS
BRANCHES	COTTON BUSHES
BARK	FRUIT PRESSING RESIDUES
SAWDUST	
WASTES	

Tab. X GAS ANALYSIS FOR MOVING BED GASIFICATION

	DOWNDRAFT SERI		UPDRAFT SFW-FUNK
	1 BAR	7 BAR	1 BAR
CO	48.15	53.56	32
H_2	31.58	16.36	27
CO_2	14.72	25.04	21
CH_4	2.23	1.98	5
REST	3.32	N. R.	15
STOICHIOMETRIC RATIO	0.27	-	0.11
	0.57	0.26	0.71

N. R.: NOT REPORTED

Tab. XI GAS ANALYSIS FOR FLUID BED GASIFICATION

	BATTELLE		MINO (20 bar)[2]		LURGI
	CATALYTIC	NON-CATALYTIC	BEFORE CATALYTIC METHANE CONVERTER	AFTER CATALYTIC METHANE CONVERTER	CFB
CO	21.4	29.0	20.0	34.7	33.50
H_2	55.7	31.6	32.5	42.2	33.36
CO_2	20.7	23.1	30.2	21.5	26.62
CH_4	2.0	13.6	17.0	0.3	4.86
REST	0.14[1]	2.7[1]	0.3	0.3	1.66[3]
STOICHIOMETRIC	0.83	0.16	0.03	0.37	0.11
RATIO	2.23	0.93	1.37	1.07	0.87

[1] $C_2H_4 - C_3H_{6/8}$ [2] EXPECTED FIGURES [3] WITHOUT N_2

draft Gasifier shows some similarities to the Entrained Bed coal gasification. The high CO content is a result of high temperature treatment of the gases. Accordingly, a CO shift conversion stage is needed in this particular case.

Obviously, only limited information is available about pressure gasification. However, the few reported results don't show as large an increase in CH_4 content as might be expected. Hypothetically, the CH_4 content of the gases is determined by the kinetics of thermal degradation rather than by equilibria.

Tab. XI shows clearly, that gasification in the stationary Fluid Bed requires some reforming step (oxygen addition or catalytic steam reforming) in order to reduce the CH_4 content. Only the Circulating Fluid Bed has the capability to produce a synthesis gas containing a tolerable amount of CH_4. This is a result of the unique features of the CFB-principle, some of which should be mentioned here for clarity's sake:

- Constant and high temperature throughout the whole reactor volume, allowing gasification reactions to take place all through the reactor.
- No formation of gas bubbles which would cause a major part of the devolitization products to move through the reactor without any chance to react with the surrounding gases.
- Feeding the biomass into the bottom of the reactor which leads to a longer residence and, therefore, reaction time for the gas phase.

Generally, all the gas compositions exhibit a relatively high CO_2 content which corresponds to the low operation temperature and high heat losses of the test plants which were of the laboratory or pilot scale only as shown in Tab. XII. This table lists the throughputs achieved so far in the individual plants. As the gasification results were gained from rather small plants only, some uncertainty remains with respect to the performance in bigger reactors and under pressure. This should be kept in mind when looking at the overall plant configurations and at the cost data in the following chapters.

4. METHANOL PRODUCTION ROUTES

4.1 Coal to Methanol

A general scheme for the production of methanol from coal is shown in Fig. 4. The coal is gasified with steam and oxygen in a pressurized gasifier. The H_2 and CO contents of the raw gas depend on the selected process (see Tab. VI, VII, VIII) and on the feed coal. In some cases (LURGI Dry Ash e. g.), the required H_2/CO ratio can be achieved simply by reducing the CO_2 content of the gas. In many other cases it is necessary to convert part of the CO contained in the raw synthesis gas. This is accomplished by catalytic shift conversion without prior desulfurization.

Desulfurization, which is required in any case, and CO_2 removal take place simultaneously in the Rectisol gas purification section.

Tab. XII TEST CONDITIONS OF BIOMASS GASIFICATION UNITS

	PRESSURE BAR	TEMPERATURE °C	THROUGHPUT KG/H
SERI	1 7	800 - 1 000	28
SFW-FUNK	1	900 - 1 000 EXIT: 350	516
BATTELLE	1	700 - 750	25
MINO	30	750	10
LURGI	1	750	300

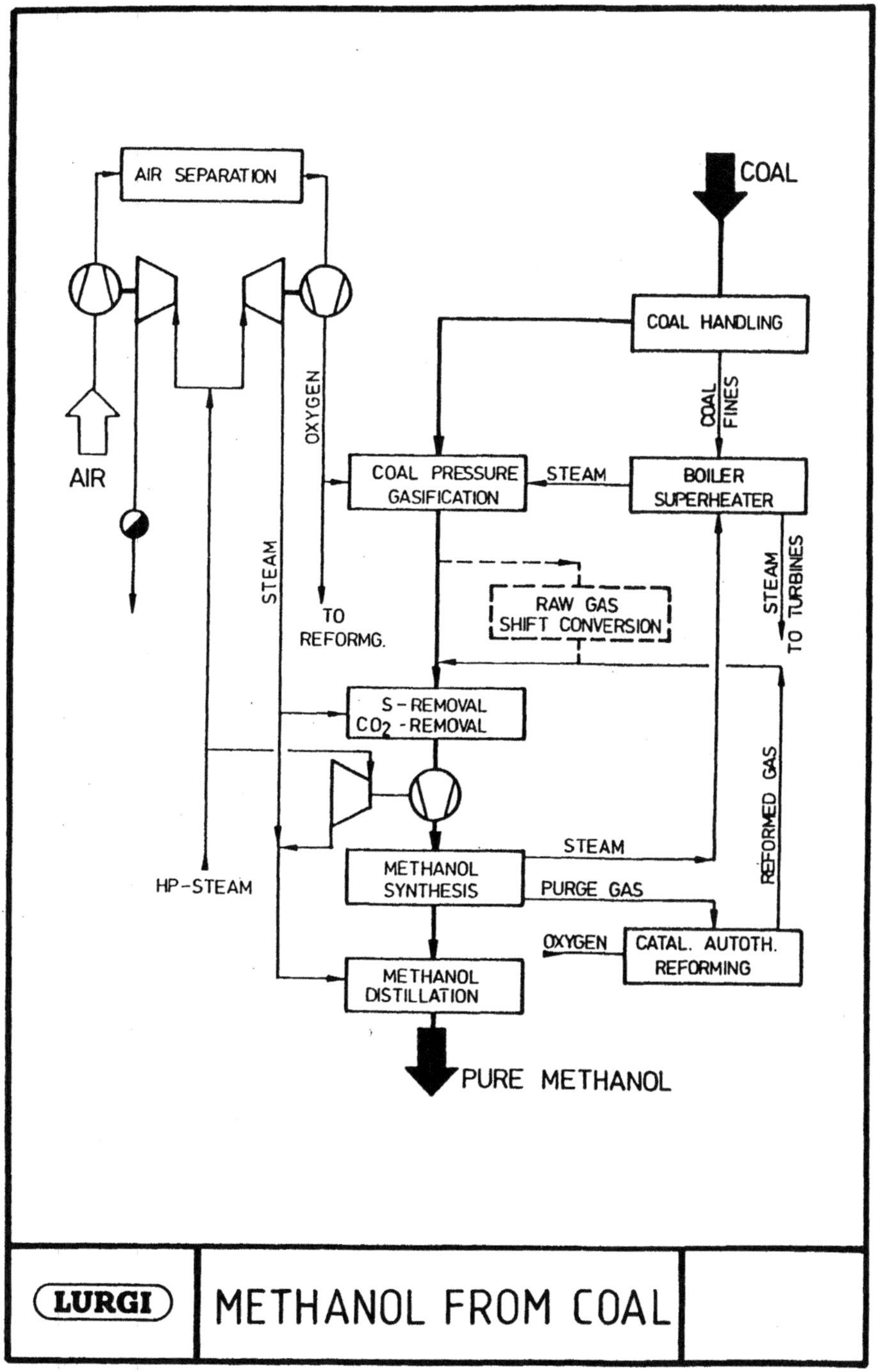
AIR SEPARATION
COAL
AIR
COAL HANDLING
OXYGEN
COAL FINES
COAL PRESSURE GASIFICATION
STEAM
BOILER SUPERHEATER
STEAM
TO REFORMG.
RAW GAS SHIFT CONVERSION
STEAM TO TURBINES
S - REMOVAL
CO_2 - REMOVAL
REFORMED GAS
STEAM
HP-STEAM
METHANOL SYNTHESIS
PURGE GAS
OXYGEN
CATAL. AUTOTH. REFORMING
METHANOL DISTILLATION
PURE METHANOL
LURGI
METHANOL FROM COAL

Figure 4

Using methanol as a solvent and applying temperatures below 0 °C, the sulfur content is reduced to less than 0,1 ppm.

Having been treated in this way, the gas can be routed to the methanol synthesis unit which works as already described in chapter 2.

A purge gas stream has to be withdrawn from the synthesis loop to maintain the level of inerts (CH_4, N_2, Ar). This purge gas can serve as a clean gaseous fuel. If the purge gas contains considerable amounts of methane, which is mostly the case when applying moving and fluidized bed gasification, this purge stream should be subjected to a reforming process. As shown in Fig. 4, this can be accomplished by autothermal catalytic reforming with oxygen. This reformer has the capability to process also the light ends recovered in the methanol distillation stage. The reformed gas is routed back into the synthesis loop. However, a purge stream is necessary also in this case to get rid for instance of N_2.

A block flow diagram in Fig. 5 shows the entire plant in greater detail as it would look like if a moving bed gasifier were used. The description of the relevant units can be found elsewhere /12/.

Considering the fact that the purge gas may contain a remarkable amount of methane, the idea was born to establish a co-production route of methanol and SNG as shown in the block flow diagram in Fig. 6. Of course, this route is applicable only where consumers are available for both products. Then, this route offers some economic advantages.

4.2 Wood to Methanol

The general scheme for methanol production from wood is very similar to that for methanol from coal. The main difference lies in the low sulfur content of the wood and, hence, of the synthesis raw gas. The respective levels are approx. 0.5-1 Vol. % H_2S for coal and less than 100 ppm for wood. This low sulfur content in general no longer allows the application of a crude gas shift conversion stage, but requires a clean gas shift, which tolerates this small amount of sulfur. The removal of sulfur might be accomplished over a fixed bed of ZnO or in a washing unit (HP, Purisol, Rectisol). The washing unit would simultaneously serve to remove the CO_2.

Whereas in the case of methanol from coal large-scale plants could be designed and built right away, because each single process step represents commercially proven technology, the situation is different for wood as a feedstock. Here, all the processes downstream of the wood gasifier can be regarded as proven technology. A commercial-scale wood gasifier under pressure has, however, not yet been operated. It will need some further test work in existing pilot facilities or even need to build new pilot facilities before a demonstration unit will be operable.

However, that methanol can doubtless be produced from wood indirectly, i. e. via gasification, was recently proven in the LURGI laboratories under a contract with C.E.C. A synthesis gas collected from a pilot plant based on the Circulating Fluid Bed principle was routed through shift conversion, gas purification and methanol synthesis units on line /22/, yielding roughly 0.4 kg methanol per kg of dry wood.

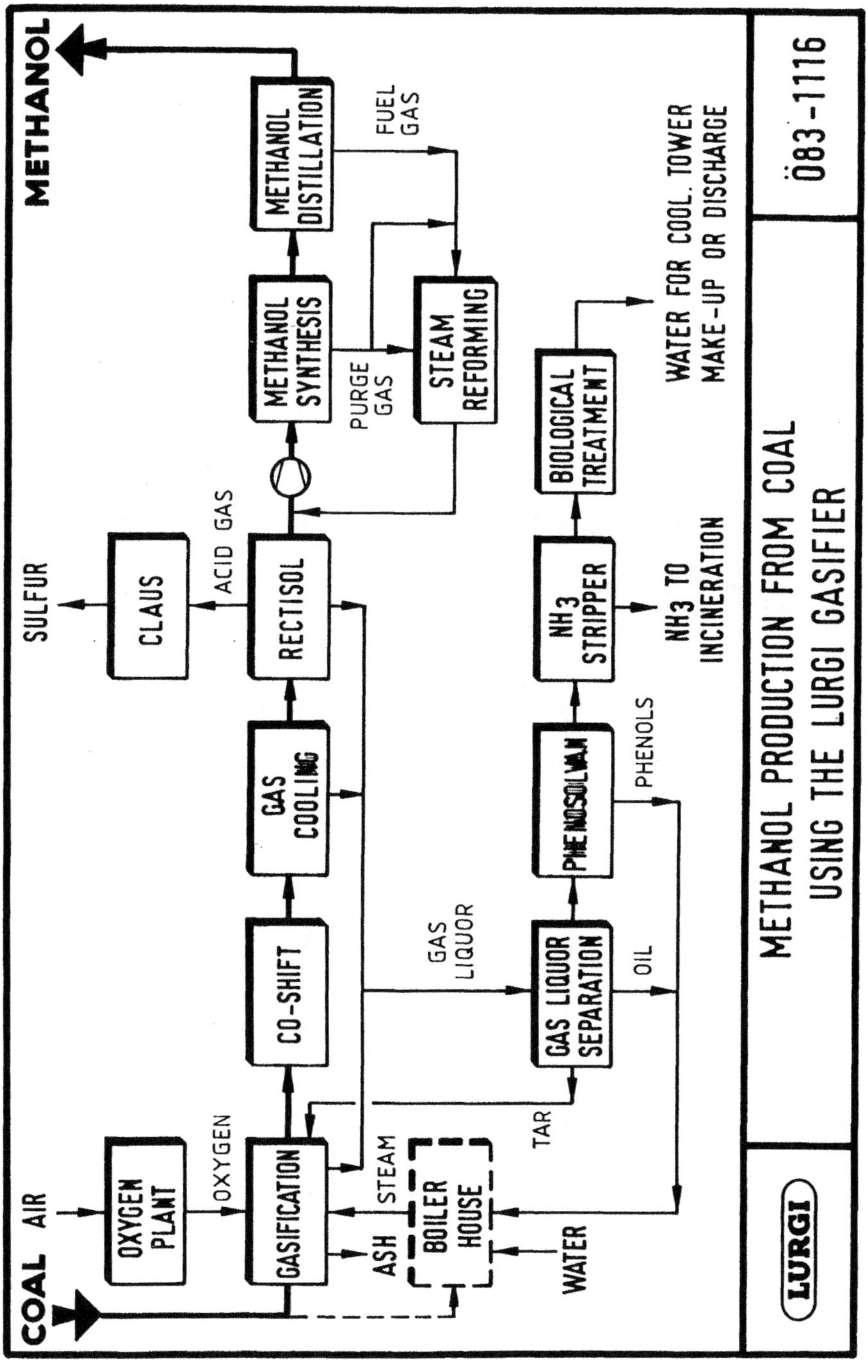
COAL
AIR
OXYGEN PLANT
OXYGEN
GASIFICATION
ASH
STEAM
BOILER HOUSE
WATER
CO-SHIFT
GAS COOLING
RECTISOL
ACID GAS
CLAUS
SULFUR
METHANOL SYNTHESIS
PURGE GAS
STEAM REFORMING
METHANOL DISTILLATION
FUEL GAS
METHANOL
TAR
GAS LIQUOR
GAS LIQUOR SEPARATION
OIL
PHENOSOLVAN
PHENOLS
NH3 STRIPPER
NH3 TO INCINERATION
BIOLOGICAL TREATMENT
WATER FOR COOL. TOWER MAKE-UP OR DISCHARGE
LURGI
METHANOL PRODUCTION FROM COAL USING THE LURGI GASIFIER
Ö83-1116

Figure 5

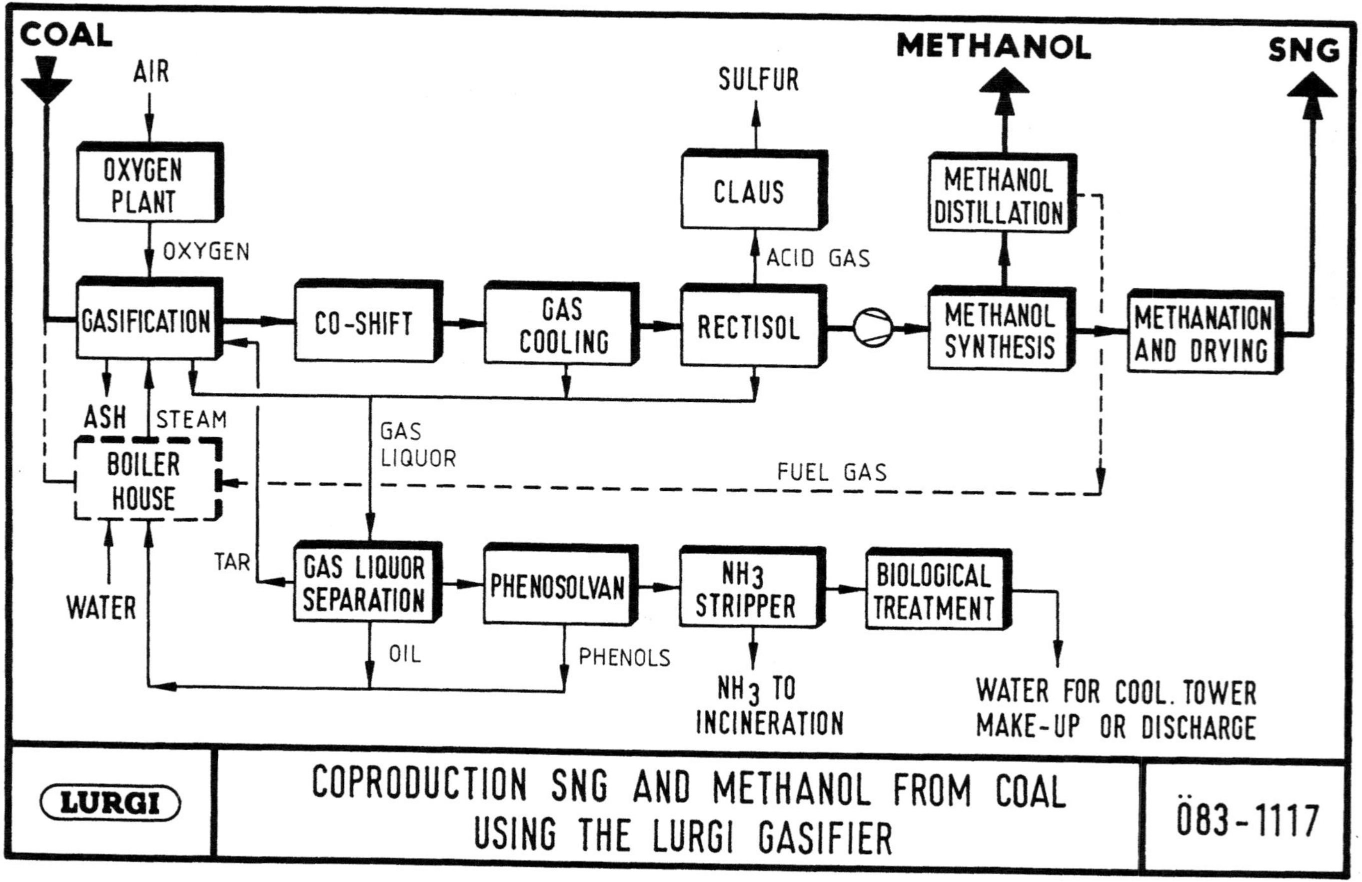
COAL
AIR
OXYGEN PLANT
OXYGEN
GASIFICATION
CO-SHIFT
GAS COOLING
RECTISOL
ACID GAS
CLAUS
SULFUR
METHANOL
METHANOL DISTILLATION
METHANOL SYNTHESIS
METHANATION AND DRYING
SNG
ASH
STEAM
BOILER HOUSE
WATER
GAS LIQUOR
FUEL GAS
TAR
GAS LIQUOR SEPARATION
PHENOSOLVAN
NH3 STRIPPER
BIOLOGICAL TREATMENT
OIL
PHENOLS
NH3 TO INCINERATION
WATER FOR COOL. TOWER MAKE-UP OR DISCHARGE
LURGI
COPRODUCTION SNG AND METHANOL FROM COAL USING THE LURGI GASIFIER
Ö83-1117

Figure 6

5. COST CONSIDERATIONS

Depending on the coal (wood) prices, the production costs for methanol are more or less significantly determined by the investment costs; e. g. to an extent of

- 85 % at low coal prices of 1 US \$/10^6 BTU
- 55 % at coal prices of 3 US \$/10^6 BTU.

The high investment costs reflect the fact that handling of solid materials and cleaning the raw synthesis gas requires expensive equipment. Besides, these particular feedstocks result in a poorer conversion efficiency, which drops from 70-75 % for natural gas or naphtha to approx. 50 % for coal. However, the specific investment costs for a coal-to-methanol plant are comparable with those of an electric power plant based on coal /23/. Practically the same can be stated for wood as feedstock.

When reporting and comparing cost data, all the assumptions, inclusions and exclusions must be taken into consideration. For a better comparison, a self-supporting plant has been investigated to derive the next figures. The only energy import to such a plant is the basic fuel for the final product, i. e. coal in this particular case.

The production costs of methanol shown in Fig. 7 have been calculated under the following assumptions /12/:

- self-supporting plant
- 10 % contingency
- 25 % of the investment for interest during construction
- 25 MM US \$ working capital
- 22 %/a interest on total capital cost
- 3,5 %/a of total capital cost for maintenance
- 3 %/a of total capital cost for labor
- 1,5 %/a of total capital cost for catalysts, chemicals and raw water.

The graph distinguishes between Western and Eastern US coals because they sell at different prices. Moreover, the LURGI Dry Ash Gasifier was selected for the Western Coal, which is a lignite whereas the British Gas/LURGI Slagging Gasifier is applied for the Eastern Coal (High Volatile Subbituminous - Pittsburgh No. 8). It can be seen from the diagram that for low-priced Western Coals, methanol production is close to being competitive.

To demonstrate the above-mentioned advantage lying in the coproduction of methanol and SNG and to show where methanol stands in comparison with other synfuels, the graph in Fig. 8 summarizes a bundle of information /23/. It is worth noting that the increase in efficiency for the co-production route from approx. 50 % to approx. 60 % does not only reduce the slope of the cost line in Fig. 8 (the slope represents the conversion efficiency) but also lowers the investment costs. The data in Fig. 8 are also based on a LURGI Moving Bed Gasifier.

To finalize the part dealing with coal, it should be repeated that each process step has already been installed and operated under commercial conditions and that therefore cost information can be collected very reliably.

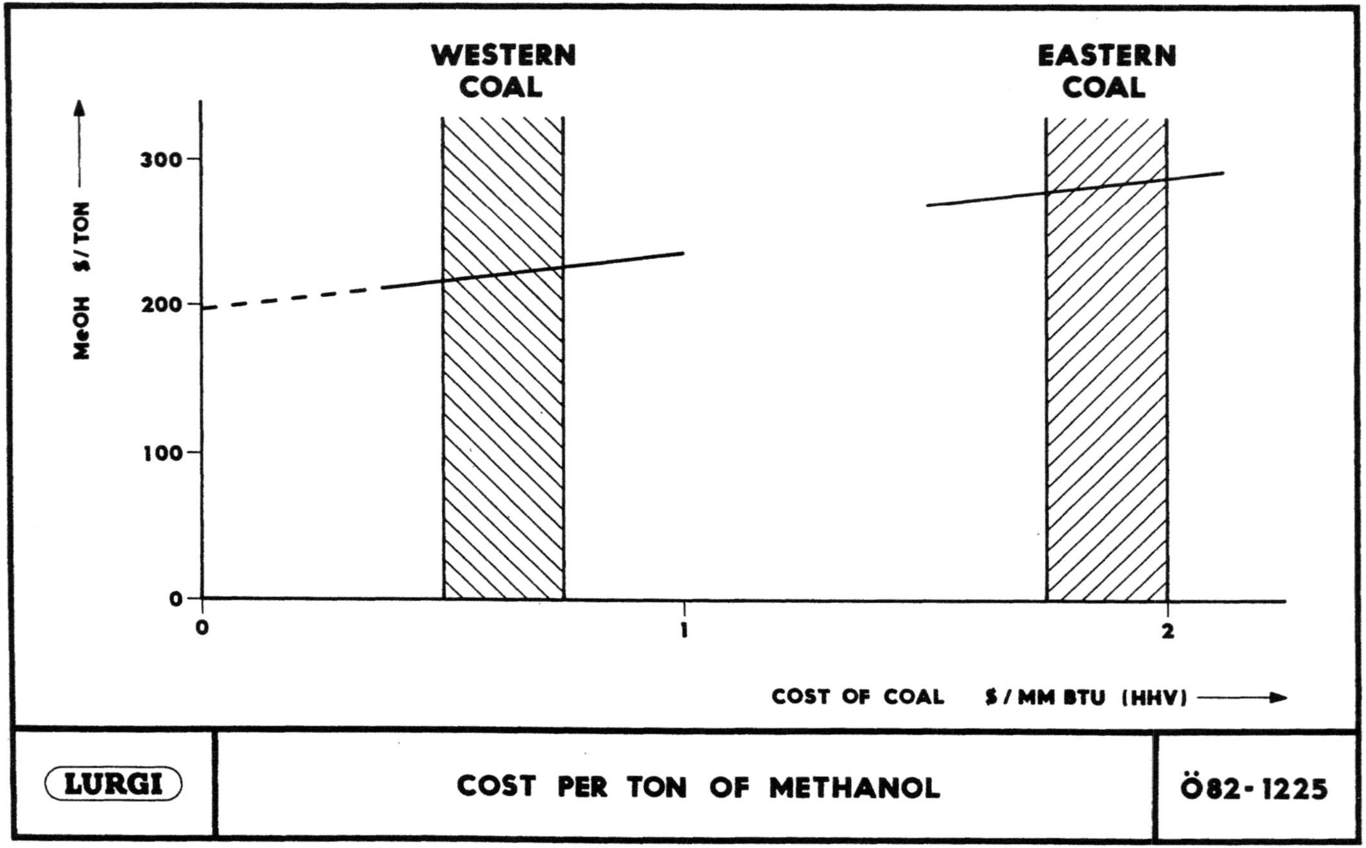
WESTERN
COAL
EASTERN
COAL
MeOH $/TON
300
200
100
0
0
1
2
COST OF COAL $/MM BTU (HHV)
LURGI
COST PER TON OF METHANOL
Ö82-1225

Figure 7

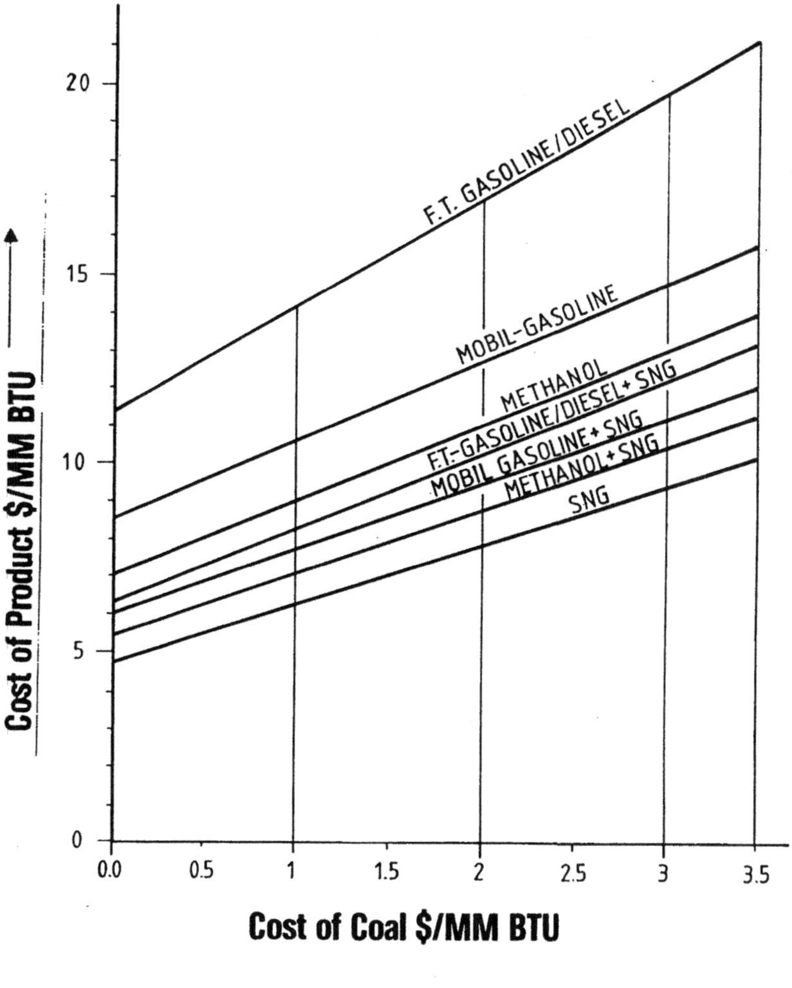

LURGI

Cost of Product versus Cost of Coal (based upon LHV)

Figure 8

The situation is different if wood is used as a feedstock for methanol production. Here, only little information is available. Most data are generated by extrapolation over a wide range of capacities. This leads to problems of scaling down and of corresponding specific cost increases. (The problem is not as severe for coal-based plants where maximum-size units or even multiple trains are considered.)

A detailed cost study is reported in /21/ for the Battelle catalytic gasification process. For zero wood price and for the most common price of 30 US $/t dry wood, the following production costs (converted) are obtained:

Methanol Production Costs

Wood price \ Capacity	200	2 000	t/d dry wood
0	531	188	US $/t
30 $/t	604	270	US $/t

Financing is based on private investors.

These data indicate that wood is competitive with coals costing approx. 2 US $/MM Btu when compared with Fig. 7.

Similar projected methanol production costs of 270 US $/t (wood at 30 US $/t) were calculated using a modified Texaco Coal Gasification process, i. e. an entrained bed gasifier /24/ for a plant exceeding the high capacity above by about 20 %.

Another cost estimate was established for an atmospheric fixed bed gasifier in Brazil /25/. The production costs vary between 305 US $/t and 474 US $/t, depending on the wood price (20 $/t; 40 US $/t) and the specific investment costs (2,5 US $/gal/a; 4 US $/gal/a). Taking the lower specific investment costs and a wood price of 30 US $/t for the sake of comparison, one ends up with production costs of 328 US $/t.

Preliminary calculations based on atmospheric pilot plant tests (s. Tab. XI) with a Circulating Fluid Bed gasifier result in production costs of 250 US $/t. This includes a contingency of 25 % on the total investment, 25 % capital interest and depreciation, 3 % labor and 4 % maintenance.

When reviewing the uncertainty of the cost estimates for methanol-from-wood plants it should be kept in mind

- that only the wood pretreatment (regarding its capacity) and the wood gasification (regarding its overall performance) units are uncertain,
- that these units represent roughly 25 % of the total investment,
- that the downstream processes are well-known and need only to be downscaled.

6. SUMMARY

The production of synthesis gas from coal via gasification and subsequent gas cleaning is well-known, commercially established practice. Gasification of coal under pressure is performed in the LURGI Moving Bed gasifier in many commercial units. Pressure gasification applying Fluid Bed (HTW) and Entrained Bed Gasification (Texaco) is at or near the demonstration stage.

Methanol production from synthesis gas has an installed capacity of more than ten million tons per year already. Plants built during the last decade or under construction now apply the economically favourable Low Pressure Methanol Synthesis.

Due to current price relations, methanol from coal is competitive in case of cheap coals (less than 1 US $/MM BTU) only. However, such coals are available.

Up to now, the production of synthesis gas from wood has been proven in pilot or bench scale units only. This process still needs demonstration on a larger scale. The development of wood gasification is pursued in many places in North America and Europe.

Reflecting the current status of the technology, cost estimates for methanol from wood have a higher level of uncertainty. Keeping this restriction in mind, forecasts have methanol production costs based on wood close to the costs calculated for cheap coals. This statement is true for large-size plants only (approx. 1 000 t/d of methanol). In so far, restrictions might origin from the size of the plant rather than from the technology. Such large-size plants require a forest industry and forest areas of approx. 800-1 000 km^2.

7. REFERENCES

/1/ Marschner, F.; Möller, F.-W.; Gelbke, H. P.: 'Methanol'; Ullmanns Enzyklopädie der technischen Chemie, 4. Aufl., Bd. 16, S. 621-633

/2/ Supp, E.: 'Technology of LURGI's low pressure methanol process'; Chemtech, 1973, 7, S. 430-435

/3/ Methanol; 'The LURGI Low Pressure Methanol-Process'; Lurgi Express Information Ö 1414

/4/ Emerging Energy and Chemical Applications of Methanol: Opportunities for Developing Countries. World Bank, Washington D.C., 1982

/5/ Yearbook of Industrial Statistics, 1979

/6/ Patel, J. G.: 'The U-Gas Process', Energy Res., Vol. 4, (1980), S. 149-165

/7/ Flesch, W.; Yagneswaran, S.: 'New Experience with Winkler Coal Gasification at Neyveli', Chem. Age of India, Vol. 27, (1976), S. 825-831

/8/ Schlinger, W. G.: Energy Res. 4, (1980), S. 127-136

/9/ Oorsouw, F. v.: in Kohlevergasung in der Energietechnik, VGB-TB 201, Essen 1979

/10/ Staege, H.: 'Energiesysteme bei der Anwendung der Flugstromvergasung von Kohle', Chem.-Tech., Vol. 9, (1980), S. 129-134

/11/ Schäfer, W. et al: 'Ruhr 100 - Neue Ergebnisse von der Weiterentwicklung der LURGI-Druckvergasung', Vortrag 27. DGMK-Tagung, 06.-08.10.1982, Aachen

/12/ Cornelius, G.; Vierrath, H.: 'Methanol and Electric Power from U.S. Coals Using LURGI Gasification', 5th Int. Coal Utilization Conf. Houston, 1982

/13/ Reed, T. B.; Graboski, N., Editors: Proc., Biomass-to-Methanol Specialists Workshop, Tamorron Durango, Colorado 1982

/14/ Zerbin: 'Vergasung von Biomasse', in /18/

/15/ Schiefelbein, G. F. et al: 'Thermochemical Conversion of Biomass: A Summary of R & D Activities Sponsored by the Biomass Energy Technology Division of DOE', Pacific North West Laboratory, Richland Wash., 1981

/16/ Van Swaaij, W. P. M.: 'Gasification - The Process and the Technology', Proc. of the Int. Conf. on Biomass, Brighton, 1980

/17/ Reed, T. B., Markson, M.; Graboski, M.: 'The SERI High Pressure Oxygen Gasifier', in /13/

/18/ Gehrmann, J., Editor: '<u>Thermochemische Gaserzeugung aus Biomasse - Vergasung und Entgasung -</u>, Jül.-Conf.-46, Jülich 1981

/19/ Himmelsiep, H.: 'SFW-Funk-Verfahren', in /18/

/20/ Beenackers, A. A. C. M.; van Swaaij, W. P. M.: 'Methanol from Wood', a State of the Art Review, Proc. of the Int. Conf. on Biomass, Berlin, 1982

/21/ Mudge, L. K.; Baker, E. G.; Mitchel, D. H.: 'Catalytic Generation of Methanol Synthesis Gas from Wood', in /13/

/22/ Lindner, Chr.: 'Gasification of Self-Renewing Raw Materials Particularly of Wood', Proc. of the Int. Conf. on Biomass, Berlin, 1982

/23/ Rudolph, P.: 'Synfuels from Coal - How and at what costs', 7th Energy Technol. Conf. Washington, 1980

/24/ Stevenson, W. A.: 'Large Wood to Methanol Plants', in /13/

/25/ Yang, V.: 'The Outlook for Methanol in Brazil', in /13/

HYDROCARBONS FROM METHANOL

K.-H. Keim, W. Dolkemeyer, H.-H. Gierlich, M. Schmidt
Union Rheinische Braunkohlen Kraftstoff AG
Postfach 8
5047 Wesseling
W. Germany

INTRODUCTION

Methanol Availability and Potential Use

In the past, methanol has mainly been considered as a chemical only, produced in limited amounts with high purity for chemical purposes such as formaldehyde synthesis. Only since the first oil crisis, in 1973, has the use of methanol as a petrochemical feedstock or fuel, for instance as motor fuel, and as an energy source for boilers and gas turbines, been seriously taken into consideration again and intensively investigated.

Methanol can be produced with proven technology from virtually any carbonaceous material such as heavy oil residues, flare gases, natural gas, coal, peat, shale-oil or even garbage, by steam reforming or partial oxidation respectively of these materials and subsequent conversion of the resulting synthesis gas to methanol (Fig. 1). Thus the use of methanol as a fuel would make the fuel and energy market less dependent on the presently predominating crude oil.

With respect to environmental aspects, methanol fuels yield several advantages compared with fuels derived from crude oil. They are sulfur-free, unleaded if used as motor fuel, and have a high burning efficiency. Thus the emission of detrimental products is decisively lower from methanol-powered engines or power plants.

However, the introduction of methanol fuel onto the market is problematic, as the necessary methanol capacities will only be built if marketing of the methanol fuel can be guaranteed by simultaneous development of the necessary infrastructure, mass production of methanol-powered automobiles, etc. These problems could be overcome, however, following the discovery made by MOBIL workers in the early 1970's that methanol can be converted with relative ease to high octane gasoline, which in its properties is identical to gasoline derived from crude oil. This Methanol to Gasoline (MTG) process permits the installation of the methanol capacities necessary for the subsequent introduction of methanol fuel, as it provides an outlet for these methanol capacities until such time as the requisite infrastructure changes for the distri-

G. E. Beghi (ed.), Synthetic Fuels, 165–185.

bution and use of methanol fuel have been effected. Here the European Community will have to play an important role in coordinating the joint introduction of methanol fuel onto the European market, as the independent action by only one of the members would further exacerbate the difficulties involved, particularly with infrastructure.

As soon as economics allow the introduction of methanol onto the market, whether in the form of methanol fuel or as MTG gasoline, methanol will also become an economically very interesting alternative petrochemical for aromatics or light olefins.

It is essentially because of these aspects that the conversion of methanol to hydrocarbons has been extensively investigated, especially in industrial research laboratories, within the past ten years.

In this paper an attempt will be made to review the activities in this field and to give an outlook with respect to potential commercial application.

FUNDAMENTALS OF METHANOL CONVERSION TO HYDROCARBONS

Stoichiometry and Thermodynamics

Reports of hydrocarbon formation from methanol can be found in the literature dating back as far as 1880, when Le Bel and Greene first described the decomposition of methanol in molten $ZnCl_2$ to hexamethylbenzene and methane.[1]

Hydrocarbons formation from methanol can essentially be considered as a dehydration of methanol in accordance with the following equation:

$$n\ CH_3OH \longrightarrow (CH_2)_n + n\ H_2O \qquad \text{Eq. 1}$$

where $(CH_2)_n$ either represents an olefin, cycloparaffin or a balanced mixture of paraffins and aromatics.

If paraffins were the sole reaction product, then according to stoichiometry the following equations could be written, which account for the hydrogen deficit:

$$(n+1)\ CH_3OH \longrightarrow C_nH_{2n+2} + C + (n+1)\ H_2O \qquad \text{Eq. 2}$$

$$(2n+1)\ CH_3OH \longrightarrow 2C_nH_{2n+2} + CO + 2n\ H_2O \qquad \text{Eq. 3}$$

whereas for the formation of mononuclear aromatics the coformation of H_2 has to be assumed:

$$n\ CH_3OH \longrightarrow C_nH_{2n-6} + 3\ H_2 + n\ H_2O\ (n \geq 6) \qquad \text{Eq. 4}$$

The conversion of methanol to hydrocarbons is an exothermic reaction. The heat of reaction released depends strongly on the product distribution, which can be influenced by the catalyst employed and by the reaction conditions used. All the promising processes for the conversion of methanol to hydrocarbons known so far are heterogeneously catalyzed over zeolite catalysts.

METHANOL CONVERSION TO HYDROCARBONS OVER ZEOLITE CATALYSTS

Zeolite Catalyst

Zeolites are porous crystalline alumina-silicates composed of AlO_4 and SiO_4 Tetrahedra. They are interconnected through shared oxygen atoms forming a three-dimensional framework with cavities, pores and channels of molecular dimensions. The chemical composition of zeolites follows the general formula:

$$M_n^+ \left[(AlO_4)_n \quad x \quad (SiO_4)_m \right]^{n-} \quad x \quad z\, H_2O$$

Crudely, a zeolite may be pictured as a modified silica, in which some silicon atoms have been replaced by aluminium. Each time such a substitution occurs, one exchangeable monovalent cation (or half a bivalent one) must enter the structure to preserve electron neutrality. If the cations are exchanged by protons, the zeolites become a strong Brönstedt acid. The protons can be easily detached from the "active sites" on the inner walls of the zeolite pores to which they are initially anchored, so that they can induce acid catalyzed reactions inside the pore system. It is thus the dehydrated H-form of the zeolites which is catalytically active

$$H_n^+ \left[(AlO_4)_n \quad x \quad (SiO_4)_m \right]^{n-}$$

The catalytically most significant zeolites are those having pore-openings characterized by 8-, 10- and 12 rings of oxygen atoms. These zeolites have a well-defined geometry of pores, channels and cages of molecular dimensions. This enables these zeolite catalysts to be unique in their ability to discriminate between reactant molecules depending on molecular size and shape, thus controlling product selectivity. The ZSM-5 catalyst, a synthetic zeolite of the pentasil group first synthesized by Mobil workers (Formerly Socony-Mobil), for example, only allows hydrocarbon formation of up to C_{10} - C_{12} within its pore system, thus being highly selective for the conversion of methanol to hydrocarbons in the gasoline boiling range.[2,3]

Reaction Mechanism

The catalytic conversion of methanol over zeolites has been intensively investigated, especially with zeolites of the ZSM-family. The reaction mechanism has still not been completely elucidated. Several mechanisms are presently being discussed:[4, 5, 6]

a) Carbene Mechanism This mechanism postulates the formation of carbene by dehydration of methanol and/or dimethylether over the acidic sites of the zeolite, as the initiating reaction. Carbene combines to form ethylene. Further addition of carbene to ethylene leads to cyclopropene, which is consequently rearranged to propylene.

$$CH_3OH \xrightarrow{H^+} :CH_2 + H_2O$$

$$CH_3\text{-}O\text{-}CH_3 \xrightarrow{H^+} :CH_2 + CH_3OH$$

$$2\ :CH_2 \longrightarrow C_2H_4$$

$$:CH_2 + C_2H \longrightarrow CH_2 — CH_2 \text{ (ring with } CH_2\text{)} \xrightarrow{H^+} CH_3—CH = CH_2$$

b) Carbonium-ion Mechanism According to this mechanism, a carbonium-ion is formed from dimethyl-ether as a first step on the acidic sites of the zeolite. This carbonium-ion then induces a series of consecutive reactions, also leading to lower olefins as primary reaction product:

$$CH_3\text{-}O\text{-}CH_3 + H^+ \longrightarrow CH_3OH + CH_3^+$$

$$CH_3^+ + CH_3OCH_3 \longrightarrow CH_3OH + CH_3CH_2^+$$

$$CH_3CH_2^+ \longrightarrow CH_2 = CH_2 + H^+$$

$$CH_3OCH_3 + CH_3CH_2^+ \longrightarrow CH_3CH_2CH_2^+ + CH_3OH$$

$$CH_3CH_2CH_2^+ \longrightarrow CH_3CH = CH_2 + H^+$$

c) Methylethylether Mechanism The third mechanism suggested assumes the formation of methylethylether from methanol and dimethylether as the primary reaction step.

The ether is cracked into methanol and ethylene. The ethylene then forms methylpropylether with dimethylether, which is split into methanol and propylene:

$$CH_3OH + CH_3\text{-}O\text{-}CH_3 \longrightarrow CH_3\text{-}CH_2\text{-}O\text{-}CH_3 + H_2O$$

$$CH_3\text{-}CH_2\text{-}O\text{-}CH_3 \longrightarrow CH_2 = CH_2 + CH_3OH$$

$$CH_2 = CH_2 + CH_3OCH_3 \xrightarrow{H^+} CH_3\text{-}O\text{-}CH_2\text{-}CH_2CH_3$$

$$CH_3\text{-}O\text{-}CH_2\text{-}CH_2\text{-}CH_3 \longrightarrow CH_3OH + CH_3\text{-}CH = CH_2$$

General Reaction Pathway

In spite of the question as to which reaction mechanism is actually valid, there is no doubt that the primary products are low molecular weight olefins, which can react further by oligomerisation, hydrogen transfer and cyclisation reactions to naphthenes, which are finally dehydrogenated to aromatics.

This yields the following general reaction pathway postulated by

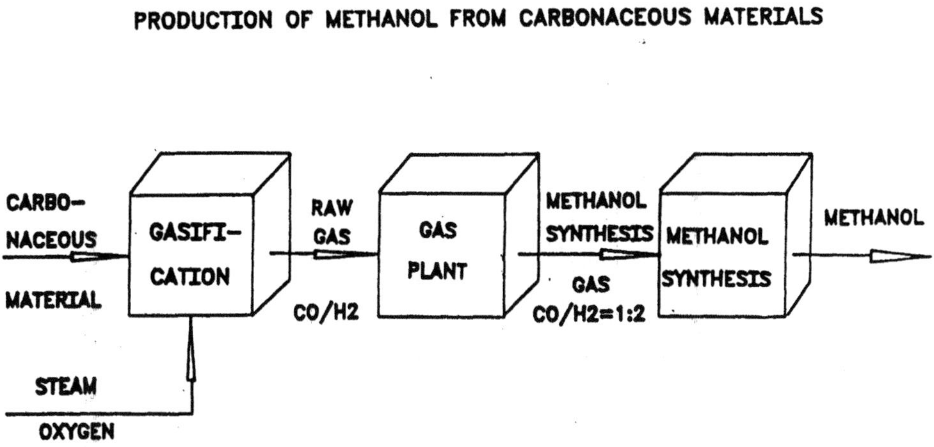

Figure 1: Production of Methanol from Carbonaceous Materials

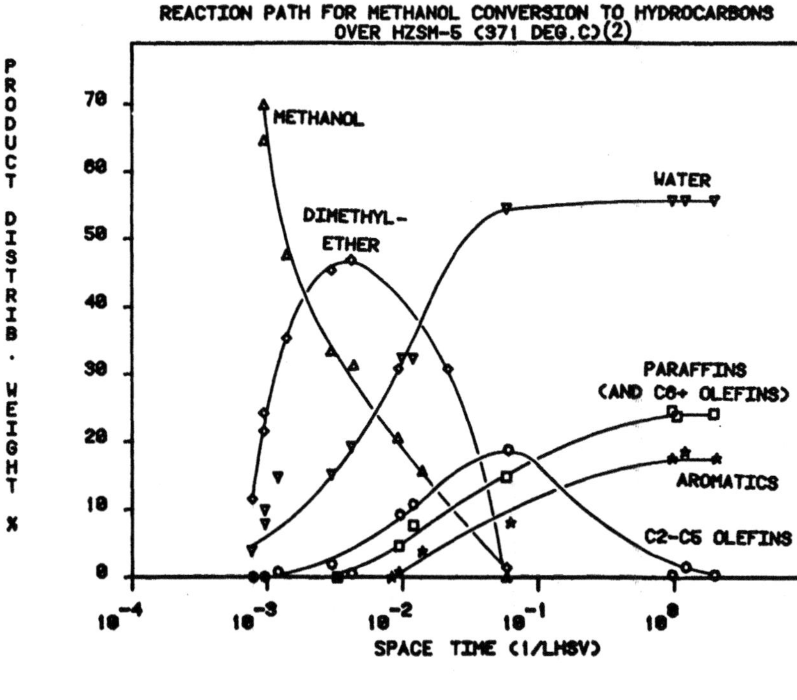

Figure 2: Reaction Path for Methanol Conversion to Hydrocarbons over HZSM-5 (371 Deg. C)

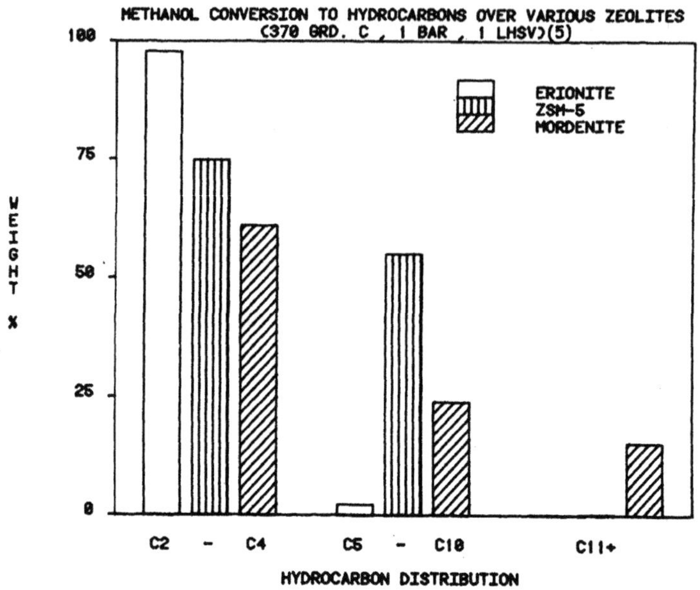

Figure 3: Methanol Conversion to Hydrocarbons over various Zeolites (370 Deg. C, 1 Bar, 1 LHSV)

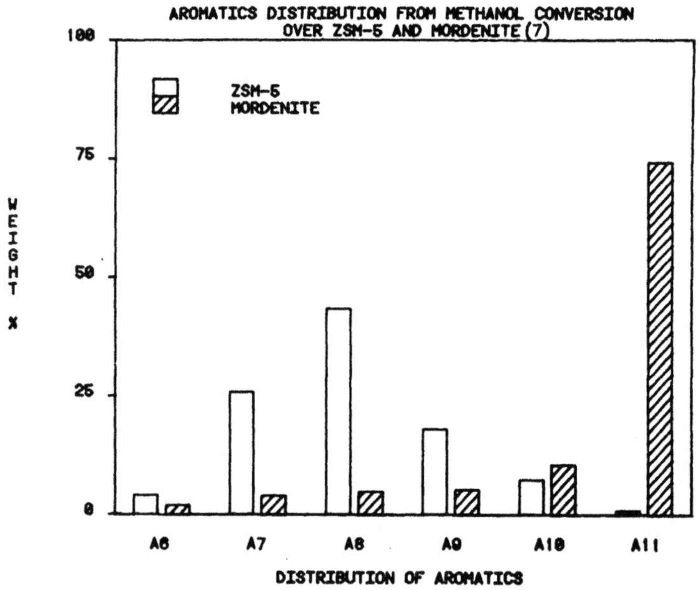

Figure 4: Aromatics Distribution from Methanol Conversion over ZSM-5 and Mordenite

Chang and Silvestri[2]:

$$x\ CH_3OH \xrightarrow{-\ x\ H_2O} C_2 - C_5\ \text{Olefins} \rightarrow \begin{array}{l}\text{Paraffins}\\ \text{Aromatics}\\ \text{Cycloparaffins}\\ C_6+\ \text{Olefins}\end{array}$$

The validity of this general reaction pathway is demonstrated in Fig. 2, where the product distribution for the conversion of methanol over ZSM-5 zeolite as a function of residence time is shown.[2] At short space times, high selectivities for C_2 - C_5 olefins are obtained, whereas with longer space times aromatics, paraffins and C_6+ olefins increase in yield.

"Shape Selective Catalysis"

Zeolite catalysts can control product selectivity because of their well-defined pore structures, which enable them to discriminate between reactant molecules depending on molecular size and shape. This phenomenon - a molecular sieve effect - has become known as "shape selective catalysis" in the literature.[5]

This effect is exemplified in Fig. 3, where the hydrocarbon distribution from methanol conversion over various zeolites at otherwise identical conditions is shown.[5] The comparison demonstrates that small pore zeolites like erionite (mean pore-diameter < 4.3 Å) yield the highest selectivity for C_2 - C_4 olefins, whereas large pore zeolites like mordenite (> 8 Å), which can accommodate molecules as bulky as hexamethylbenzene within their pore system, produce substantial amounts of C_{11}+ hydrocarbons.

The "intermediate" pore size zeolite ZSM-5, on the other hand, can only accommodate molecules as bulky as the symmetrical tetramethylbenzene (Durene) within its pore structure. Thus the formation of C_{11}+ hydrocarbons is negligible. This is also exemplified in Fig. 4, where the aromatics distribution obtained from methanol conversion over ZSM-5 and mordenite is compared.[7] The production of substituted benzenes for methanol conversion over ZSM-5 peaks at C_8, whereas with mordenite preferentially C_{11}+ aromatics (e.g. hexamethylbenzene) are produced.

Unfortunately, coke formation in zeolites, affecting the activity of the zeolite catalyst negatively, seems to be also a shape selective reaction controlled by the pore system geometry. It has been general experience that catalyst deactivation due to coking is rather rapid for small pore and large pore zeolites, whereas intermediate pore zeolites such as ZSM-5 have a high coke tolerance.[5]

METHANOL CONVERSION TO CHEMICALS

Results from bench-scale units

As the formation of the light olefins ethene and propene from methanol is a very promising concept commercially, it has been investigated very

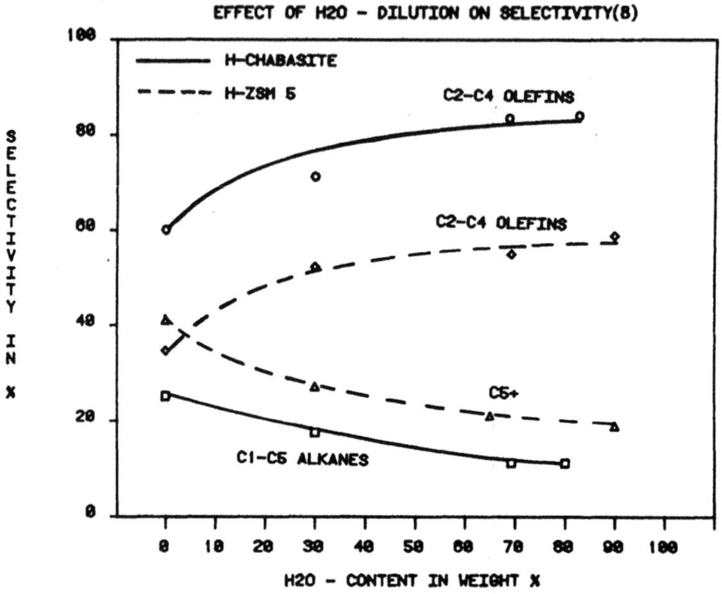

Figure 5: Effect of H_2O - Dilution on Selectivity

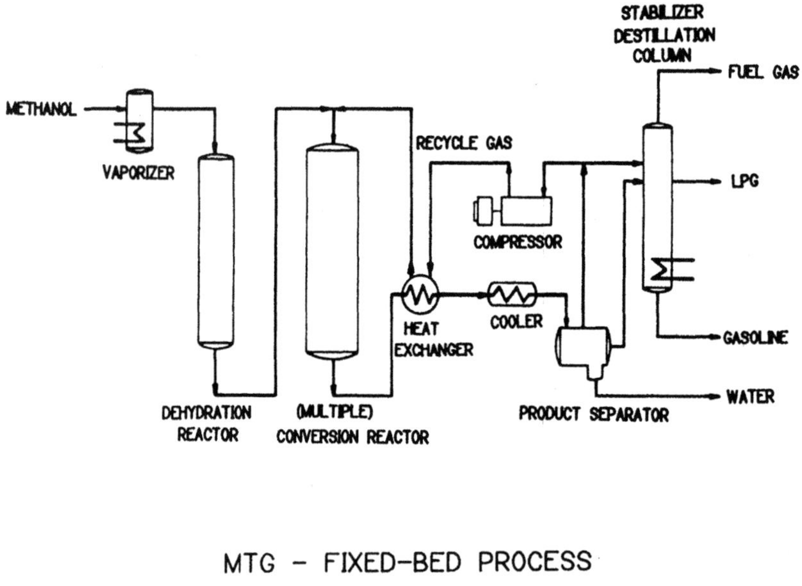

Figure 6: MTG - Fixed-Bed Process

intensively in the past few years. Because of the shape selectivity of zeolite catalysts, light olefins are preferentially produced over small-pore zeolites, like erionite and chabazite. Unfortunately, these zeolites also show the highest coking rates and, as a result, very short regeneration cycles of only several hours. Medium-pore zeolites, on the other hand, which have a higher coke tolerance, show a distinctly lower selectivity for light olefins. This is exemplified in Table I, where ethylene selectivity and lifetime for different pore size zeolites are compared.[8]

Table I: Comparison of Pore Structure, Ethene Selectivity and Lifetime

Zeolite	Pore Opening Å	Ethene Selectivity %	Lifetime
ZSM-5	5.4 x 5.6	15 - 25	Days
Chabazite-Erionite	5.3 x 3.5	35 - 45	Hours
Ferrierite (modified)	4.3	70 - 80	Minutes

According to the literature, essentially three routes for the selective conversion of methanol to ethene/propene over zeolite seem to be presently pursued:

- optimization of process conditions
- modification of medium-pore pentasil zeolites (ZSM-5, ZSM-8, ZSM-11, ZSM-23), in order to reduce accessibility of the pore structure for molecules larger than C_4
- optimization of small-pore zeolites (chabazite, erionite), to reduce coking and stabilize the catalyst framework against hydrothermal degradation.

The fact that light olefins are intermediates in the methanol conversion pathway can be utilized to optimize process conditions for higher light olefin yields. Essentially the main measures are:

- reduction of methanol partial pressure, e.g. by dilution of the methanol feed with water
- operation at partial conversion, e.g. by reduction in residence time.

The influence of residence time on light olefin selectivity has already been described (see Fig. 2).

The effect of water dilution of the methanol feed on the product selectivity for ZSM-5 and chabazite is shown in Fig. 5.[8] In both cases, a distinct increase in light olefin selectivity is observed with increasing water content in the feed.

Table II summarizes the results of a fluid bed study carried out with ZSM-5 catalyst with partial conversion.[9] Here, too, the methanol feed was diluted with various amounts of water. The C_2 - C_4 olefin selectivities obtained in this study with ZSM-5 are in the same range as those shown in Fig. 5.

Table II: Conversion of Methanol to Light Olefins over ZSM-5: Hydrocarbon Distribution

Reactant Composition	83 % CH_3OH 17 % H_2O	16 % CH_3OH 84 % H_2O
Conditions:		
Temperature, ^oC	299	324
Pressure, bar	1.42	2.9
Methanol, WHSV	0.4	0.2
Per pass conversion, %	52	62
Hydrocarbon product, wt. %		
C_2 - C_4 Olefins	45.6	51.2
C_5 - Olefins	2.1	1.3
C_1 - C_5 Paraffins	18.3	24.7
C_6+- Hydrocarbons	34.0	22.8
	100.0	100.0

Modification of medium-pore pentasil zeolites (especially ZSM-5) for higher C_2 - C_4 olefin selectivity involves impregnation and ion-exchange techniques as well as the synthesis of novel zeolites which have other oxides than silica and alumina (e.g. Fe, B, Cr, Ti, Zr) incorporated into the zeolite structure.

Table III shows an example of the effect on C_2 - C_4 olefin selectivity claimed if Zr- or Ti-oxide is incorporated into the ZSM-5 silica-alumina structure.[8)]

Table III: Effect of ZSM-5 Modification on Light Olefin Selectivity

Catalyst	Al-ZSM-5	Zr-Al-ZSM-5	Ti-Al-ZSM-5
C_2 - C_4 Olefin Selectivity %	56	65	64
C_5+-Selectivity	23	14	13
Feed: 33 Weight % Methanol, 66 % Water			
WHSV: 2			
Temp.:350 ^oC			
Pressure: 1 bar			

Further examples of ZSM-5 modifications which are claimed to increase the C_2 - C_4 olefin slectivity distinctly are impregnation with silica or phosphorous compounds[10)] (reported selectivity to C_2 - C_4 olefins 70 %) and ion-exchange with e.g. Zn, Pd, Mg-compounds[11)] (reported selectivity to C_2 - C_4 olefins > 70 % at partial conversion).

The optimization work performed on the small pore zeolites in order to increase C_2 - C_4 olefin selectivity is comparable with that described for the medium-pore zeolites, i.e.

- modification of the pore structure by incorporation of additional oxides other than silica and alumina into the zeolite framework

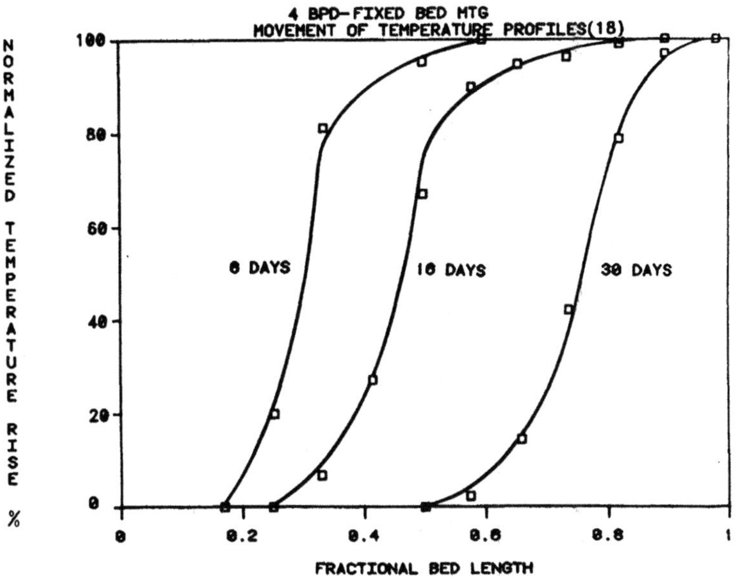

Figure 7: 4 BPD Fixed Bed MTG - Movement of Temperature Profiles

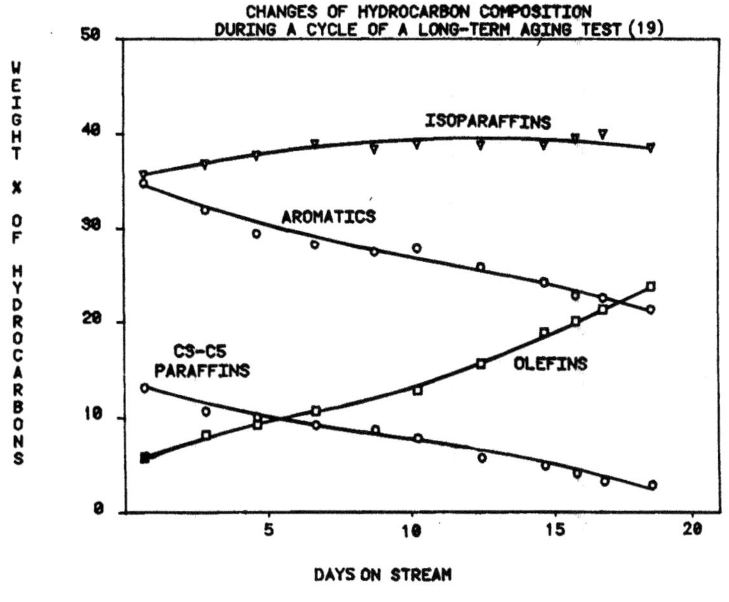

Figure 8: Changes of Hydrocarbon Composition during a Cycle of a long-term Aging Test

- modification of the catalyst properties by ion-exchange and/ or impregnation.

In addition, the small-pore zeolites with a relatively low SiO_2/Al_2O_3 ratio need stabilization against hydrothermal degradation which occurs at elevated temperatures in the presence of water. Thus the optimization work on small-pore zeolites also has to involve stabilization against hydrothermal degradation. The following options are presently being pursued:

- defined hydrothermal pretreatment
- stabilization by impregnation with decomposable silica-compounds and dealumination by, e.g., treatment with complexing agents.

These measures increase the silica/alumina ratio of these zeolites and make them more stable against hydrothermal degradation.

Examples for the improvements achieved in the conversion of methanol over modified small-pore zeolites are mostly reported in the patent literature.

Table IV shows as an example the product distribution obtained over a silanized chabazite/erionite mixture which was ion-exchanged with Mn. The patent cited reports unchanged activity after 53 regeneration cycles.[12)]

Table IV: Methanol Conversion to Light Olefins over Mn-exchanged, silanized Chabazite/Erionite

Temperature:	400 °C
Pressure:	atmospheric
WHSV:	0.75 h^{-1}
Feed:	65 % Methanol, 35 % H_2O
Conversion:	98.2 %
HC-Product Distribution:	
Ethene	39.3 Weight %
Propene	29.7 " "
Butene	7.8 " "
Methane	6.7 " "
C_2 - C_4-Paraffins	12.0 " "
Others:	0.8 " "
Dimethyl-ether	3.7 " "

Status of the Processes for Methanol Conversion to Chemicals

As stated earlier, methanol conversion to chemicals is a commercially very promising concept. Thus all major chemical companies like e.g. BASF, HOECHST, ICI and MOBIL perform extensive research work in this field.

As can be concluded from the description of the present activities performed, the processes for the conversion of methanol to olefins are still at the research stage. Speculations about potential industrial application are, thus, invalid as long as the questions with respect to long-term steady state selectivity to ethene and propene and the amount and price of necessary catalyst make-up cannot be answered.

METHANOL CONVERSION TO GASOLINE (MTG)

Results from bench-scale units and pilot plants

The basis of the MTG process is the selective conversion of methanol over ZSM-5 into hydrocarbons in the gasoline boiling range rich in aromatics but with a very low benzene content. It has already been shown that this can essentially be done by increasing the residence time of the reactant in the reactor to allow the secondary reactions, i.e. olefin-oligomerisation, cyclisation and aromatization, to proceed (see Fig. 2).

Two versions of the MTG process are applicable, the fixed-bed and the fluid-bed process. Both have been investigated extensively since 1975, in bench-scale units and in 4 BPD pilot plants,[13, 14] mainly by MOBIL workers, who discovered this process. The aim of the studies was to collect sufficient data for the definition of potential operating conditions yielding maximum gasoline with optimum quality.

Based on all collected data, acceptable process ranges guaranteeing excellent gasoline yield and quality conditions could be determined. For the 4 BPD fixed-bed and fluid-bed units they are as shown in Table 5.[15] As can be seen from the comparison, the gasoline yield of the fluid-bed is about 3 percent higher, while the product composition is nearly identical. In both cases alkylation of the C_3/C_4-olefins is assumed.

The comparison of the gasoline properties is again in favour of the fluid-bed process, which exhibits a distinctly higher Research and Motor Octane Number.

Although many synfuel projects have been stopped in the past few years, studies of the MTG-process have been continued, due to the promising results obtained. They have culminated in the construction of a commercial fixed-bed plant in New Zealand and the operation of a 100 BPD fluid-bed demonstration plant at URBK in Wesseling, Germany. The latter project is financially supported by the German Bundesminister für Forschung und Technologie and the U.S. Department of Energy.

Table V: Process Conditions and Product Yields/Properties

	Fixed Bed	Fluid Bed
Temperature °C		
Dehydration Reactor	304	-
Conversion Reactor	415	413
Pressure bar	21.7	2.75
Recycle Ratio (mole)	9.0	-
Space Velocity WHSV	2.0	1.0
MeOH Conversion %	>99.9	>99.9
Hydrocarbon Product WT %		
Light gas	1.4	5.6
Propane	5.5	5.9
Propylene	0.2	5.0
i-Butane	8.6	14.5
n-Butane	3.3	1.7
Butenes	1.1	7.3
C_5+-Hydrocarbons	79.9	60.0
	100.0	100.0
Aromatics Vol. %	32	27
Research Octane Number	93	97
Motor Octane Number	83	87
Distillation °C		
10 % Evap.	46	46
50 % Evap.	99	102
90 % Evap.	166	168
Gasoline (including Alkylate) (RVP = 0.63 bar ((9 psi))	85.0	88.0
LPG	13.6	6.4
Fuel Gas	1.4	5.6
	100.0	100.0

Status of the Fixed-Bed Process

When the New Zealand Government gave top priority to maximum self-sufficiency in liquid fuels, it decided to exploit the nation's natural gas reserves in order to produce methanol and partly convert it into gasoline. After a comprehensive analysis of alternative synthetic fuel technologies, the government selected the MOBIL MTG process, providing a more economic and energy-efficient route than the Fischer-

Tropsch process.[16] When the decision was made in the early eighties to build a unit for the production of 12,000 - 14,000 bbl/d of gasoline, only the fixed-bed MTG process was commercially available. A schematic diagram of the fixed-bed process is shown in Figure 6.

Methanol is dehydrated to an equilibrium mixture of methanol, dimethylether and water in the dehydration reactor. The equilibrium mixture is then diluted with recycle gas and converted to hydrocarbons and water over the zeolite catalyst in the adiabatic conversion reactor. The recycle gas, which is comprised primarily of light hydrocarbon products (methane, ethane, propane), provides mass to absorb the heat of reaction and thereby reduces the temperature rise in the conversion reactor to a tolerable level. The recycle gas lowers the partial pressure of the reactant, which has to be compensated by increasing the reactor pressure to about 20 bar. About 20 and 80 % of the total heat of reaction are released in the dehydration and conversion reactors, respectively. The reactor effluent is cooled and partially condensed; the aqueous, liquid hydrocarbon and gaseous phases are separated; most of the gas is recycled.

In the New Zealand plant the MTG hydrocarbon product is refined in three distillation columns. The first removes dissolved gases and some of the more volatile hydrocarbons, which are used as fuel in the plant. The second removes the LPG components, and also recovers high-vapour-pressure gasoline components. The third splits the remaining product into "light" and "heavy" gasoline. The heavy gasoline from the splitter contains durene, which is an unwanted high-melting-point compound. Durene concentration is reduced by passing the heavy gasoline mixed with hydrogen-bearing recycle gas over a special catalyst at about 35 bars and 350 ^{o}C. Blending of light and heavy treated MTG-gasoline and a high vapour pressure gasoline fraction along with additives forms a premium gasoline with a quality as required by the N.Z. standards.[17]

The adiabatic fixed-bed process operates at essentially 100 % conversion of methanol to hydrocarbons and water.

Most of the conversion reaction occurs over a narrow band of the catalyst bed. As the catalyst is deactivated by coking, the band moves down the reactor. Finally, methanol breakthrough occurs and unconverted methanol appears in the aqueous phase product, and the cycle is terminated.

Typical temperature profiles of the bench and pilot-plant reactors, as shown in Fig. 7, indicate aging patterns similar to those expected in a commercial unit.[18] A corresponding change in product distribution through a cycle is plotted in Fig. 8[19] In order to maintain a reather even product distribution, the catalyst in the conversion reactor requires intermittent regeneration by burning the coke off the catalyst with dilute oxygen.

According to information from literature, the output of the New Zealand fixed-bed unit will be about 500,000 - 600,000 t/a of gasoline, which is supposed to be roughly 80 weight % of the hydrocarbon produced. An alkylation unit is not planned because the LPG will be used as such. Including alkylation, the gasoline yield could be increased up to 85 %, which is still more than 3 percentage points lower than in the fluid-bed process. This and the changes in hydrocarbon com-

Figure 9: 100 BPD MTG Demonstration Plant

position between regeneration cycles are some of the disadvantages of the fixed-bed process compared with the fluid-bed process. Table VI outlines the advantages and disadvantages of the fixed-bed process.[20]

Table VI: MTG-Fixed-Bed Process

Advantages	Disadvantages
- No catalyst losses - Simple technology	- Multiple reactors - Losses of thermal efficiency - Recycle gas - High system pressure - Intermittent catalyst regeneration - Changing product composition

Status of the Fluid-Bed Process

Because of its complexity, scaling-up of a fluid-bed process in general needs more steps before commercialization than a fixed-bed process does. This is one of the reasons for the MTG fixed-bed process being already commercially available today and why a fixed-bed unit will go on-stream already in New Zealand in 1985.

The fluid-bed process for the conversion of methanol to gasoline was developed stage by stage. Laboratory tests with small bench-scale units were started in the mid-seventies. They were followed by intensive investigations in a 4 BPD fluid-bed pilot plant.[14]

Sufficient data for commercial scale-up will finally be received from a 100 BPD unit, presently being operated at URBK in Wesseling, Germany.

The 100 BPD fluid-bed demonstration plant project was initiated nearly four years ago, when the three industrial partners, Mobil Research and Development Corporation, Uhde GmbH and Union Rheinische Braunkohlen Kraftstoff AG (URBK) signed an agreement regulating the cooperation in the joint venture. Mobil supplied the basic design package and the catalyst, Uhde provided the detail engineering and built the plant, whereas URBK, as project manager, is responsible for plant operation.[21, 22]

The construction of the plant was completed within the schedule, in August/September 1982.

First methanol feed to the reactor was on December 15, 1982, after pretests with a non-reactive fluid-bed catalyst and nitrogen had ascertained the expected hydrodynamics in the system.

Figure 9 shows the 100 BPD MTG plant at Wesseling. The height of the unit is about 52 m.

Since methanol-to-gasoline conversion is a highly exothermic reaction, the design of the heat removal system may have an important impact on commercial scale-up. About 1700 kJ/kg of methanol converted have to be removed from the reactor. Heat removal in a fluid-bed sys-

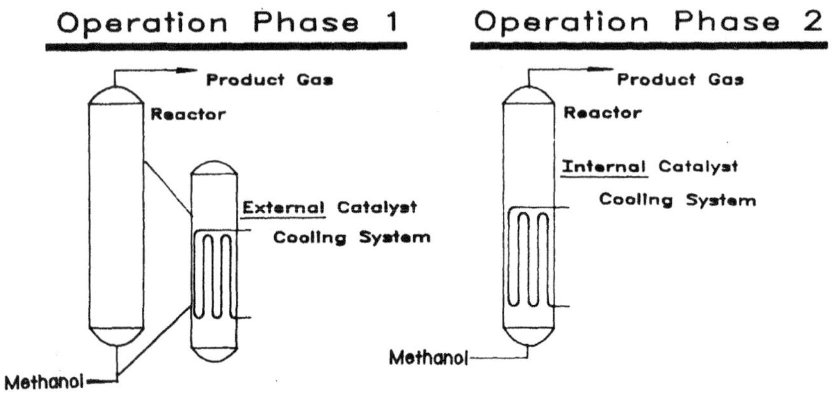

Heat removal can be carried out

Phase 1: by circulating catalyst through an external cooler

Phase 2: by means of internal heat exchanger pipes immersed in the dense bed reactor

Figure 10: Operation Phase I versus Operation Phase II

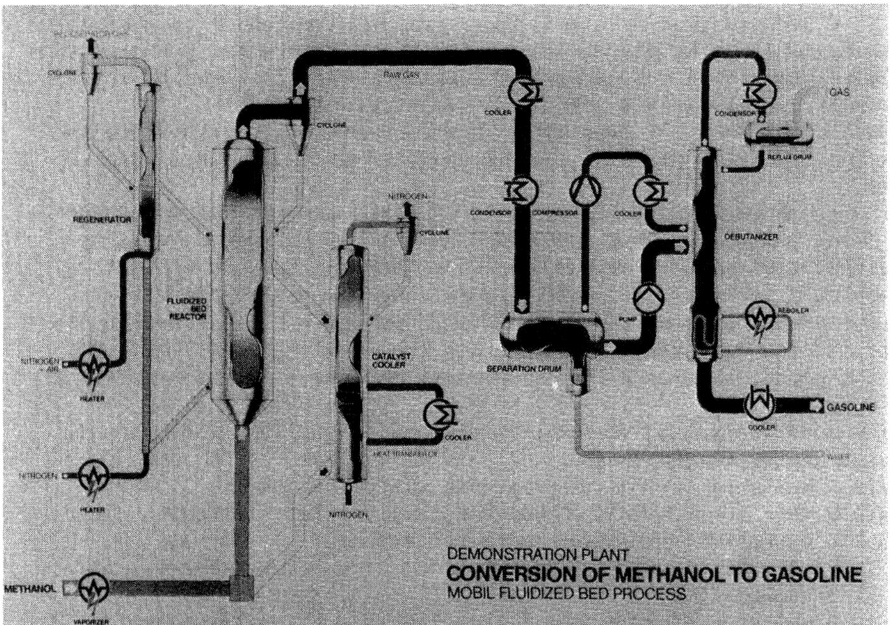

Figure 11: Demonstration Plant - Conversion of Methanol to Gasoline Mobil Fluidized Bed Process

tem as shown in Fig. 10 can be carried out according to two modes:

- by circulating catalyst through an external cooler or
- by means of internal heat exchanger pipes immersed in the dense bed reactor.

Both modes, each having its advantages, will be tested in the 100 BPD demonstration plant at Wesseling. Fig. 11 shows a simplified flow diagram of the Phase I operation mode, with external catalyst cooling, which was in operation from December 1982 until the end of September 1983. The feedstock is methanol, blended with water to simulate crude methanol. Vaporized methanol is charged to the reactor riser. Together with catalyst from the cooler, it rises into the reactor, where the methanol-to-gasoline conversion takes place. The heat of the exothermic reaction is removed by the catalyst circulating to the external cooler. To maintain a constant catalyst activity in the reactor, a small fraction of coked catalyst is circulated continuously through the regenerator, where the coke is burnt off.

The product gas leaves the reactor, passes through a cyclone and a filter, where entrained catalyst particles are removed, and goes into the separation section of the plant. Water is separated and withdrawn. Both the hydrocarbon liquid phase and the gaseous phase are sent to a stabilizer distillation column. The overhead product of the column is light hydrocarbons (C_4 and lighter), whereas the gasoline is withdrawn at the bottom and pumped to the storage facilities within the battery limits.

Table VII summarizes the ranges of process conditions at which the plant has been operated and the major highlights of the first operation phase with external cooler.[22] As can be concluded from these figures, operation with the external cooler was extremely successful. All goals were met or even exceeded. A summary of the major results obtained reads as follows:

- heat removal with external catalyst cooler successfully demonstrated
- successful scale-up of the 4 BPD pilot plant to demonstration unit (scale-up factor: 40)
- operation at steady-state conditions with continuous regeneration demonstrated
- constant gasoline yield and quality at steady-state conditions obtained
- necessary data for the conceptual design of a commercial size plant with external cooler acquired.

In the last quarter of 1983 the plant was modified and has been back on stream with the internal reactor cooling mode in operation (see Fig. 10) since January 1984. The results obtained so far indicate identical performance of the internal reactor cooling mode with respect to methanol conversion, product yield and quality. The heat removal scheme proved excellent, providing an isothermal temperature profile in the reactor. Thus both versions of the fluid-bed MTG process might be a commercially viable alternative to the fixed-bed process.

Table VII: 100 BPD Demonstration Plant Ranges of Operating Conditions and Major Highlights
Phase I, Dec. 15, 1982 - Sept. 28, 1983

Reactor Pressure	2.7 -	4.5	bar	
Reactor Temperature	380 -	430	°C	
WHSV	0.5 -	1.5	h^{-1}	
MeOH Feed	100 -	200	BPD (1050 kg/h)	
Methanol Conversion	> 99.9	%		
Max. Gasoline yield	90	%		
Time on stream (with methanol)	4148	h		
Total Service Factor (incl. scheduled shutdown)	60	%		
Service Factor for scheduled test runs March - Sept. 1983	100	%		

ECONOMIC CONSIDERATIONS

In order to produce 1 t of gasoline from methanol, approximately 2.5 t of methanol have to be converted in the fluid-bed MTG process and 2.7 to 2.85 t in the fixed-bed process. The present price level for methanol in West Germany is ca. 430 - 450 DM/t. Thus the costs for the methanol feed needed per ton of MTG-gasoline produced in the case of the fluid-bed MTG process, for example, amount to 1075 - 1125 DM. This is far in excess of the present market price of premium gasoline of ca. 800 DM/t, which does not, however, cover the costs involved. This rather simple calculation demonstrates that the price of synthetic MTG-gasoline is strongly dependent on the price of methanol and thus dependent on the price of the feedstock used. Presently only extremely cheap gas resources are competitive, as long as no governmental aids, such as a reduction in taxation on synthetic fuels, are granted.

A similar calculation could be made for the conversion of methanol to light olefins. Assuming the validity of the best results reported so far, the economics show cost levels of > 2000 DM/t ethylene and so still do not reach by far the breakeven point, today. Such calculations do not account for all the imponderabilities, however, which might be encountered before the stage of commercialisation, as all of these processes are still at the development stage.

REFERENCES

1) J. E. Le Bel, W. H. Greene, Am. Chem. J. 2, 20 (1880)

2) C. D. Chang and A. J. Silvestri, J. Catal., 47, 249 (1977)

3) S. L. Meisel, J. P. McCullough, C. H. Lechthaler, P. B. Weisz, Chemtech., 6, 86 (1976)

4) F. Asinger, Erdöl und Kohle, Erdgas, Petrochemie, 36, 130 (1983) and literature cited there

5) C. D. Chang, Catal. Rev. Sci. Eng., 25 (1), 1 (1983) and literature cited there

6) F. X. Cormerais, G. Perot, F. Chevalier, M. Guisnet, J. Chem. Research (5), 362 (1980)

7) C. D. Chang, W. H. Lang and W. K. Bell, Catalysis of Organic Reactions (W. R. Moser, ed.), Dekker, New York, 1981, p. 73

8) U. Dettmeier, E. J. Leupold, H. Litterer, H. Baltes, W. Herzog, F. A. Wunder in 'C_1-Chemie, Expertengespräch und Statusseminar 25./26.05.1982', ed. Projektleitung Rohstoffforschung, Kernforschungsanlage Jülich GmbH, Jülich 1982 and: Erdöl und Kohle, Erdgas, Petrochemie, 36, 365 (1983)

9) J. A. Brannan, W. E. Garwood, S. Yurchak, W. Lee, 'Proc. of Int. Seminar on Alternative Fuels', Liege, Belgium, May 25 - 27, 1981

10) W. W. Kaeding and S. A. Butter, J. Catal., 61, 155 (1980)

11) DOS 29 12 068 of 25.10.1979

12) DE-PA 30 08 146.0 (04.03.1980)

13) S. E. Voltz, J. J. Wise, 'Final Report, Nov. 1976, Contract No. E (49-18)-1773, U.S. Energy Research and Development Administration'

14) A. Y. Kam and W. Lee, 'Final Report, Contract No. Ex-76-C-01-2490, U.S. Department of Energy'

15) F. B. Fitch, W. Lee, 'Proc. of the Int. Pacific Conf. on Automotive Engineering, Hawaii, Nov. 1981'

16) 'N.Z. Pioneers Fuel Process', Special Bulletin, November 1979, Ministry of Foreign Affairs, Wellington, N.Z.

17) A. L. Titchner, Chemistry and Industry, 6 Nov. 1982

18) D. Liedermann, S. Yurchak, J. C. W. Kuo, W. Lee, 'Proc. 15th Intersociety Energy Conversion Eng. Conf. 1980', p. 1573

19) S. Yurchak, S. E. Voltz, J. P. Warner, Ind. Eng. Chem. Process Des. Dev., 18, 527 (1979)

20) W. Dolkemeyer, D. Flatow, N. Thiagarajan, R. Gould, 'Vortragsveröffentlichung 453', Haus der Technik, Essen, Herbst 1981

21) H.-H. Gierlich, N. Daviduk, H. Heiart, Compendium 82/83, Ergänzungsband Erdöl und Kohle, Erdgas, Petrochemie, p. 99, 1982

22) K.-H. Keim, J. Maziuk, A. Tönnesmann, 'Proc. VI Symposium on Alcohol Fuels, Ottawa, May 21 - 25, 1984', II, pp. 160/166

PYROLYSIS OF COAL AND PYROLYSIS UNDER HYDROGEN PRESSURE

R. CYPRÈS
Free University of Brussels (C.P. 165)
50 Av. F.D. Roosevelt
1050 Brussels
Belgium

ABSTRACT. The pyrolysis of coal must be considered in terms of its condensed polycyclic and mainly aromatic structure. Considering the bonding energies, the order of the different stages of devolatilisation can be predicted. The experimental parameters such as heating rate, influence of mineral impurities, reaction pressure and reactive atmosphere play an important role in the kinetics of the formation of volatile matter. The influence of the coal mass and of the gaseous atmosphere on its gasification rate will be considered. A mass effect has been shown which implies to work in a fluidised or in an entrained bed to obtain high gasification rates.
Hydropyrolysis means carbonisation under hydrogen pressure, allowing a considerable increase in the percentage of volatile matter which can be extracted from coal by increasing the partial pressure of hydrogen. The constituents of the oils are different in nature from those of the tars because of the chemical reaction which takes place between the hydrogen and the free radicals produced by the intramolecular bonds breaking within the coal 'macromolecule'.
Hydropyrolysis is a third possible way of coal gasification and liquefaction having the great advantage of requiring a relatively small investment if the installation is integrated in a large coal-consuming plant, such as a thermal power station or a gasification plant.

PART I. STRUCTURE OF COAL AND PYROLYSIS PROCESS

1. INTRODUCTION

Pyrolysis is the thermal decomposition of coal or other organic compounds. It may take place in a confined atmosphere (coke-oven), in inert gas or in the presence of a reactive gas. This is the case with hydropyrolysis or carbonisation in the presence of hydrogen, of pyrolysis under pressurised steam, in the case of steam gasification, which are processes where the pyrolysis of coal always occurs before its gasification.
The primary stage of the thermal decomposition of the coal macromolecule

G. E. Beghi (ed.), Synthetic Fuels, 187–218.

has to be distinguished from the secondary reactions of post-cracking and association of the volatile matter produced.

2. COAL STRUCTURE

The macromolecular structure of coal has given rise to a large number of studies. Several models of the coal molecule have been developed (1). Figures 1, 2 and 3 present the models of Given, Ladner and Wiser respectively.

The structure of coals is not uniform. From all these studies one can draw the conclusion that coals have a macromolecular structure, which is strongly aromatic and are formed from condensed polycyclic clusters of 4 to 6 cycles which are linked together by bridges (- CH_2 - etc.) and carry side chains. The aromaticity increases with the rank of the coal. There is therefore no monomer making up the elementary building block, whose repetition forms a polymer. The problems posed are those of determining the distribution of the clusters, the number of cycles of which it is formed, how they are bridged together, their degree of aromaticity, the number of alkyl and hydroxyl groups and the distribution of the O, N and S heteratoms in the macromolecular structure.

Many physical and chemical experimental methods have been used to establish a distribution of the molecular masses of the groups forming the molecule. One of these methods is that developed by Sternberg and his collaborators in 1971 (2), which consists of performing the alkylation of the coal so that 'coal polyanions' are produced. These authors have given a distribution of the molecular mass of the fragments, which is represented in figure 4. This is not the molecular mass of the coal, but that of the fragments produced by the breakdown of the most fragile bridges.

3. PYROLYSIS MECHANISMS

The characteristics phenomena of the pyrolysis of coal should be mentioned here, before other problems are dealt with. The three main phenomena which occur when the temperature of the coal is increased are :

a) a weight loss due to the release of volatile matter ;

b) a variation in volume during carbonisation which can be described by a contraction due to coal melting, then the dilatation and finally by a contraction during postcarbonisation which corresponds to the transformation of semi-coke into coke.

We shall only mention the aspect of the diagrams which may be either isothermal, i.e. the phenomenon is measured at constant temperature as a function of time, or with progressive increase in temperature giving the dynamic aspect of the phenomenon (figures 5 and 6).

The isothermal diagram shows that at each temperature the coal devolatilisation rate tends asymptotically towards equilibrium. This corresponds to the different bond energies in the coal molecule given in table I.

TABLE I. Bond energies in the coal molecule

	Bond energy (kcal/mole)
C - C aliphatic	77
C - H aliphatic	87 to 96
C - H hydroaromatic	92
C - H aromatic	102
C - C aromatic	125

The order in which the coal molecule fragments are liberated corresponds to the growing bond energy.
At low temperatures only the weakest bonds are broken.
A certain quantity of residual volatile matter will remain in the semi-coke which will be eliminated at higher temperatures. One may speak of 'extractible' volatile matter of a coal or a char.

3.1. Isothermal kinetics of the pyrolysis

The pyrolysis rate of the organic compound obeys a law of type

$$-\frac{d\,[A]}{d\,t} = k\,[A]^n$$

when there is no reaction between the primary products formed.
In this expression : [A] is the concentration of the substance which can be decomposed at time t ; k is the rate constant ; n is the order of the reaction which in general is equal to 1.
The rate constant of the reaction of decomposition varies with the temperature according to Arrhenius' general law

$$k_T = k_o . e^{-\frac{E}{RT}}$$

in which k_T is the rate constant at a temperature T ; k_o is the frequency factor ; E the activation energy of the pyrolysis reaction ; T the pyrolysis temperature in K.
For the gaseous primary compounds formed (the volatile matter) if V is the volume of gas which can be given off present in the coal at time t, the devolatilisation rate is given by relation

$$-\frac{dV}{dt} = kV \qquad (1)$$

or

$$\frac{dV}{V} = -\,k.dt$$

Integrating between the initial time t_o and the time t_1, one has

$$\int_{t_o}^{t_1} \frac{dV}{V} = -\,k.dt$$

$$\ell n \frac{V_1}{V_o} = -\,k.t \qquad \text{or} \quad V = V_o.e^{-kt}$$

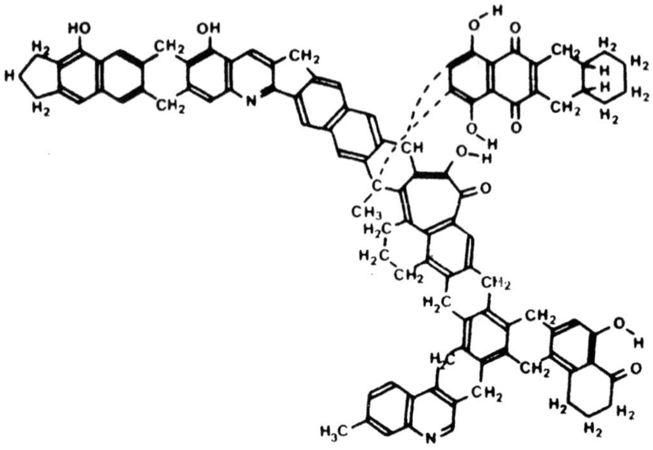

FIG. 1 : PROPOSED STRUCTURAL ELEMENTS OF COAL
(GIVEN, 1960)

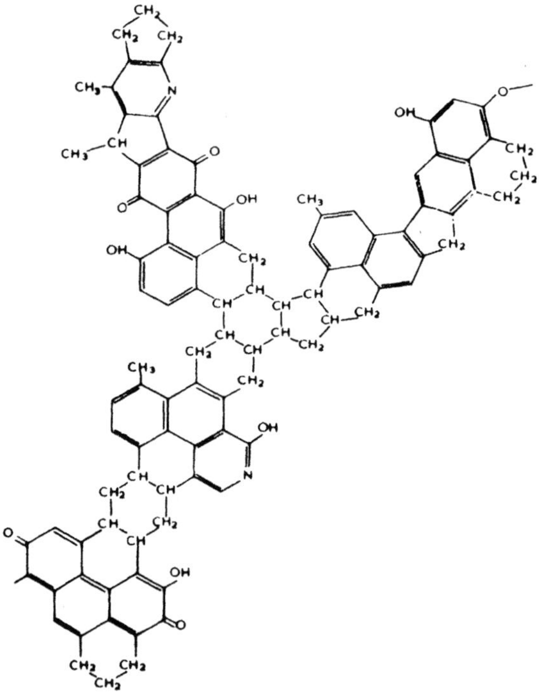

FIG. 2 : PROPOSED MOLECULAR MODEL·OF AN 82%C VITRINITE
(by W.R. LADNER, 1963 IN GIBSON, 1978)

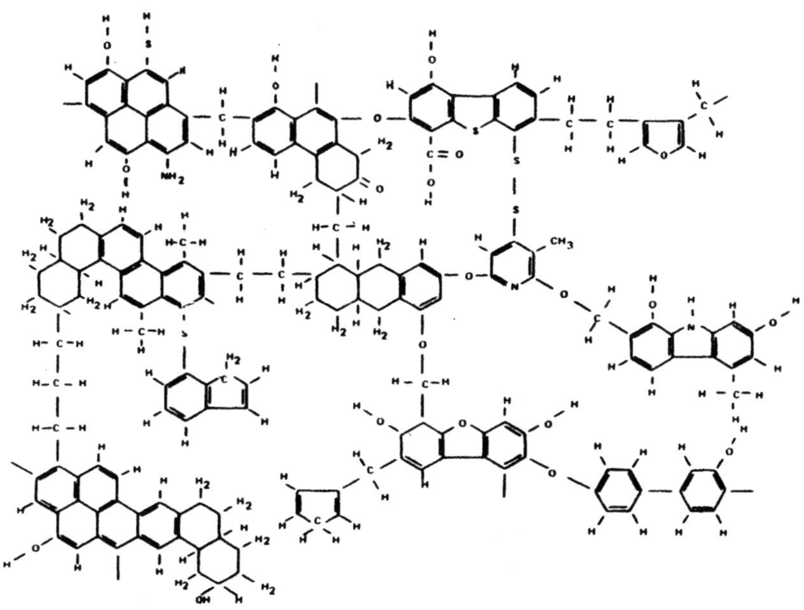

FIG. 3 : SCHEMATIC REPRESENTATION OF STRUCTURAL GROUPS AND CONNECTING BRIDGES IN BITUMINOUS COAL (BY W.H. WISER, IN STERNBERG, 1975)

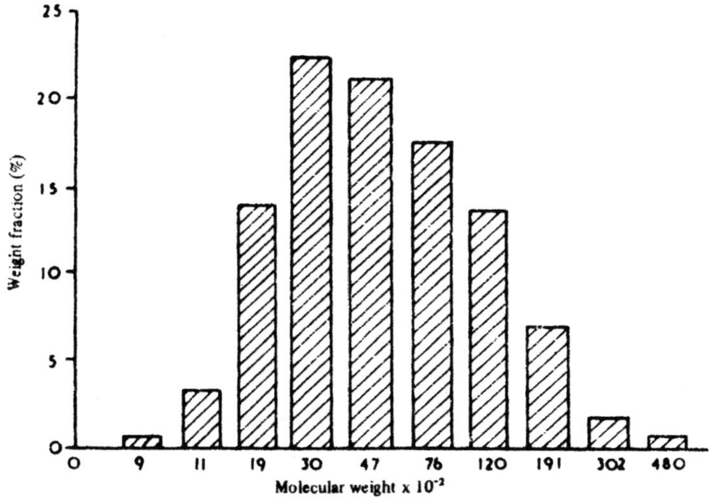

FIG. 4 : MOLECULAR WEIGHT DISTRIBUTION OF ETHYLATED POCAHONTAS COAL. (STERNBERG and others, 1971)

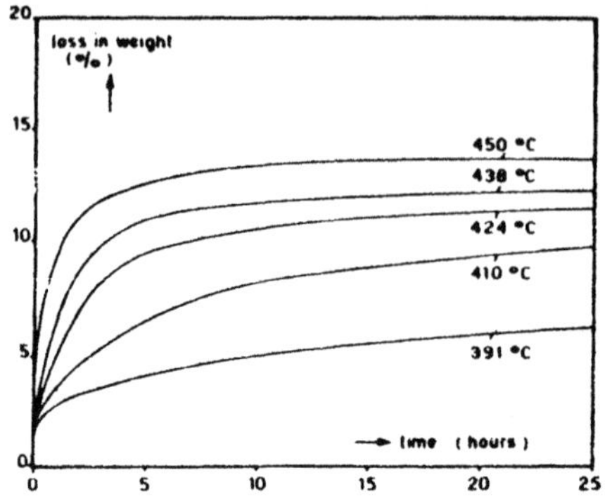

FIG. 5 : DECOMPOSITION (LOSS IN WEIGHT) OF A COKING COAL AT CONSTANT TEMPERATURES (23.3 PER CENT VM) (FROM VAN KREVELEN)

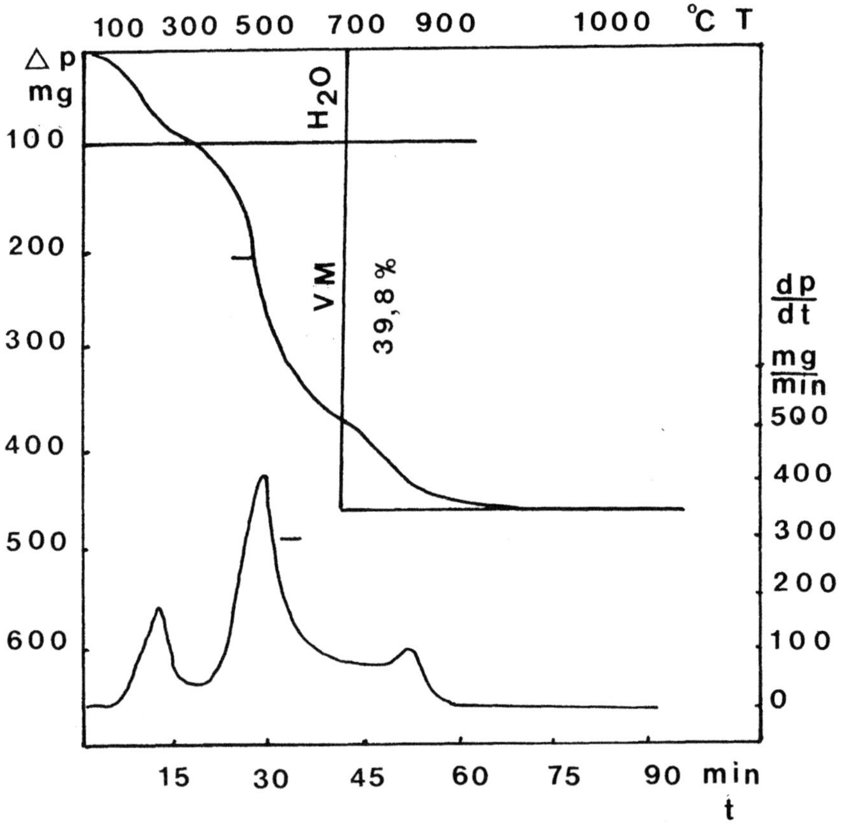

FIG. 6 : THERMOGRAVIMETRY OF A HIGH VOLATILE COAL (SULCIS)

from which, replacing V by its value of (1), one obtains :

$$-\frac{dV}{dt} = k.V_o.e^{-kt}$$

For a series of parallel reactions one writes :

$$-\frac{dV}{dt} = \Sigma_i\ k.V_{i,\ o}.e^{-k_i.t}$$

The validity of this approach is illustrated by the diagram of figure 7 which gives the rate as a function of $\frac{1}{T}$ (graphic representation of the Arrhenius equation).

Thermogravimetric studies allow to represent the gasification rates, by the weight loss per unit time, as a fraction of the weight present at the moment considered. The devolatilisation reaction rate is given by the coal weight loss

$$v = -\frac{dP}{dt} = k.P$$

The rate constant k, at a given temperature is

$$k = -\frac{dP}{dt} \cdot \frac{1}{P}$$

expressed in g of gasified coal per minute and per g of coal present at time t.

3.2. Description of the pyrolysis

The carbonisation of coal is the result of two simultaneous and opposite processes : cracking and polymerisation. Cracking produces free radicals which are unstable and which can either continue to decompose or can join together to form heavier molecules. Foch and Loison (3) have given the description of this process represented in figure 8. Account should be taken of the postcracking of compounds produced during the initial stage.

We have shown in a paper published already in 1965 (4) that starting from almost any simple model compound (benzene, toluene, phenol, cresols, etc.) thermal cracking in an inert gas atmosphere gives a wide range of compounds, from light gases (CH_4, CO, H_2, etc.) to heavy polycyclic compounds such as naphthalene and phenanthrene. There is always an association of molecular fragments with the formation of the most stable compounds. This also explains the difference in composition between low temperature tars and coke-oven tars. We have also shown that the aliphatic fraction of low temperature tars leads above 750°C to the formation of aromatic by cyclisation of the short chain olefines (C_2 to C_4) formed by the cracking and the dehydrogenation of long chain alkanes such as decane (5, 6).

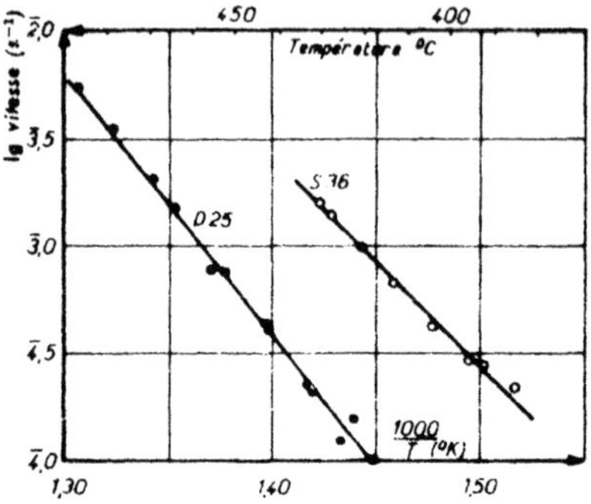

FIG. 7 : INITIAL RATE OF ISOTHERMAL WEIGHT LOSS FOR TWO COALS WITH 25 AND 36 V.M. ARRHENIUS PLOT LOG V = f(1/T) (LOISON R, and others 1970)

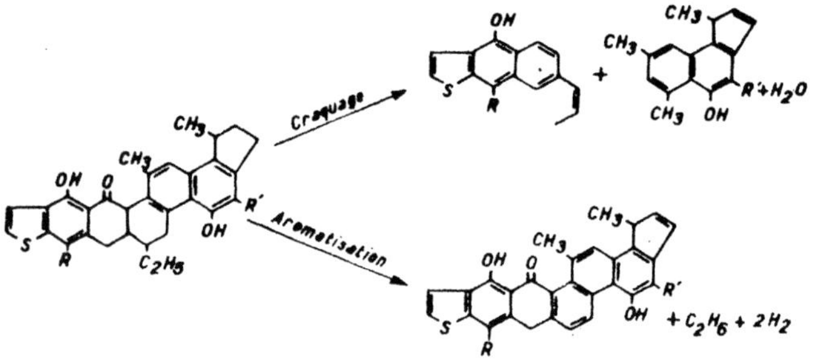

FIG. 8 : SCHEMATIC REPRESENTATION OF THE TWO MAIN CARBONIZATION PATHWAYS (LOISON R. and others, 1970)

PART II. PYROLYSIS UNDER HYDROGEN PRESSURE ⁎

1. INTRODUCTION

Within the framework of the research programme 'Chemical and physical valorisation of coal' of the Commission of the European Communities, research on the carbonisation of coals under reactive gas pressure have been carried out in our department of carbochemistry at the Université libre de Bruxelles since 1975.
Hydropyrolysis is the carbonisation of coal under hydrogen pressure. Hydrogen reacts (even at atmospheric pressure but with more efficiency under pressure) with the fragments produced by the thermal decomposition of the coal macromolecule.
Devolatilisation is always the first step in coal gasification. Whatever the process is - synthesis gas production or direct hydrogenation - the primary stage of the reaction always proceeds through coal devolatilisation. If this devolatilisation is carried out under hydrogen pressure, there are chemical reactions between hydrogen and the free radicals formed. The equilibrium between the two antagonist processes occurring during the carbonisation - namely cracking and polymerisation - is shifted to the cracking. All the parameters affecting this equilibrium can be exploited to increase the devolatilisation rate : increase in hydrogen partial pressure, rapid heating rate at high temperature and very fine particle diameter (7).
Coal devolatilisation occurs very quickly : it is completed within a few minutes if the coal is brought rapidly to the reaction temperature (8). At the same pressure and temperature hydrogen is able to extract more volatile matter than can an inert gas. The products of the hydropyrolysis have a composition which is different to those obtained by carbonisation, because hydrogenation occurs during the postcracking.
Among the numerous studies carried out on coal liquefaction and gasification in the last 150 years, hydropyrolysis is a relatively recent field. In 1937, Dent put forward the idea of carbonising coal in a water gas flow (H_2 + CO) under atmospheric pressure (9). In 1954, Union Carbide published results on the exploitation over several years of a mild hydrogenation unit of 300 t/d of coal (10, 11).
It was not until the seventies that hydropyrolysis returned to prominence and this occurred mainly in the USA. Furfari has recently reviewed the hydropyrolysis of coal (7).

This paper deals with the influence of coal rank, hydrogen partial pressure and temperature on the devolatilisation yields and on the composition of gases and oil produced in fixed-bed semi-batch carbonisation under hydrogen between 0.1 and 5.0 MPa. Gasification yields were established by thermogravimetry. A thermogravimetric installation capable of working up to 10 MPa and 950°C was constructed. A fixed-bed

---

⁎ Paper entitled 'Hydropyrolysis, a third way for coal conversion' presented by R. Cyprès, S. Furfari, C. Tilte-Neumann and M. Ghodsi at the 27-Deutsche Gesellschaft für Mineralölwissenschaft und Kohlechemie Haupttagung, Aachen, October 1982.

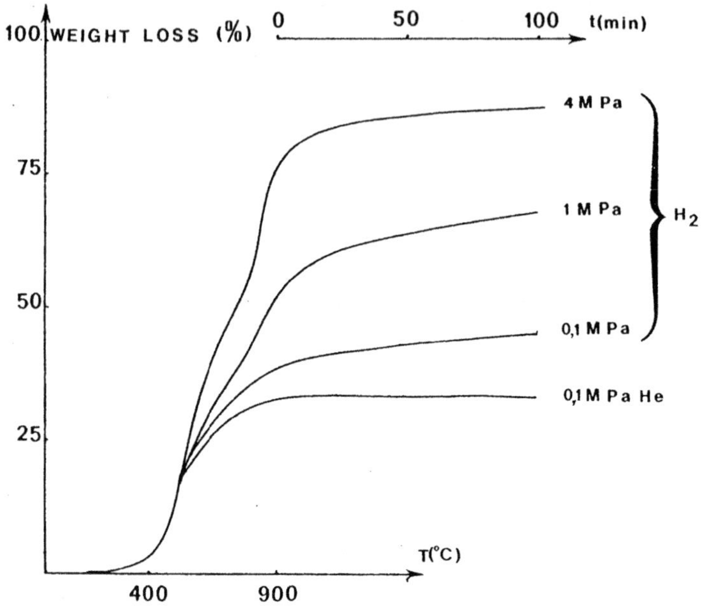

FIG. 9 : GASIFICATION OF BERINGEN F COAL (mf)

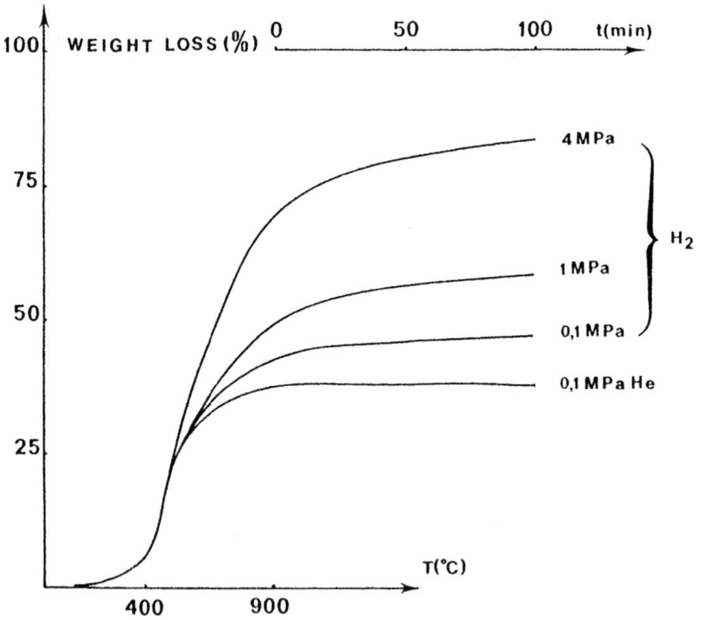

FIG. 10 : GASIFICATION OF ILLINOIS N°6 COAL (mf)

reactor with collection of oil and gas permits the carbonisation of 100 g of coal under the same conditions. Both apparatus have been described in earlier papers (12, 13).

2. INFLUENCE OF COAL RANK ON DEVOLATILISATION YIELD

A series of experiments was performed with three bituminous coals (Belgian Beringen, Illinois n°6, English Limby), one subbituminous coal (Italian Sulcis) and one lignite (German Rheinbraun).
Table II gives the analysis of these coals.

TABLE II. Coal analysis

a) Proximate analysis (wt %)					
	Sulcis	Illinois	Beringen	Limby	Lithotype n°8
Moisture	6.0	6.5	3.7	3.7	2.7
Ash (mf)	12.9	10.7	7.3	2.5	3.3
VM (mf)	49.1	35.8	35.5	37.0	52.0
b) Ultimate analysis (wt % mf)					
	Sulcis	Illinois	Beringen	Limby	Lithotype n°8
C	63.2	71.3	74.4	79.9	61.0
H	3.7	4.6	4.3	5.1	4.5
N	1.8	1.3	1.9	1.8	0.7
O	14.1 [a]	8.9 [b]	10.9 [a]	9.5 [a]	26.8 [a]
S	4.3	3.6	0.6	0.98	1.6

[a] measured ; [b] by difference

Figures 9 to 11 show the results for some of these coals, of the devolatilisation yields versus temperature and pressure obtained by thermogravimetry ; 100 mg of coal ground to < 106 μm, compacted, were heated up to 900°C at a heating rate of 10°C min^{-1} and maintained at the experimental temperature for 100 minutes. Even for the Beringen coal containing 35.5 wt % V.M. a devolatilisation yield higher than 75 % can be achieved at 4 MPa and 900°C. All the thermogravimetric curves show qualitatively the same behaviour of the different coals. Devolatilisation is rapid. Up to 500°C there is no significant difference in the gasification rate under 0.1 to 4 MPa, nor in comparing the kinetics and the rate of devolatilisation under hydrogen and helium at 0.1 MPa. Hydropyrolysis is related only to the first part of each curve. The second part is due to the hydrogasification process, if complete gasification is performed.

3. 'EXTRACTIBLE' VOLATILE MATTER

The aptitude of different cokes, prepared by preliminary devolatilisation under helium, under the same experimental conditions than for hy-

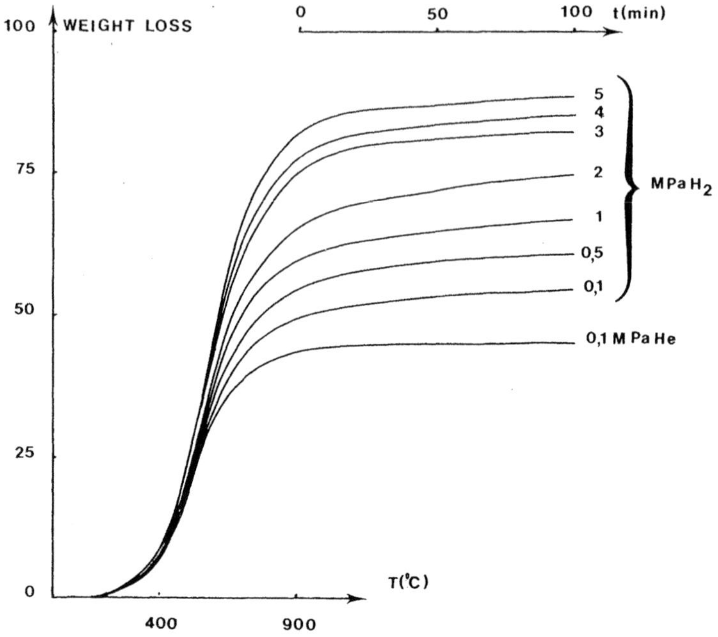

FIG. 11 : GASIFICATION OF SULCIS COAL (mf)

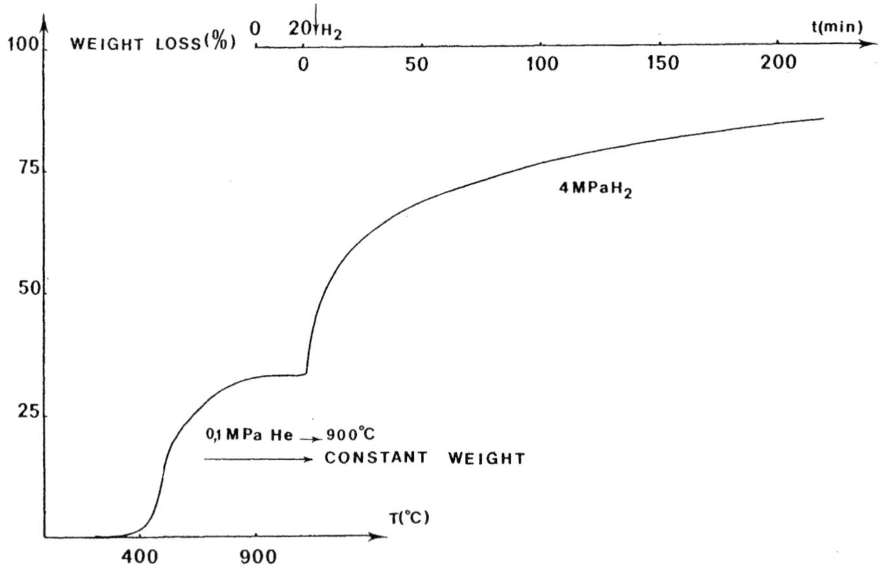

FIG. 12 : TWO STEPS GASIFICATION OF BERINGEN F COAL (mf)

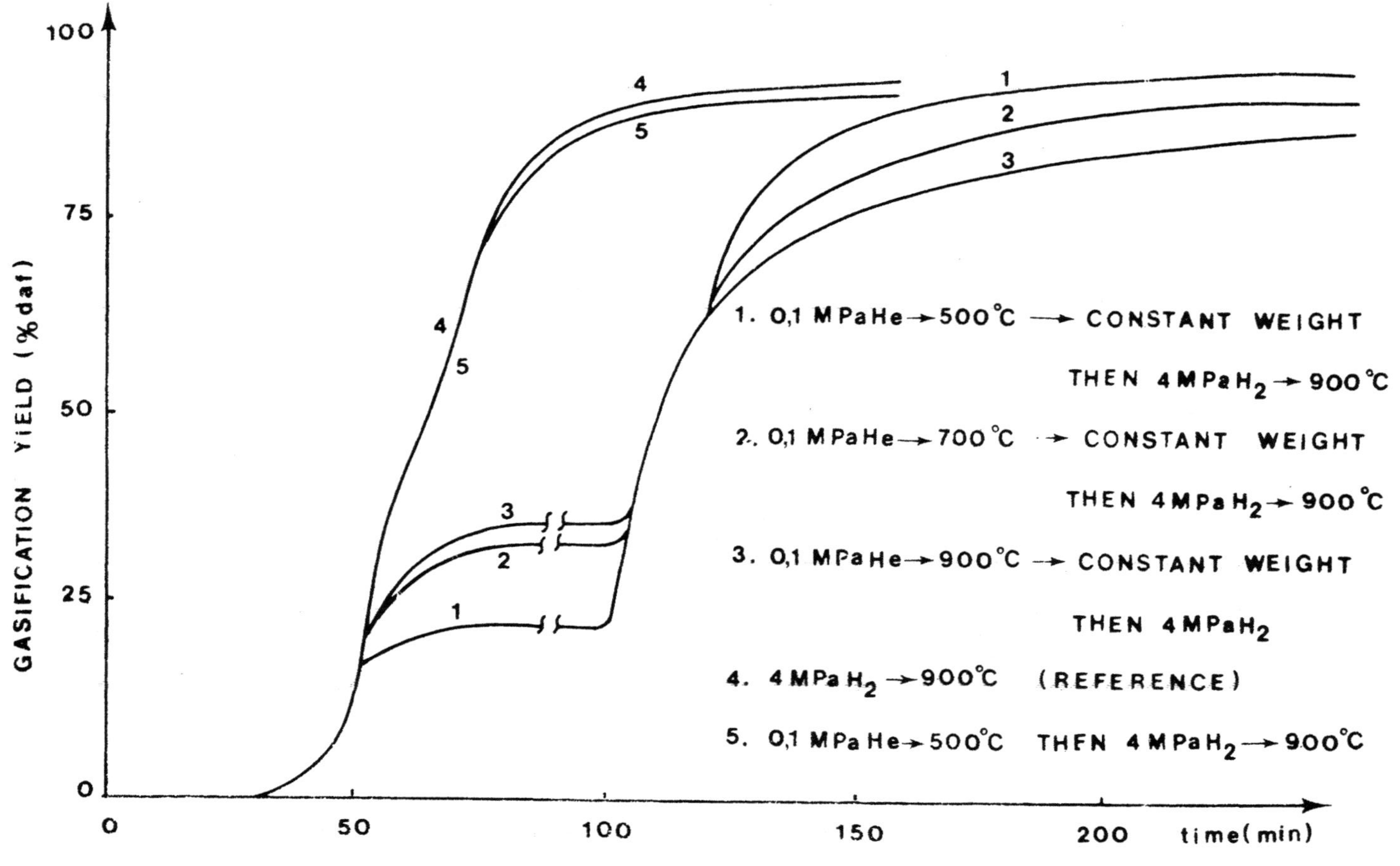

FIG. 13 : INFLUENCE OF PRECARBONISATION TEMPERATURE.

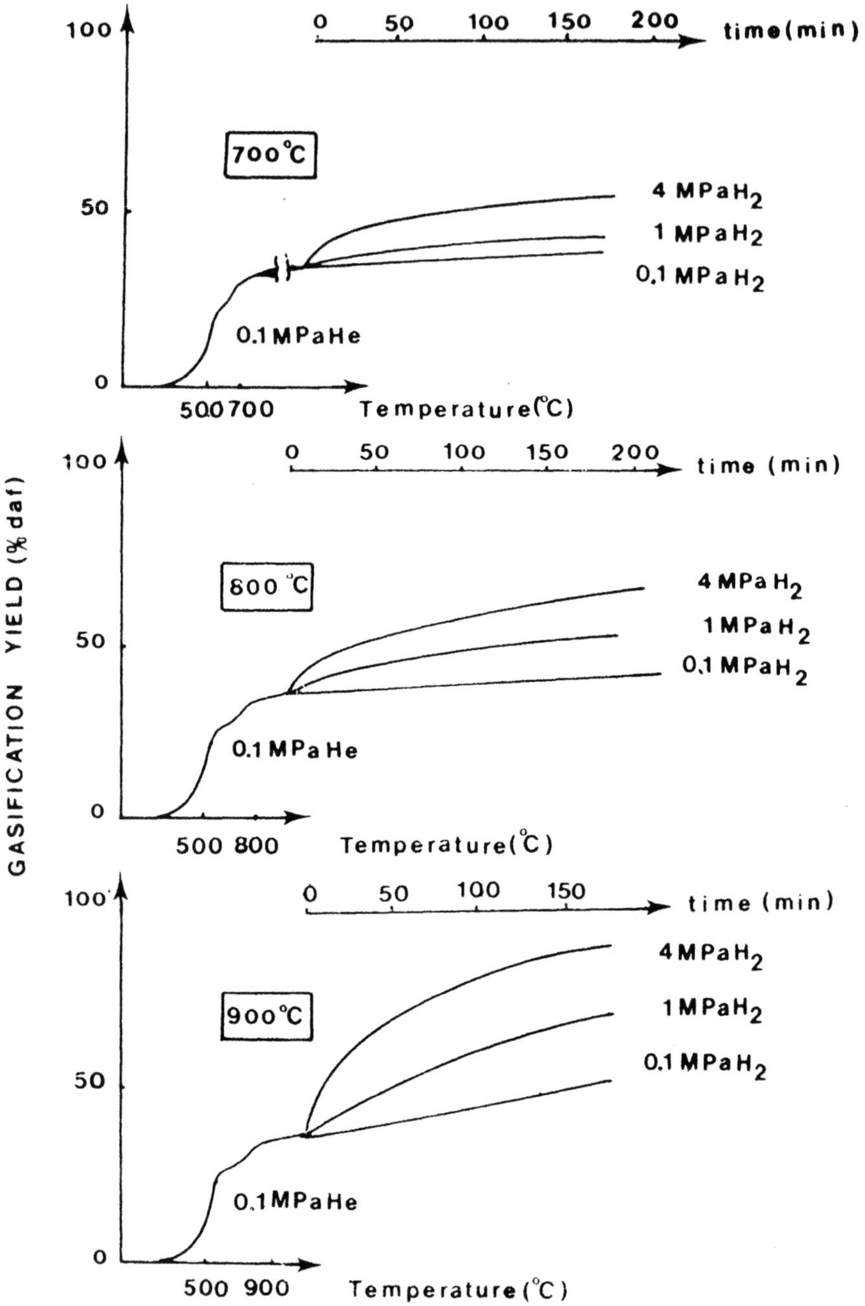

FIG. 14 : ISOTHERMAL TWO STEPS GASIFICATION

drogen, to undergo secundary devolatilisation under hydrogen pressure, has been studied.
Figure 12 shows Beringen coal heated under helium at 0.1 MPa up to 900°C and maintained at this temperature for 20 min in order to reach weight stabilisation ; hydrogen was then introduced and the pressure was set at 4 MPa. The devolatilisation started again very quickly. The devolatilisation kinetics beyond a rate of 70 % (figure 12) is more rapid than that of the residual char formed in direct hydrogenation under hydrogen at 4 MPa. The same phenomena are observed with the other coals but the kinetics of devolatilisation vary from coal to coal. Figure 13 shows the influence of precarbonisation temperature and figure 14 results on isothermal two steps gasification.
These experiments lead to the conclusion that the fraction of coal which is rapidly volatilisable is not completely destroyed during carbonisation in an inert atmosphere. The fraction of this coke which is extractible by hydrogen under pressure at 900°C depends on the hydrogen partial pressure.
The practical significance of this is that chars produced at atmospheric pressure of inert gas and at low temperature are suitable for devolatilisation under hydrogen pressure at higher temperature, with only a slight loss in devolatilisation yield, as compared with what is obtained in one-stage hydrodevolatilisation at the same pressure and final temperature. We are studying what variations in the composition of the volatile matter occur in this case.

4. INFLUENCE OF THE COAL SAMPLE MASS

The thermogravimetric devolatilisation yields are very high especially for bituminous coals. With the same coals, but in a hydrocarbonisation fixed-bed reactor described previously, Furfari has only obtained yields not exceeding 50 wt % by treating 100 g of coal (12). For this reason, in the same reactor and in the same conditions 1 to 100 g of Beringen coal have been processed. Results are given in figure 15 : there is a negative mass effect. With small samples of coal, the devolatilisation yields obtained are the same as that obtained in the thermobalance with 100 mg, whereas with 100 g the yield of conversion is again not higher than 45 wt %. Coal behaves as a solid with a great effect of retention of his own volatile matter. A small bed of coal particles is sufficient to fix part of the volatile matter which has to cross this bed. Probably a part of this decrease in the devolatilisation yield is also due to a postcracking of volatile matter inside the coal mass with a carbon formation. Thermogravimetric measurements on the devolatilisation yield can be considered as representing the maximum potentialities of a coal. Experimental conditions will have to be determined, in order to achieve, in fluidised or entrained bed, on an industrial scale, the same gasification yields as those demonstrated to be possible with one particular coal, in a thermogravimetric experiment.
It has also been demonstrated that high heating rates of the coal are definitely not necessary to achieve a high devolatilisation yield, contrary to what is asserted in numerous papers.

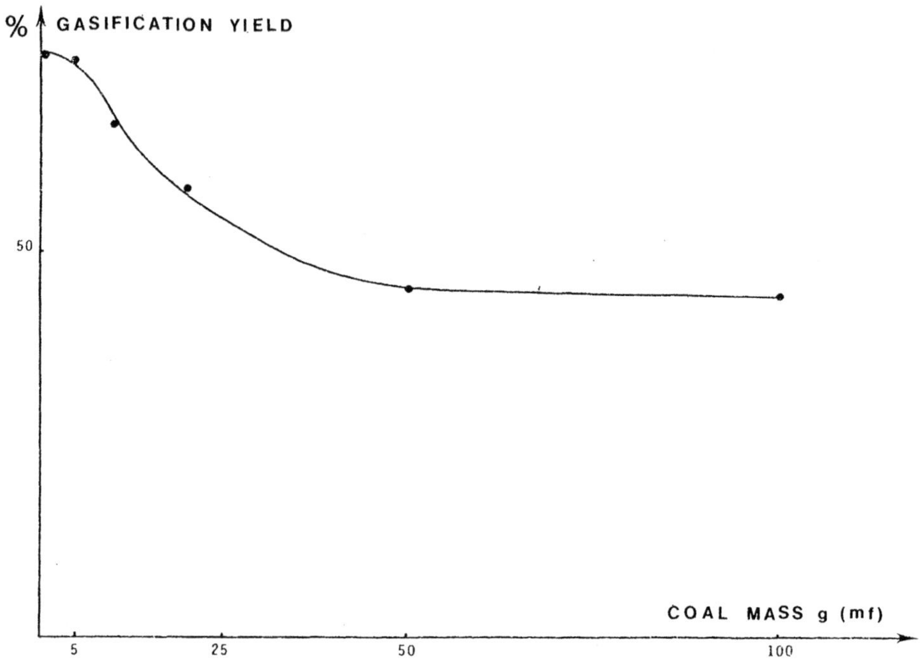

FIG. 15 : INFLUENCE OF COAL MASS ON THE GASIFICATION YIELD.

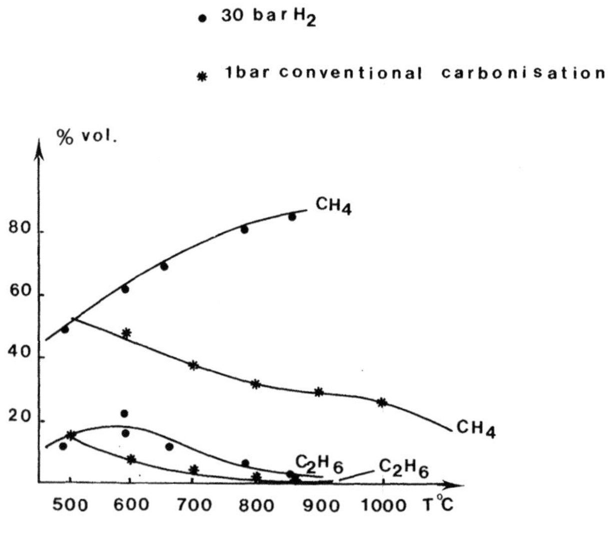

FIG. 16 : HyPy AND CONVENTIONAL CARBONISATION
GAS COMPOSITION AS A FUNCTION OF TEMPERATURE

The essential factor is that the volatile matter liberated by a coal particle is not trapped on another particle, as to avoid cracking with carbon deposition.

5. CHEMICAL REACTION OF THE HYDROGEN DURING THE PYROLYSIS

Hydrogen acts not only as an extractor of coal constituents, transferring them from the solid to the gas phase. Hydrocracking occurs in the postcracking phase of hydrodevolatilisation. Gases and oils are of different composition compared to those of inert gas carbonisation.
Figure 16 shows the evolution of methane content in the gas in both cases.
It is the same with the evolution of the quality and quantity of the light aromatic hydrocarbons and the phenols. Without going into a detailed discussion of these results, which have been published elsewhere (12), we can say that hydropyrolysis allows an increase in their yield and a degradation at the same time of the heavy hydrocarbons in energetic compounds of higher commercial value.
Our results and those obtained elsewhere, particularly in the USA, show that the quantity and quality of the gaseous and liquid hydrocarbons can vary by changing the following parameters : pyrolysis temperature, coal residence time, volatile matter residence time, hydrogen partial pressure and total pressure (7, 12).
Figure 17 shows the evolution of the composition of light aromatic hydrocarbons and phenols in the oil produced under 3 MPa hydrogen, versus temperature.

6. HYDROGEN CONSUMPTION

Different authors have shown that the hydrogen consumption is low or even nil at low temperature and increases with this parameter (14, 16). Coal liquefaction is obtained by incorporating hydrogen in the clusters formed by the coal molecule depolymerisation. The higher the hydrogen content of the coal, the higher will be the oil yield of the carbonisation, provided that the devolatilisation products are not postcracked due to a too long residence time in the hot zone of the reactor.
Figure 18 represents the variation of the C, H and O content of coals according to their range (17). The ultimate analysis of raw coal liquid and gasoline have been also plotted. It is claimed that the amount of hydrogen consumed is represented by the 'hydrogen distance' between the point representing the coal composition and that of the desired product. This is not exact, as account has not been taken of the oxygenated compounds which are not hydrogenated during the hydropyrolysis. It is well known that the gases produced by lignite hydrogasification contain CO_2 and CO. It has been shown, also in our investigation that the release of CO_2 from a lignite always starts at low temperature (18). Figure 19 gives the evolution of the gas yields versus the hydrogen pressure during the carbonisation of a Rhineland lignite at 550°C. The CO_2 production decreases slowly with the pressure from 15.2 wt % at 0.1 MPa to

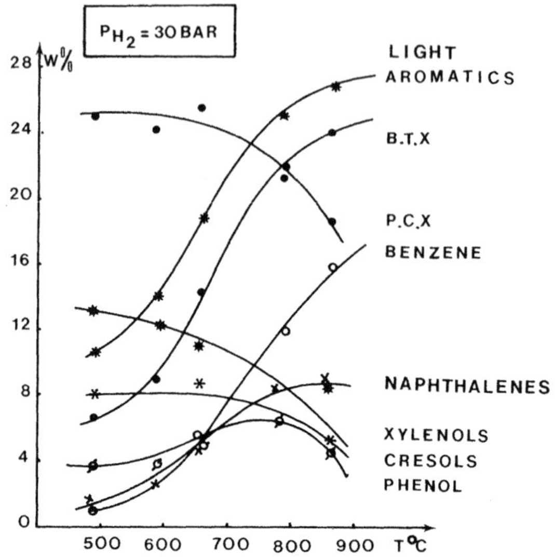

FIG. 17 : HYDROPYROLYSIS OF BERINGEN COAL
OIL COMPOSITION AS A FUNCTION OF TEMPERATURE
t = 60 s, < 500 μm

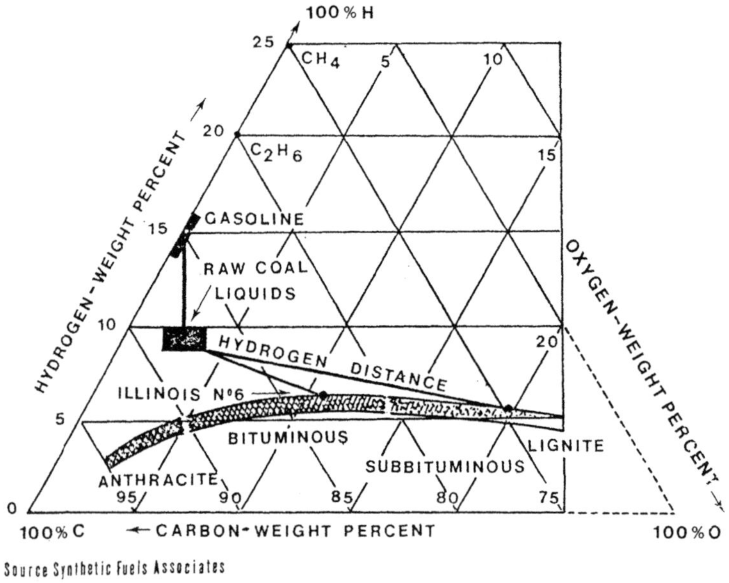

FIG. 18 : C, H, O COMPOSITION OF COAL AND COAL HYDROGENATES

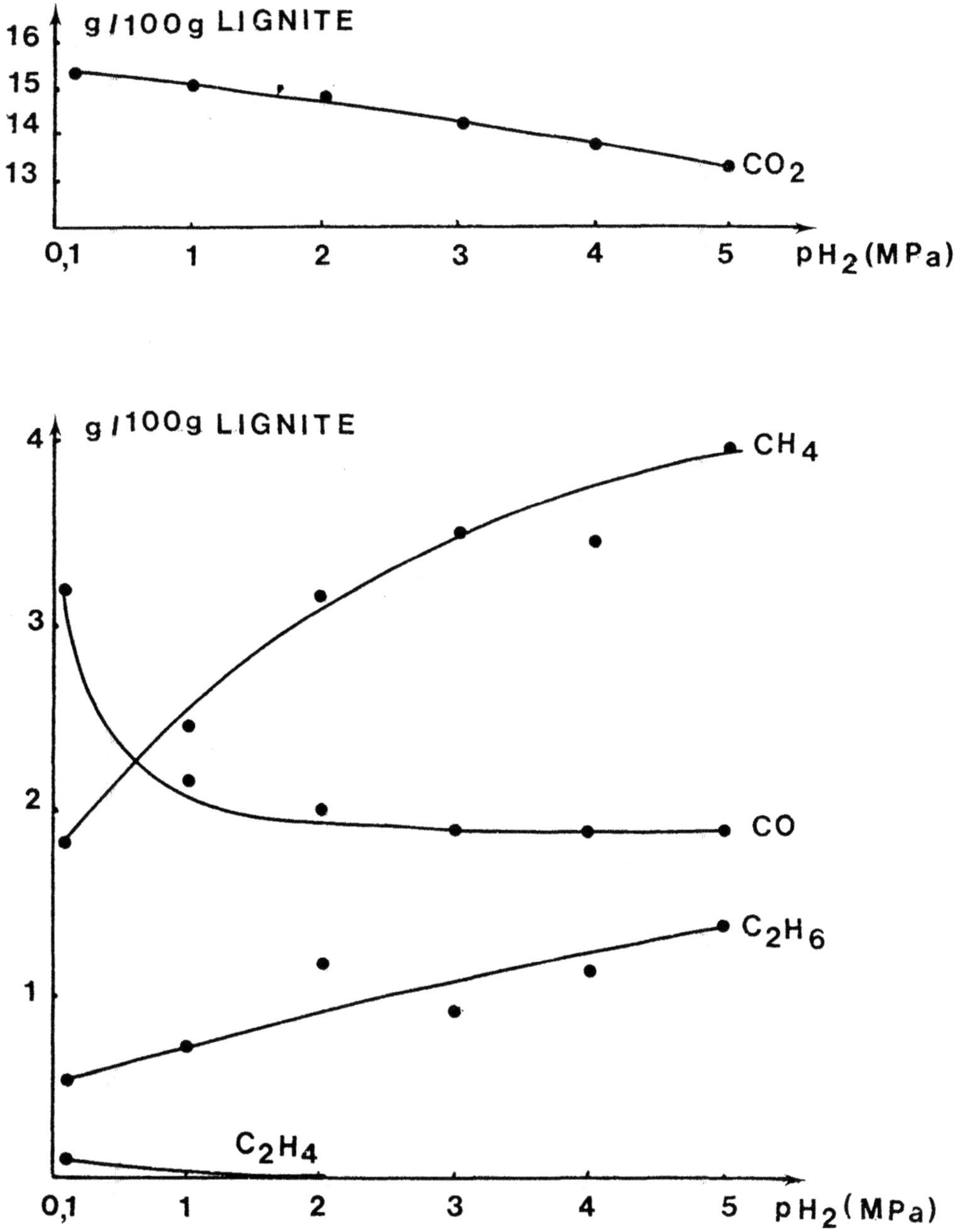

FIG. 19 : HYDROPYROLYSIS OF LIGNITE
INFLUENCE OF H_2 PRESSURE ON THE GAS YIELDS
(T° : 580°C)

13.3 wt % at 5 MPa. The CO yield stabilizes at ≈ 2 wt % at 1 MPa. Therefore, a large fraction of the oxygen content of the lignite is eliminated up to 5 MPa as CO_2, without hydrogen consumption. Figure 9 also shows that the cracking of the phenols is only substantial above 750°C. This demonstrates that it is possible to remove a large part of the total oxygen of the coal under hydrogen pressure without hydrogenating it to water which should lead to a high consumption of hydrogen.

7. HYDROPYROLYSIS COAL DESULFURISATION

7.1. Importance of the mineral matter composition

In previous publications we have shown that the sulfur elimination yield by hydropyrolysis depends on the mineral matter composition of the coal (19 - 22). This important aspect will be only summarised here.
By SEM with simultaneous EDAX XR analyses and XR diffraction, it has been confirmed that in the coal rich in calcite and dolomite, the sulfur liberated during the hydropyrolysis as hydrogen sulfide, is trapped by the alkaline-earth minerals. Under hydropyrolysis conditions the mineral matter of the Sulcis coal have the same effect as dolomite or lime is added, during combustion, to sulfur containing coals, as to reduce atmospheric SO_2 pollution.
Figure 20 shows a SEM microphotograph and XR patterns of untreated Sulcis coal. It can be seen that the untreated coal is rich in sulfur and in calcium. There is also some iron. The XR images show that sulfur is uniformly distributed in the sample. This is because Sulcis coal is rich in organic sulfur and relatively poor in mineral sulfur. The calcium and iron are well localised. The Sulcis coal is rich in calcite. The increase of sulfur signal in some area is due to the pyrite as confirmed by the iron XR image.
The microphotographs of the char formed at 780°C under hydrogen at 3 MPa show (figure 21) that there is now a simultaneous localisation of calcium and sulfur and also of magnesium and sulfur (20). There is also a change in the proportion between sulfur and iron. The areas rich in iron do not contain any more sulfur. The pyrrhotite formed at ≈ 500°C is completely reduced to iron under hydrogen at 3 MPa at 780°C, while under helium at 3 MPa at 540°C FeS_2 is transformed only into $FeS_{1.5}$ (21). Thus, hydrogen reacts with pyrite and with organic sulfur to form hydrogen sulfide :

$$FeS_2 + H_2 \rightarrow FeS + H_2S$$

$$FeS + H_2 \rightarrow Fe + H_2S$$

and

$$R-S-R' + 2\,H_2 \rightarrow H_2S + RH + R'H$$

Hydrogen sulfide reacts then with $CaCO_3$ and/or CaO and $MgCO_3$ as follows :

$$CaCO_3 + H_2S \rightarrow CaS + CO_2 + H_2O$$

$$CaO + H_2S \rightarrow CaS + H_2O$$

$$MgCO_3 + H_2S \rightarrow MgS + CO_2 + H_2O$$

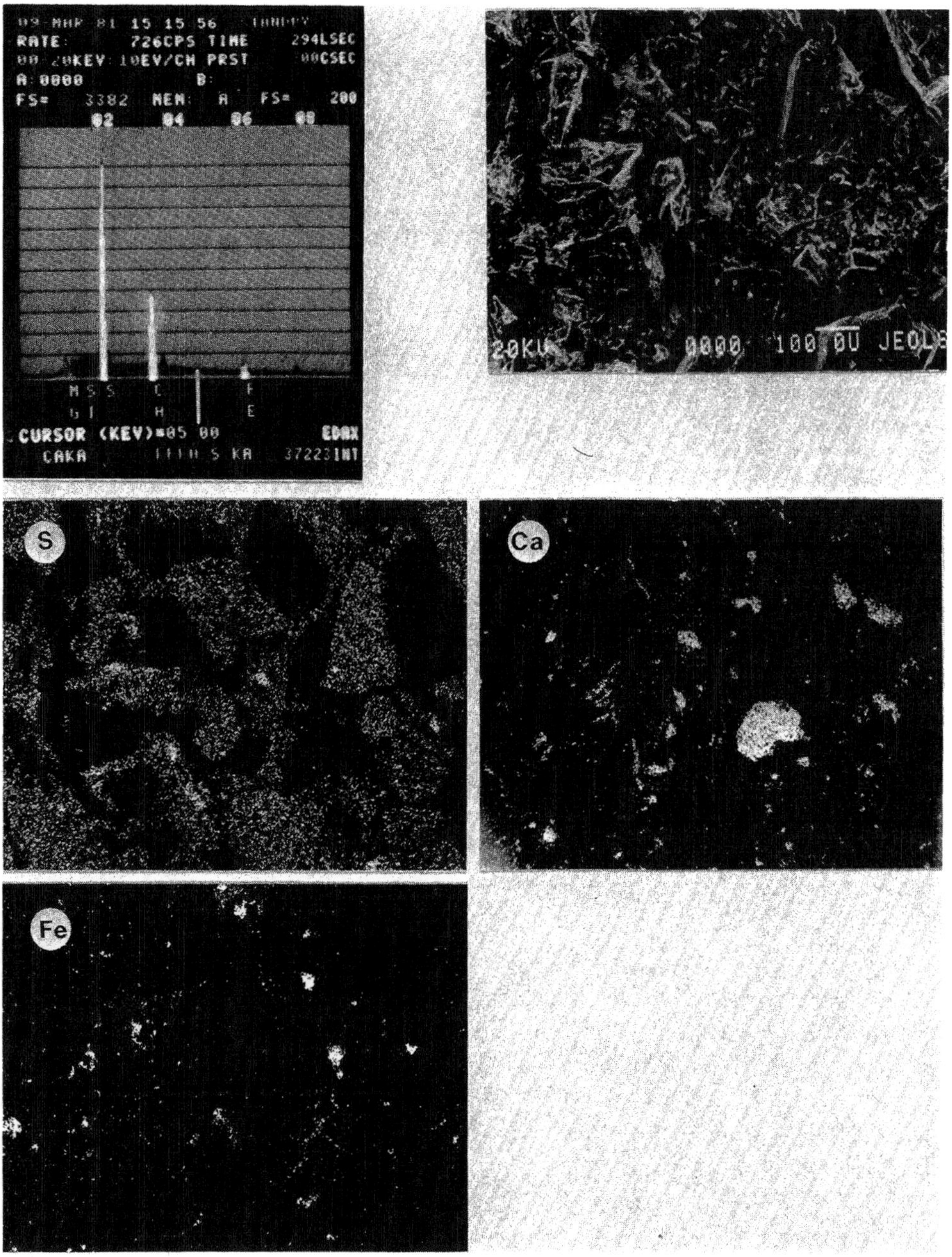

FIG. 20 : SEM MICROGRAPHY WITH S, Fe AND Ca DISTRIBUTION UNTREATED SULCIS COAL.

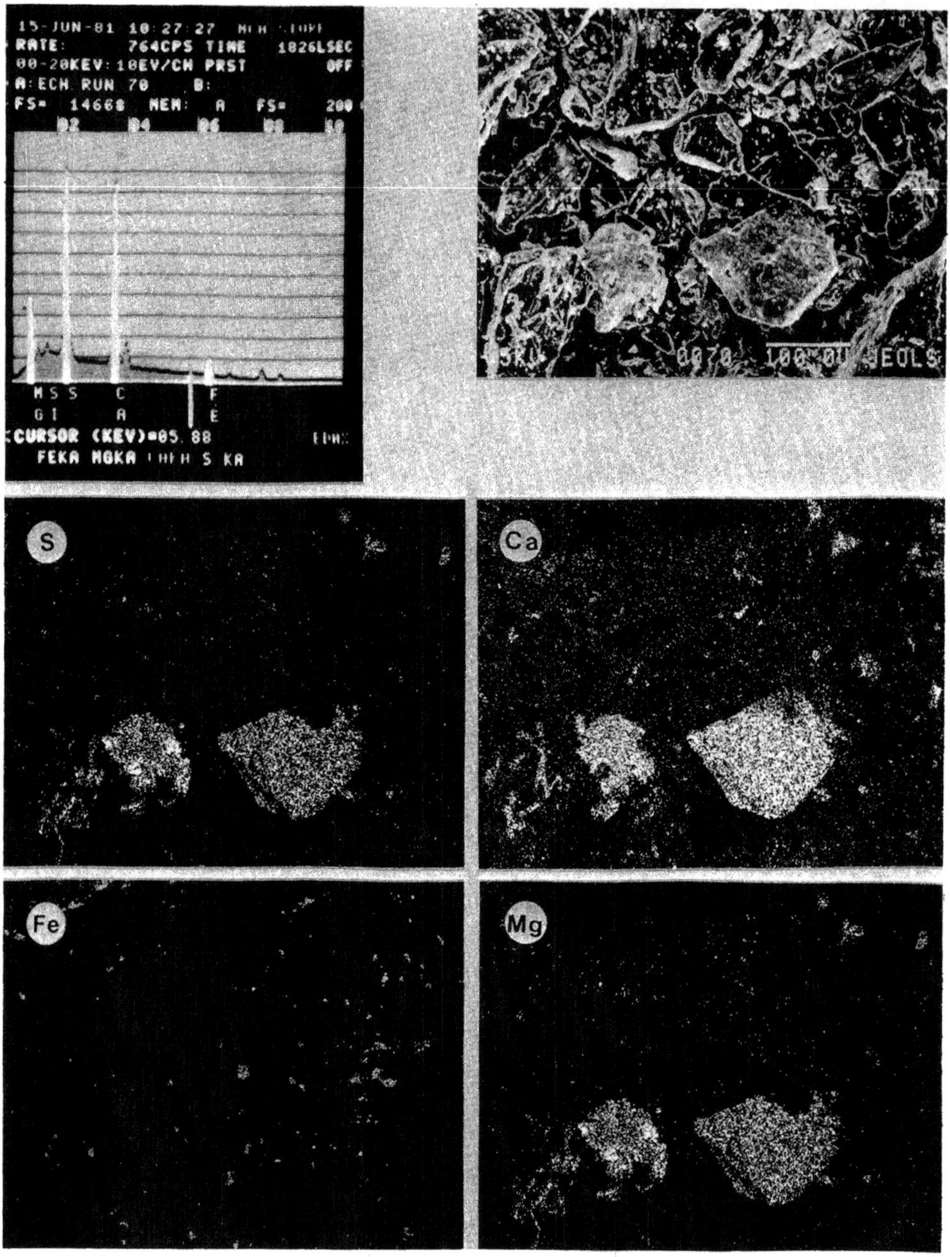

FIG. 21 : SEM MICROGRAPHY WITH S, Fe, Ca AND Mg DISTRIBUTION SULCIS CHAR FORMED AT 780°C UNDER 3 MPa H_2

The reaction

$$MgO + H_2S \rightarrow MgS + H_2O$$

is not possible thermodynamically.
Hydropyrolysis of Illinois bituminous coal containing 3.6 % sulfur but low in calcite content confirms the general interpretation of the mechanism of sulfur retention in the char. Results obtained in our laboratory by Tilte-Neumann and Bertau are briefly summarised.
Sulfur content of the char, obtained in the same experimental conditions as for Sulcis coal, was reduced to 1.5 %. Scanning electron microscopy and Edax analysis of S, Al, Si of Illinois coal and char obtained by hydropyrolysis at 900°C and 1 MPa are presented in figures 22 and 23. Figure 22 shows the importance of S peak relative to Al and Si and the coïncidence of Al and Si due to kaolin 2 $SiO_2.Al_2O_3$ of the clay. There is also an excess of SiO_2. S is well dispersed throughout the sample, due to organic sulfur. Figure 23 shows a general aspect of the char. There is an important decrease in the relative height of the peaks of S to Si and Al. Al and Si are again localised at the same place. Remaining S is well dispersed, suggesting its heterocyclic origin.

7.2. Combustible and non-combustible sulfur

The fractions of combustible and non-combustible sulfur have been determined as to evaluate if the chars containing important amount of sulfur are able to be used as a clean fuel. The samples were placed in an electrical furnace and burned in an air flow at atmospheric pressure by progressive heating up to a final temperature of 1000°C. The results are summarised in figure 24 where the hatched areas represent the combustible sulfur (sulfur evolved as sulfur dioxide) and the white ones the non-combustible sulfur. The combustible sulfur in all the chars is very low with regard to the total sulfur. In all the chars obtained, the combustible sulfur is less than 0.8 wt % mf ; on the contrary, in the untreated coal this amount is 2.8 wt % mf. In the chars the combustible sulfur represents only 14 to 22 % of the total sulfur while in the coal it reaches 64 %. The combustible sulfur content varies little with the pressure and temperature (20). The char formed under helium does not give a worse result than hydrogen, as far as combustible sulfur in char is concerned. This drastic decrease in combustible sulfur is due to the nature of the mineral matter and not specifically to the hydrogen presence. Pyrolysis allows the sulfur to be eliminated as hydrogen sulfide or sulfur-containing tar species. Part of the hydrogen sulfide reacts with alkaline-earth mineral matter to form alkaline-earth sulfides. If the temperature is too low, this reaction does not occur.
During the combustion, the oxidisation of the sulfides occurs :

$$CaS + 2\ O_2 \rightarrow CaSO_4$$

$$MgS + 2\ O_2 \rightarrow MgSO_4$$

Those compounds are detected in the ashes by XR diffraction (figure 25). Figure 26 is an oversimplified representation of the sulfur reactions during hydropyrolysis of coal and combustion of the hydropyrolysis char.

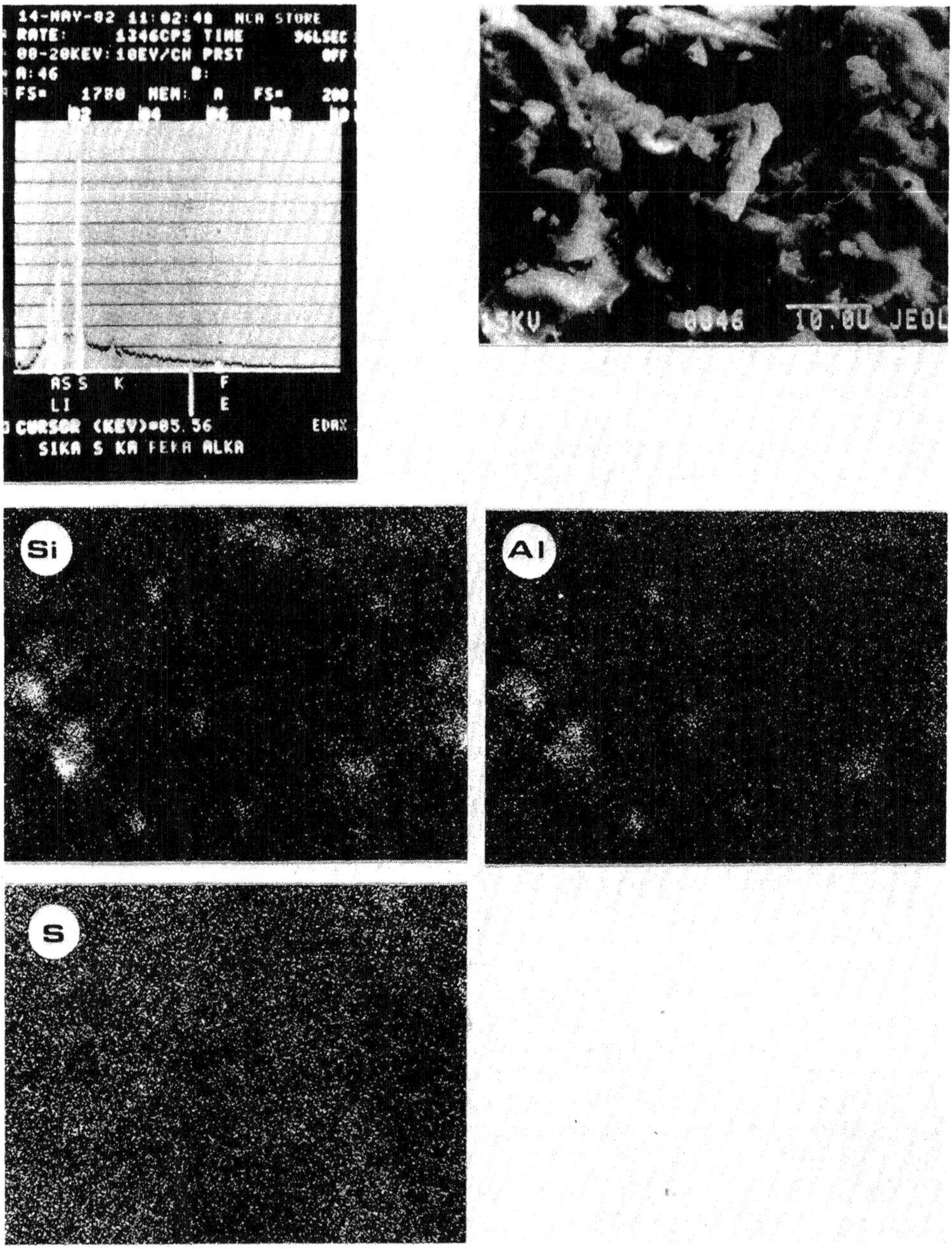

FIG. 22 : SEM MICROGRAPHY WITH Si, Al AND S DISTRIBUTION
UNTREATED ILLINOIS COAL.

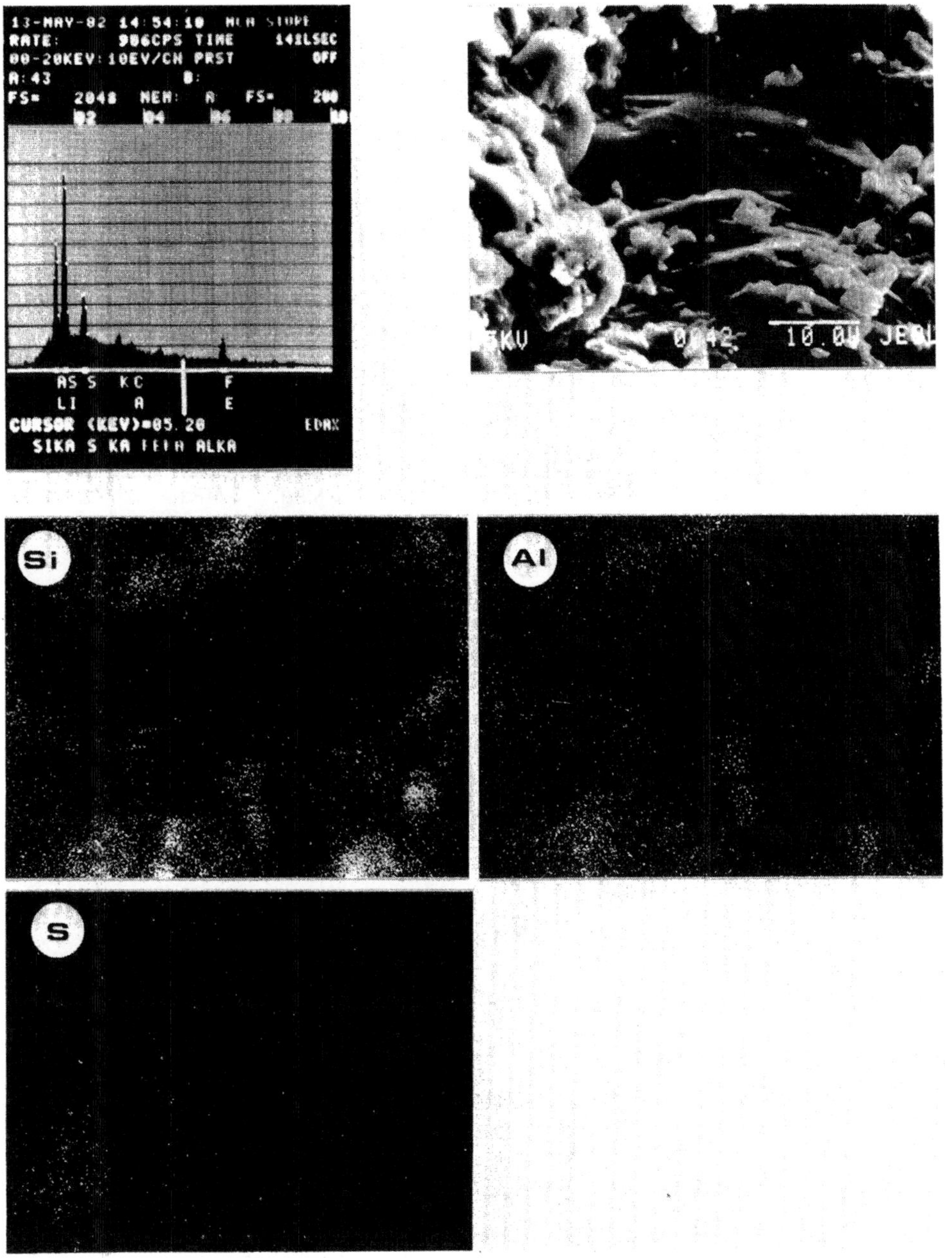

FIG. 23 : SEM MICROGRAPHY WITH Si, Al AND S DISTRIBUTION
ILLINOIS CHAR FORMED AT 900°C UNDER 1 MPa H_2

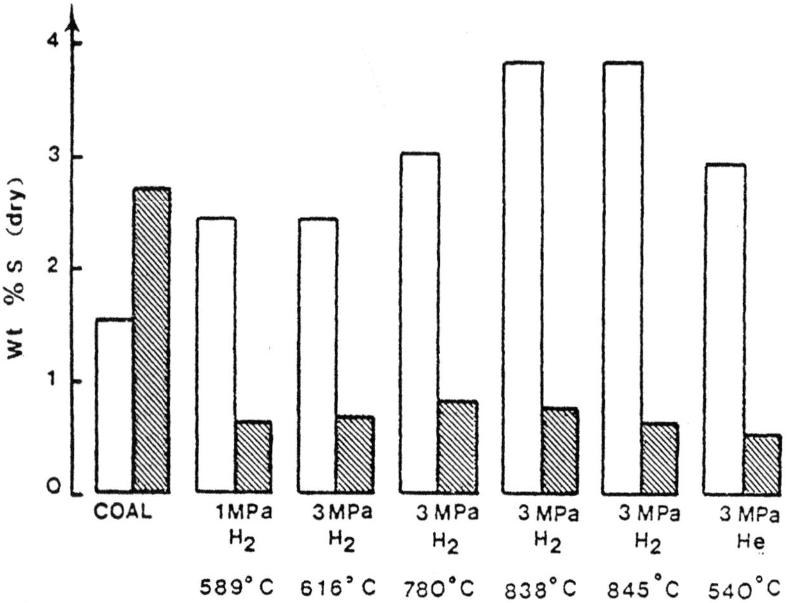

FIG. 24 : DISTRIBUTION OF THE TOTAL SULFUR IN COMBUSTIBLE SULFUR (HATCHED AREA) AND NONCOMBUSTIBLE SULFUR (WHITE AREA) AT 1000°C

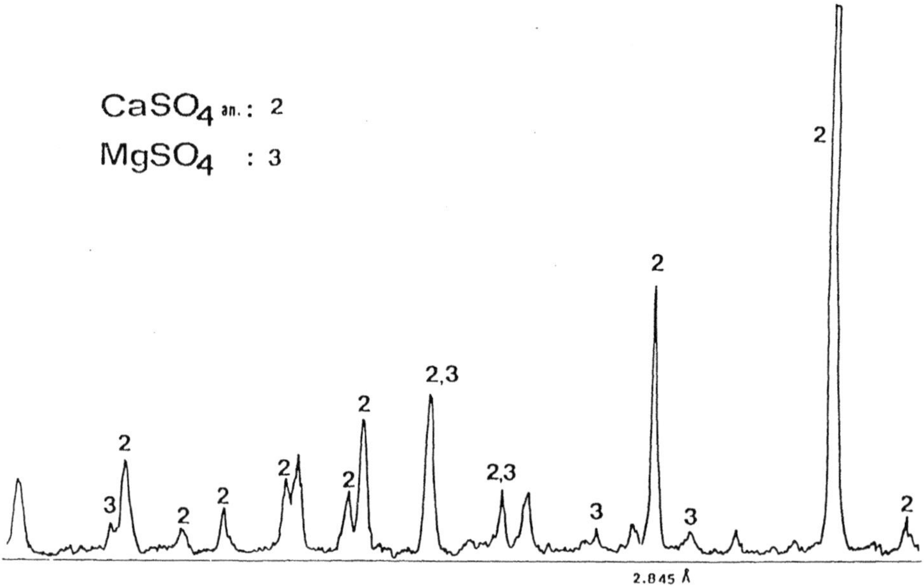

FIG. 25 : X.R. DIFFRACTION SPECTRUM. SULCIS CHAR (FORMED AT 836°C AND 3 MPa H_2) ASHES.

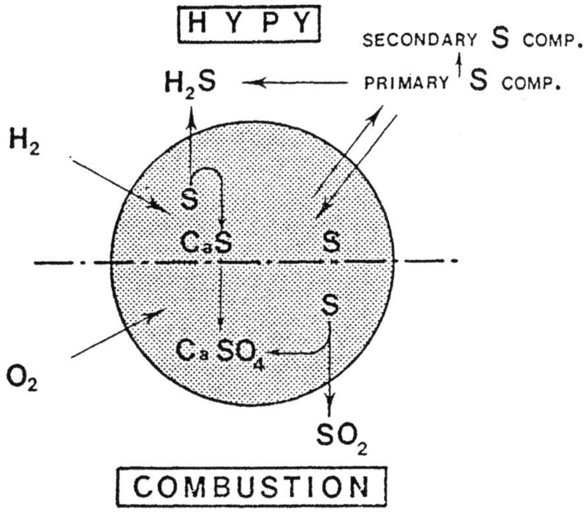

FIG. 26 : SCHEMATIC OF THE PRINCIPAL MODES OF SULFUR REACTION DURING HyPy AND COMBUSTION

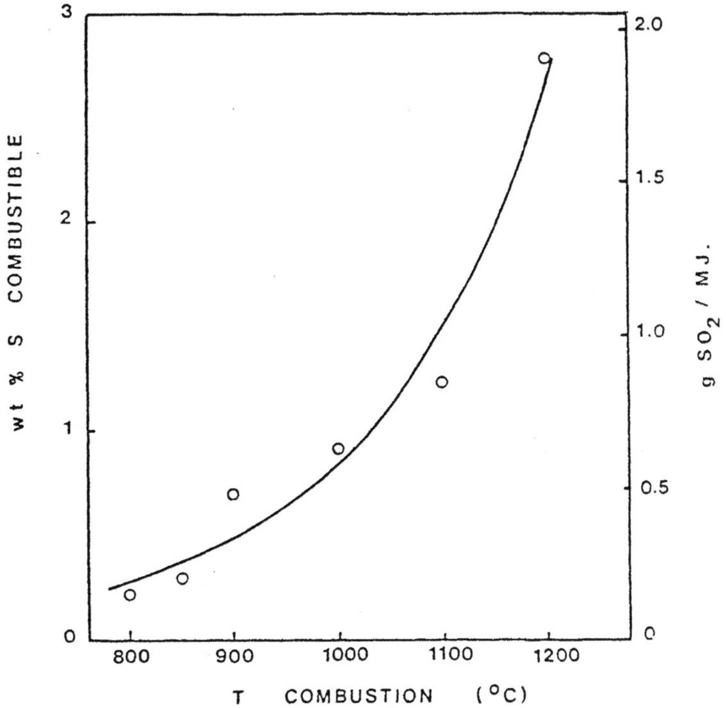

FIG. 27 : PRODUCTION OF COMBUSTIBLE S AS A FUNCTION OF COMBUSTION TEMPERATURE.

An important contribution to the problem was made by the research group of the Rheinische Braunkohle of West-Germany.
H. Teggers, L. Schrader, K.A. Theis and R. Fladerer (22, 24, 25) have studied sulfur retention in the chars produced in their lignite hydrogasification pilot-plant. Lignite submitted to hydrogasification has less than 0.5 % sulfur and is poor in ashes (1.5 to 2.5 %). Sulfur content in the char was determined as a function of the hydrogasification parameters. Residence time of the coal appeared to be the most important factor. The authors have shown that it is possible to produce a residual char whose sulfur had become fixed by the alkaline constituents of the ashes, in particular by CaO as CaS. It was demonstrated that compared with the SO_2 released during combustion of the lignite, the SO_2 emission by combustion of the hydrogasification residual char of the same lignite, was reduced by 79 % to 96 %. This results of the oxidisation of CaS to $CaSO_4$ which remains in the ashes.
This mechanism was given by H. Teggers et al. (22 - 25) to explain in the case of hydrogasification of a low sulfur and low calcite lignite, that the remaining sulfur in the char reacts with CaO to CaS as the ratio of Ca/S is increased by a factor 2 to 5 depending gasification yield.
CaS is oxidised to $CaSO_4$ during combustion. It has been demonstrated that this mechanism proposed by Teggers et al. for low sulfur and low calcite containing lignites is also the one that explains why a 4 % S containing char produced by hydropyrolysis of a more than 4 % S coal is a SO_2 non polluting fuel. A given char will be a more or less polluting fuel when burned, depending on the combustion temperature ; the amount of SO_2 liberated by combustion as a function of combustion temperature was determined between 800 and 1200°C. Experiments were made on 780°C and 3 MPa H_2 char. Results are plotted on figure 27. SO_2 vapor pressure, from $CaSO_4$ thermal dissociation, increases with temperature. This is an argument in favour of burning such type of chars in fluidised bed combustors, well adapted for low temperature combustion.

8. THE THIRD WAY

The research performed in our laboratory on coal carbonisation under hydrogen pressure and the work carried out in the USA on rapid hydropyrolysis, allow the development of a third route of partial gasification and liquefaction of coal which offers good economic prospects for application in industrialised countries, which have neither oil nor gas reserves. It is founded on the coupling of hydropyrolysis with the residual char combustion, in power-plant stations. Another alternative of the process is to gasify the residual char, instead of burning it. The advantages of hydropyrolysis can be summarised as follows :

8.1. Relatively low investment

Compared to the investments required by the known gasification or liquefaction processes, those which are necessary for the hydropyrolysis are much more modest. This is explained by the process characteristics themselves. First of all, reaction time is very short, in the order of

one minute, whereas 30 to 60 minutes are necessary to achieve complete gasification. Consequently, the mass flow for the same volume of reactor is multiplied by 30 to 60. The hydrogen pressure is not too great. The pressure effect is marked mainly between 0.1 and 2 MPa. While gasification is achieved at 5-7 MPa and liquefaction at 10-40 MPa, hydropyrolysis is already efficient at 2 MPa. This obviously brings about a reduction in the price of the apparatus.
Rapid hydropyrolysis does not require, which is the case for gasification with steam and oxygen under pressure, an oxygen production plant.

8.2. Good energetic yields

Hydropyrolysis reaction is exothermic. The reaction heat contributes towards bringing the coal to reaction temperature. The contrary is true for the gasification and liquefaction processes based on synthesis gas. The reaction between coal and steam is strongly endothermic. Some 30 % of the coal is consumed to produce the energy necessary to the process. At Sasol, for example, the energetic yield is 40 to 45 %.
Table III compares the material necessary to produce one ton hydrocarbons by the Sasol process and by the hydropyrolysis process, even if for the latter lower figures are taken for the hydrocarbon production yield as what is achieved in many experimental installations.

TABLE III. Requirements for one ton hydrocarbons production by Sasol and hydropyrolysis processes.

Sasol			Hydropyrolysis	
		Requirements :		
Coal (pure)	2.8 t		Coal	3.3
Oxygen	0.35 t		Hydrogen (1 to 2 %)	0.06
Steam	1.25 t			
		Production :		
Hydrocarbons	1.0 t		Hydrocarbons	1.0
			Clean char	1.97

8.3. Low hydrogen consumption

The hydrogen consumption depends on the degree of hydrogenation of the products produced by devolatilisation. At high temperature, all the volatile matter is converted into methane. In such a case, the hydrogen consumption will obviously be high. On the other hand, at low temperature (≃ 600 - 700°C), the liquid hydrocarbons yield is high and the methane one is low. Light aromatic hydrocarbons (benzene type) are mainly produced. The H to C ratio then passes from 4 (CH_4) to 1 (C_6H_6). Hydropyrolysis is then nearly self-sufficient for its hydrogen needs. The coal pyrolysis evolved enough hydrogen to hydrogenate the volatile matter in liquid hydrocarbons. The external hydrogen required is very small or nil.

It has been demonstrated that it is wrong to consider that all the oxygen in coal is hydrogenated to water. The carbon dioxide produced by the decomposition of the carboxylic function of humid acids, evolves at low temperature and is present in the gases. The phenols hydroxyl functions are not hydrogenated if the temperature is lower than ≃ 750°C since the oils contain up to 25 wt % of phenols.

8.4. The residual char is a non polluting solid fuel

Hydropyrolysis always strongly reduces the combustible sulfur content of the char. If the mineral constituents of the coal are acid, the major part of the sulfur is eliminated as H_2S or organic sulfur compounds in the oil and gas. If they are alkaline, part of the sulfur is trapped on $CaCO_3$ and $MgCO_3$, as CaS and MgS. This sulfur is non polluting because it is non combustible. It remains as sulfates in the ashes.
The investments required for providing, in coal power plant stations, environmental protection against SO_2 pollution, could thus be suppressed.

9. CONCLUSIONS

The Belgian Government decided in 1981 to sponsor our project. It has granted $ 2 million for the construction and the experimentation of a hydropyrolysis pilot plant at the 'Institut National des Industries Extractives' (INIEX) in Liège. It will be an original conception and the result of a collaboration between INIEX, industry and our carbochemistry department of the Université libre de Bruxelles. It is planned, if the pilot experimentation is conclusive, to construct a demonstration plant in Kempen to upgrade the high volatile bituminous coal and to fire a small electric power-station with the produced smokeless char.
If these industrial experiments are conclusive, every power-station could be coupled to a module of hydropyrolysis pretreatment of the coal consumed. Instead of constructing a completely new coal gasification or liquefaction facility, the integration of the rapid hydropyrolysis in the present line of work, of the coal power-stations, requires only the specific investment of this coal pretreatment before its combustion and of the combustion of the oil. The economy of all the handling and conditioning installations in the power-station is realised.
Hydrocarbons will have to be only collected where they are produced. Their upgrading will have to be performed in a central unit of great capacity. Petroleum industry could well take it up and valorise this new source of oil. The refining overcapacity in Europe permits to consider the reconversion of existing units, to this end.
Rapid devolatilisation, limited to the very reactive fraction of coal, can be ideally integrated in the developments of the new technologies of fluidised bed combustion, at atmospheric pressure or under pressure. Hydropyrolysis is performed in a fluidised bed. Consequently one can conceive that the fluidised combustor would be fed, with the hot char leaving the fluidised hydropyrolysis reactor, at a grain size and with a residual volatile matter content, required for a good combustion.
The technology now under development at INIEX which has to be settled in

a 100 kg/h semi-industrial pilot plant is not related to the upgrading of a particular coalfield. It can be integrated wherever there is a unit of high consumption of coal, such as an electric power-station, a district heating boiler or a coal gasification plant.

10. ACKNOWLEDGEMENT

This work is part of the European Community Program on the 'Chemical and Physical Valorisation of Coal'. The author thanks C.E.C. Coal Directorate, for the financial assistance that it has provided.

11. REFERENCES

1 DAVIDSON R.M., 'Molecular structure of coal', *ICTIS/TR 08 Rep.*, (1980), *IEA Coal Research*, London.

2 STERNBERG H.W., DELLE DONNE C.L. ; PANTAGES P., MORONI E.C. ; MARKBY R.E., 'Solubilization of an lvh coal by reductive alkylation', *Fuel* (1971), 50, 432.

3 R. LOISON, P. FOCH, A. BOYER, *Le Coke*, Dunod, Paris (1970).

4 R. CYPRES, C. LEJEUNE, 'Craquage thermique du m-cresol, du benzène, du toluène et du phénol entre 650 et 850°C', *Ann. Mines de Belgique* (1965), 7-8, 1001.

5 R. CYPRES, C. BRAEKMAN-DANHEUX, 'Processus d'aromatisation dans la formation des goudrons de haute température', *Ann. Mines de Belgique* (1974), 41, 1109.

6 P. BREDAEL, 'Mécanismes de l'aromatisation du n-decane', *Ann. Mines de Belgique* (1975), 11, 1045.

7 S. FURFARI, 'Hydropyrolysis of coal', *ICTIS/TR 20 Rep.*, (1982), *IEA Coal Research*, London.

8 J-L. JOHNSON, 'Kinetics of bituminous coal char gasification with gases containing steam and hydrogen', *Coal Gasification*, Ann. Chem. Soc., Adv. Chem. Ser. (Ed. Massey, L.G.) (1974), n° 131, 145.

9 DENT F.J. ; BLACKBURN W.H. and MILLET H.C., 'Synthesis of gaseous hydrocarbons at high pressure', *Inst. Gas Eng. Commun.* (1937), 167.

10 Anonymus, 'Coal for chemicals', *Coal Age* (1952), Oct., 70.

11 DONATH E.E., 'Fuels and chemicals from coal', *Ind. Eng. Chem.* (1954), 46, 2032.

12 CYPRES R. and FURFARI S., 'Fixed bed pyrolysis of coal under hydrogen pressure at low heating rates', *Fuel* (1981), 60, 768.

13 GHODSI M. ; NEUMANN-TILTE C., 'Construction of an isothermal balance with two symmetrical pans for direct hydrogenation of coals', *Thermochim. Acta* (1983), 62, 1.

14 HOLMES J.M. ; COCHRAN H.D.Jr ; EDWARDS M.S. ; JOY D.S. and LANTZ P.M. 'Evaluation of coal carbonization processes', *Coal Proc. Techn.*, A.I.Ch.E. (1977), 3, New York.

15 GREENE M.I., 'Engineering development of a short residence time coal hydropyrolysis process', *Fuel Process. Techn.* (1977-78), 1, 169.

16 OBERG C.L. ; COMBS L.P. and SILVERMAN J., 'Coal conversion by flash hydropyrolysis and hydrogasification', *Proc. Intersoc. Energy Convers. Eng. Conf.* (1978), **13**, 402.
17 SIMBECK D.R. ; DICKENSON R.L. and MOLL A.J., 'Coal liquefaction : Direct versus Indirect', *Coal technology '80.* Third International coal utilisation conference, Houston (1980), vol. V, 216.
18 NGUYEN CU QUYEN, 'Influence de la pression et de la présence de gaz réactifs sur les produits de la cokéfaction du lignite'. Ph.D. thesis (1979). Université libre de Bruxelles.
19 CYPRES R. and FURFARI S., 'Hydropyrolysis of a high sulphur - high calcite Italian Sulcis coal. 1.Hydropyrolysis yields and catalytic effect of the calcite', *Fuel* (1982), **61**, 447.
20 FURFARI S. and CYPRES R., 'Hydropyrolysis of a high sulphur - high calcite Italian Sulcis coal. 2.Importance of the mineral matter on the sulphur behaviour', *Fuel* (1982), **61**, 453.
21 FURFARI S. and CYPRES R., 'Hydropyrolysis of a high sulphur - high calcite Italian Sulcis coal. 3.Pyrite behaviour', *Fuel* (1983), **62**, 615-619.
22 CYPRES R. and FURFARI S., 'Hydropyrolysis desulphurisation of high sulphur coal', *4th Miami International Conference on Alternative Energy Sources*, Miami Beach, Flo USA (1981), 14-16 Dec.
23 SCHRADER L. ; TEGGERS H. and THEIS K.A., 'Hydrierende Vergasung von Kohle', *Chem. Ing. Tech.* (1980), **52**, 794.
24 FLADERER R. and SCHRADER L., 'Berechnung der Verbrennungswärmen von Rheinischen Braunkohle sowie von Restkoksen aus der Hydrierenden Kohlevergasung', *Braunkohle*, (Heft 1981), **5**, 129.
25 TEGGERS H. and ENGELHARD J., 'Hydrierende Vergasung von Kohle mit Prozeswärme aus Hochtemperatur - Kernreaktoren und Vernvendung des Restkokses aus der hydrierenden Braunkohlevergasung : Vereinigung der Grosskraftwerksbetreiber E.V.', *V6B Konferenz 'Kohlvergasung in der Energietechnik'* (1979).

METHANE PRODUCTION PROCESSES BY FERMENTATION OF BIOMASS

H.C. DUBOURGUIER, G. ALBAGNAC, D. VERRIER
Station de Technologie Alimentaire
INSTITUT NATIONAL DE LA RECHERCHE AGRONOMIQUE
BP 39
59651 VILLENEUVE D'ASCQ Cedex
France

ABSTRACT . Main biological characteristics of methanogenic fermentation are first presented with a peculiar emphasis on the limiting steps and the resulting technological implications. Strategies of anaerobic digestion are discussed according to the physico chemical composition of the liquid waste. Finally, the most significant designs are compared on practical examples.

INTRODUCTION

The methanogenic fermentation is an anaerobic biological process which converts organic materials to methane and inorganic end products including carbon dioxide and ammonia. This fermentation is widespread in nature (rumen, sediments, hot springs ...). Rational use of this process induces energy recovery from various agricultural or domestic refuses and industrial wastewaters. But the economic profitability depends not only on the investment and its productivity but also on the cost of competitive energies. It also depends on concurrent capacities of methanisation, i.e. commercial value of compost, depollution, decrease of energetic dependance.

In recent years, interest in anaerobic treatment has expanded considerably because methane can serve as a useful fuel to help offset a growing demand for energy. This increase in interest induced rapid progress in basic knowledges on methanisation, particularly in biochemistry and microbiology. Concurrently, development researchs resulted in some important new processes and in full-scale systems. In 1982, one can enumerate more than five hundred digesters in E.E.C. countries of which about fifty industrial full-scale systems.

In this report, biology of methanogenesis will be summarized taking in account technological consequences of limiting steps. Processes will be reviewed afterwards according to the nature of refuses or wastes to be treated.

G. E. Beghi (ed.), Synthetic Fuels, 219–233

1. BIOLOGY OF METHANE PRODUCTION

It is clearly established that methanogenesis is realised by a well defined trophic group, i.e. methanogenic bacteria. These organisms can use only mono carbon compounds (H_2/CO_2, formate, methanol, methylamines) and acetate. Consequently, conversion of organic materials to methane involves preliminary degradation into acetate, carbon dioxide and hydrogen. This conversion is due to the combined action of hydrolytic and fermentative bacteria and acetogenic bacteria (Fig. 1). In that system, syntrophic associations between acetogenic and methanogenic bacteria play a key role. In digesters and low-sulfate ecosystems, methane production by decarboxylation of acetate stands for 60 to 70% of the total carbon flow.

1.1. Hydrolytic and fermentative flora is characterized by a great diversity in constitutive species and a wide spectrum of their metabolites. Mesophilic and thermophilic species had been described. Their metabolic end products vary with the substrate and the temperature range. Generally, fermentation of polysaccharides (cellulose, xylan) provides a mixture of volatile fatty acids (VFA), ethanol and lactate. In contrast, with effluents containing large amounts of simple sugars, the VFA are mainly produced after lactic fermentation of the carbohydrates.

During hydrolysis and fermentation, two main bacterial interactions are well demonstrated. The first one leads to a shift of sugar fermentation to exclusive acetate production by interspecies hydrogen transfert to methanogens or acetogens. The second one is due to complementation of activities between hydrolytic and fermentative bacteria, this interaction leading in batch cultures to an increase in degradation yields of complex carbohydrates.

Association between *Clostridium thermocellum* and *Methanobacterium thermoautotrophicum* or *Clostridium thermocellum* and *Acetogenium kivui* demonstrates the role of interspecies transfer of hydrogen. This interaction does not modify degradation yield but increases the energy gain (ATP) of *Cl. thermocellum* by the shift to acetate production. Ethanol is no longer produced. This metabolic shift plays an important role in mixed one stage reactors. Moreover, it could be hazardous to correlate metabolic activities of pure cultures with those occuring in mixed populations.

Synergy between hydrolytic and fermentative species is illustrated by cocultures of *Clostridium thermocellum* and *Clostridium thermohydrosulfuricum*. This last organism increases the yield of cellulose conversion to ethanol by using sugars provided by the cellulases of *Cl. thermocellum*.

1.2. Acetogenesis from VFA and reduced compounds such as lactate or ethanol is a key step for complete conversion of organic materials to methane. Acetogenesis is due to the action of some poorly understood specialized bacteria which are named obligate hydrogen producing acetogens (OHPA). The thermodynamic principles involve that, in standard conditions, dehydrogenation of lactate, ethanol and VFA is

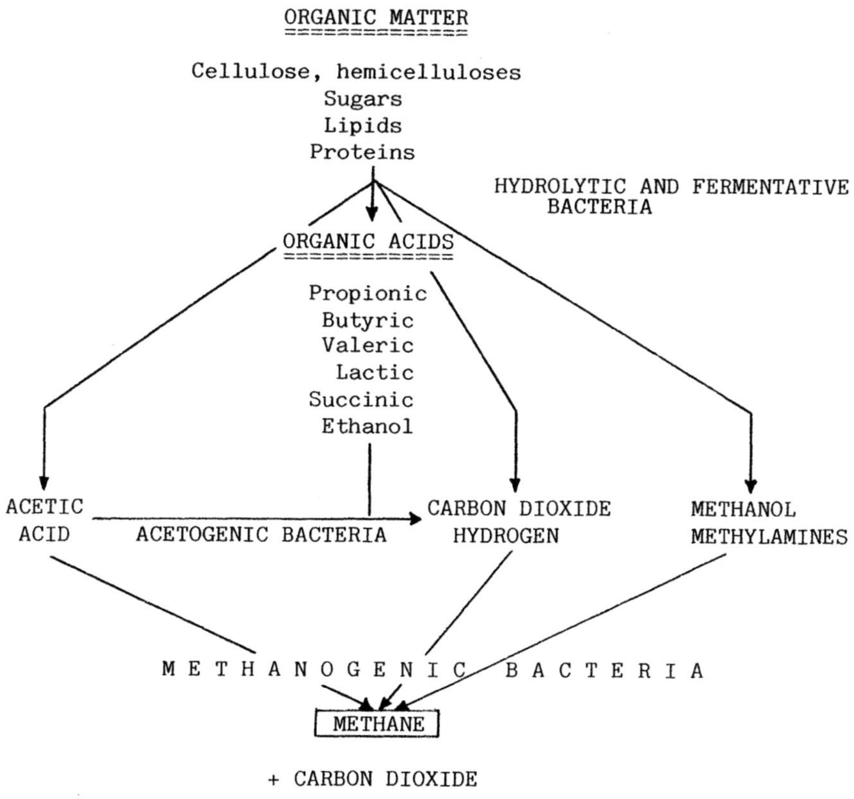

Figure 1 : Anaerobic degradation of organic matter

possible only if the hydrogen partial pressure is kept at some low levels. In this condition, energy available for OHPA will permit ATP synthesis and bacterial growth. In digesters, acetogenesis is necessarily associated with the production of methane by reduction of CO_2. According to free energy values in standard conditions, propionate is the volatil fatty acid which is the most difficult to catabolize to methane.

The OHPA bacteria could be either sulfatoreducers possessing an hydrogenase, either real syntrophic bacteria or a recently described genus Pelobacter. The role of sulfatoreducing bacteria had been demonstrated by cocultures with hydrogen-utilizing methanogens. These cocultures are able to convert lactate or ethanol to acetate and methane. Real syntrophic bacteria described by BRYANT and co-workers, Syntrophomonas, catabolize butyrate and other higher fatty acids to acetate only in coculture with a hydrogenophilic methanogen. Syntrophobacter and Syntrophus oxidize respectively propionate and benzoate in the same manner. Newly described bacteria, Pelobacter are able to catabolize ethanol to acetate only in coculture with a hydrogen-scavenger. This ethanol oxidation by cocultures was originally discovered with the isolation of "S" organism from M. omelianskii. All the OHPA bacteria are strict anaerobes, extremely sensitive to oxygen. Their doubling time is sometimes very high. For example, Syntrophobacter oxidizes propionate in coculture with a methanogen with a doubling time of 160 hrs. Thus, acetogenesis from VFA is a particularly sensitive step of methanogenic fermentation. It could even be a limiting step when large amounts of propionate are produced by fermentative bacteria.

Finally, acetate which is the main intermediate of methanogenesis in digesters is synthesized by two main metabolic ways : co-production with reduced compounds from organic materials (hydrolytic and fermentative bacteria) and homoacetogenesis with obligate production of hydrogen.

1.3. In addition to biochemical specificity, methanogenic bacteria present three main characteristics. Their cell wall lacks peptidoglycan and muramic acid. This leads to a natural resistance to antibiotics which inhibit cell wall synthesis. The methanogens have a specific methyl-carrier molecule, i.e. mercapto-2-ethane-sulfonic acid or coenzyme M. Finally, they have particular fluorescent electron-carriers of which F420 is the most famous. This fluorescence allows specific detection of methanogens under microscope and approximative quantitation of methanogenic potential of mixed populations. But some methanogens are not fluorescent.

Methanogens are divided in two trophic groups : "hydrogenophilic" and "acetoclastic". Their morphology and their kinetic properties are completely different. Their respective kinetic properties are partially due to different free energy changes of methanogenesis reactions. In standard conditions, the free energy $\Delta G'o$ of acetate decarboxylation is - 28.2 $KJ.mole^{-1}$ whereas $\Delta G'o$ of carbon dioxide reduction is - 139.2. Thus the late reaction will be more favorable to bacterial growth. Moreover, methanogenesis had been mainly studied in ecosystems (rumen, sediments, hot springs) where acetate decarboxylation is low and even absent. This fact explains the prevalence of hydrogen-utilizing methanogens in actual taxonomy. Recently, interest in acetoclastic methanogens of digesters leads to isolation of new species, Methanosarcina mazei, Methanothrix soehngenii, Methanosarcina TM1 and Methanosarcina CHTI 55. The two later are the only known thermophilic acetoclastic methanogens. All these acetoclastic methanogens are characterized by their inability to use hydrogen. The growth rates and affinity constants of hydrogenophilic species are high. Consequently, carbon dioxide-reducing methanogens fit rapidly to an increase of substrate load if there is not a concurrent decrease of pH. In these conditions, hydrogen is generally absent in biogaz.

Among acetoclastic methanogens, one can distinguish two main morphological groups. Sarcina-like bacteria (Methanosarcina) present a low apparent affinity constant for acetate and quite good growth rates. Reversely, filamentous bacteria (Methanothrix) have a very high affinity for acetate but their growth rates are very low (doubling time about 4 days). Lower will be their Ks for acetate, higher will be their numbers in acetate-limiting concentrations. So, according to the operation of the digester, different equilibrium of methanogenic populations will be observed. Methanothrix is prevalent in digesters where acetate is maintained at limiting levels, i.e. when degradation yield is to be maximum. This bacteria removes acetate even when residual concentration in the effluent is below 50 mg/l. In contrary, Methanosarcina sp. will be prevalent in digesters either when high productivity of biogaz is achieved (high levels of residual acetate) either when the digester is discontinuously fed or when methanol is provided during the first steps of degradation (pectin for example).

2. EFFECT OF SOME PARAMETERS ON METHANE PRODUCTION

Acetogenesis from volatile fatty acids and methane production from acetate are the two most sensitive steps to environmental conditions.

2.1. Redox potential and pH

Acetogens and methanogens are strict anaerobes, extremely sensitive to oxygen (E.O.S.). Normally, they grow only when redox potential of the medium is maintained below - 330 mV. Oxygen introduction but also highly oxidized mineral ions (nitrite, nitrate, sulfate) can inhibit methanogenesis. In digesters, anaerobic conditions are maintained by microaerophilic hydrolytic and fermentative bacteria.

These E.O.S. anaerobes are also very sensitive to pH. Optimal production of methane occurs only if the pH is maintained between 6.5 and 7.5. A pH below 6 adversely affects the generation of methane. One must point out that acid production is still present at this pH. Indeed, some acidogenic bacteria are active even at pH 4-4.5. Volatile fatty acids concentrations below 5 g/l do not modify methane production rates if the pH is maintained around 7.0. But acid accumulation concurrently with a pH decrease have a lethal effect on methanogens and acetogens. In accordance with these facts, all the methanogenic bacteria tested show a marked pH optimum somewhere between 6 and 8.

2.2. Temperature
Microbial methane formation occurs over a wide temperature range, namely from about 0°C to 95°C. But no acetoclastic methanogen able to grow above 65°C had been described. In this thermophilic range, methane production is only due to carbon dioxide reduction. Practically, complete degradation of organic materials in methane and CO_2 occurs in two technological temperature ranges : the mesophilic range (30-40°C) is generally used in industrial processes ; although reaction rates are higher in thermophilic range (50-55°C), economical interest is not clearly established ; moreover the later does not notably modify the yield of conversion of complex substrates. In fact, an increase of the temperature influences mainly the degradation rates of biopolymers but acetoclastic thermophilic methanogens have slightly better kinetic parameters than mesophilic ones.

Hydrolytic bacteria show a marked temperature optimum above 60°C when methane production from acetate is markedly inhibited at this temperature. These biological arguments concurrently with economical ones lead to reject thermophilic ranges for the treatment of liquid refuse containing low quantities of volatile suspended solids (VSS).

2.3. Sulfide concentration
It had been shown that sulfate inhibits methanogenesis in sediments and digesters. The hydrogen provided by oxidation of organic material (VFA included) is almost exclusively used for sulfate reduction, since this event is thermodynamically more favourable than methane formation. In this case, only carbon dioxide reduction will be affected and acetate decarboxylation still occurs. But, in some cases, sulfate reduction can occur simultaneously with carbon dioxide reduction. Total sulfide concentrations above 3 mmol/l inhibit methane formation in mixed and pure cultures. No valuable explanation is yet available. But sulfite formed by sulfate reduction could precipitate essential trace elements like iron, nickel, cobalt or molybdene. Then these ions are not easily accessible to bacteria, particularly to methanogens.

3. STOICHIOMETRY AND YIELD OF METHANE PRODUCTION

Anaerobic treatment converts efficiently the energy contained in organic materials to methane. About 90% of the calorific value of the degraded matter is recovered in methane which evolves spontaneously from the fermentation medium. Energy available for bacterial growth is thus much lower in methanogenesis than in aerobic degradation. Then mixed methanogenic populations present a low biomass productivity and low growth rates of some constitutives species (OHPA, methanogens).

Overall stoichiometry can be expressed by :

$$C_x H_y O_z + (x - \frac{y}{4} - \frac{z}{2}) H_2O \rightarrow (\frac{x}{2} - \frac{y}{8} + \frac{z}{4}) CO_2 + (\frac{x}{2} + \frac{y}{8} - \frac{z}{4}) CH_4$$

Practically, maximal yields are 443 l CH_4/kg cellulose, 370 l/kg sugar, 620 l/kg lipids and 493 l/kg proteins. However, biomass composition (CHON) could be difficult to obtain. In this case, COD (Chemical Oxygen Demand) could be used according to $P(CH_4) = 0{,}25 \times COD$

where $P(CH_4)$ is the weight of produced methane if the conversion yield is 100%. With lignocellulosic substrates, lignin cannot be degraded under anaerobic conditions. Breakdown of carbon-carbon and ether linkages of the lignin molecule is impossible by anaerobiosis. Moreover, lignin limits accessibility of cellulose to microorganisms or their cellulases. Only a portion of cellulose, namely amorphous cellulose can be fermented in a short time. Cristalline cellulose, often linked to lignin is slightly degradable and requires long term fermentations. For example, biodegradability of domestic refuse is generally between 40-50% of theoritical yield. In residual straw, about 45% of lignocellulose cannot be degraded in less than 20-40 days.

Biogas composition depends on composition of organic material. It can be calculated from COD and TOC (Total Organic Carbon) values according to :

$$COD/TOC = V(CH_4)/(V(CH_4) + V(CO_2)) \times 5{,}33$$

But theoritical biogas composition is modified by two process parameters, i.e. alcalinity of the effluent and the dilution rate. A part of produced carbon dioxide is dissolved in the liquid phase and leads to dissociation equilibrium with the cations, particularly ammonia provided by protein degradation. The carbon dioxide in form of bicarbonate is eliminated with the liquid phase, this leading to an enrichment of the biogas in methane. Similarly, high dilution rates enhance CO_2 removal in liquid phase. In anaerobic filters where high volumic loads are achieved, apparent methane ratio can be increased up to 80% because of carbon dioxide elimination in the effluent. Finally, an increase of the temperature decreases CO_2 solubility and for a same substrate, the ratio of methane in biogas and the buffering capacity of the medium will be decreased.

4. METHANISATION OF DOMESTIC REFUSE

Urban refuses contain large amounts of biodegradable organics such as paper, garbage, yard wastes ... Generally, they present high dry matter content and high amounts of minerals. But the low nitrogen and phosphorus contents can limit the refuse fermentation. Biodegradability of volatile solids ranges from 35 to 60%.

4.1. Gas production from landfill

Methanisation of domestic refuse occurs naturally in landfills. This spontaneous gas production has been estimated to 15-20 Million KWh per year for 1 Million ton landfill, i.e. a landfill equivalent to the refuse of 250 000 inhabitants. This production is slow and occurs over a long period of time (15 years). It has been pointed out that phosphorus, and to a lesser extent nitrogen, tends to disappear during degradation. This reduces the practical yield of gas production by the landfill, even though organic carbon might still remain. Collection of the produced gas is not only an economic interest. It is also a safety interest because of the fire and explosion hazards which can arise when spontaneous gas production is uncontrolled.

Research and development projects have been initiated in several countries (U.S.A., U.K., Germany, Italy, France...). In most cases, the main goals are environmental protection and promotion of gas. This last point leads to a number of researchs particularly on optimisation of gas production and on technology for safely collecting the gas.

The results of these programs are not complete. But strategies of efficient gas collection have been defined. The gas is recovered from porous wells. It is too early to predict the lifetime of the wells with confidence but some systems are now in operation for several years. In order to control the gas production and to accelerate degradation, water recirculation is under study. This technique would also provide the ability for mineral corrections (phosphorus or nitrogen).

4.2. Methanisation of refuse in digesters

In some cases, selective separation of refuse is performed to recover metals, glass and synthetics. The remaining refuse consists essentially of organics which could be fermented in digesters.

Refuse fermentation is aided by the addition of sewage sludge or other co-substrates. Early investigations of the digestion of refuse were undertaken in 1936. But detailed studies were due to PFEFFER (1979). The results lead to an industrial process, the REFCOM-Plant constructed at a disposal site in Florida Pompano Beach. The process used is based on two 1 250 m^3-completely mixed digesters working in parallel. At a 100 ton per day processing rate, the methane production rate was expected to be around 9 000 m3 per day. Start up was initiated in 1978. The plant is still under evaluation.

New processes without previous dilution of the refuse are actually under evaluation. In the VALORGA system (France) solid refuse fermentation is performed continuously at high dry matter content (30%). The medium is maintained homogenous by a pneumatic system using the produced biogas. This system is composed of an input- and output well and a fermentation vat. The substrates ply continuously between these wells. Matter fluxes are controlled by a valve using the produced biogas. At 14 days detention time, the VALORGA process could achieve 3 to 5 m^3 biogas production per m^3 of vat per day. The yield of degradation is around 40%. This system allows recovery of fermented and stabilized substrate for agricultural use.

5. METHANISATION OF ANIMAL SLURRIES

In this field of methanisation, much hopes have been created because of the energetic potential of animal refuses. For example, in France, this potential is estimated to 5 to 6 M TOE per year. This amount must be compared with the 3 M TOE provided by the use of wood for heating. But many disappointments appeared with the practical experience. Recently, a C.E.E. report pointed out some interesting results of which only significant aspects will be presented. In France, among about sixty operating plants in 1983, one half is treating solid manure, the other half liquid manure.

5.1. Solid manure

Several kinds of solid manure are treated by methanisation including bovine, pig, ovine, poultry wastes and also mixed solid manure. The most frequent process is the batch or submerged discontinuous system belonging to the concept of DUCELLIER and ISMAN. An important progress has been made with the solid state fermentation which is performed with a periodical recycling of the liquid phase. This process simplified the design of the digester which can be loaded by a farm tractor.

But regular production of gas without an extensive use of land can only be achieved by continuous feeding. Some interesting projects have been developped and they proved the feasibility of continuous fermentation of solid manure. The main problems to be solved were the handling of solid manure (grinding, pumping, hydraulic design). Thus, one can fear that development of this technology will be limited by an increase of the investment cost.

5.2. Liquid manure

Pig manure is the most methanised substrate either by semi continuous or continuous processes. It is are easily pumped and conversion of organic matter (50%) is higher than in solid manure. In many locations, depollution is the primary objective of anaerobic treatment and production of methane is an added value to the process. Bovine manure and diluted poultry manure are also treated by methanisation.

Actually, continuous mixed reactor is the most used design. But development of other processes, mainly anaerobic filters and anaerobic contact digesters are under investigations.

5.3. Results

When they are available, performances do not fit with owner's hope. The main problems are mainly technological : gas leak, homogeneisation, temperature. But also organic content of the substrate is commonly overestimated. Furthermore, purification and storage of the gas, adjustment of the production to the energetic requirements of the farm are not commonly taken in account. Main results have been obtained in pilot or demonstration plants. In these cases, they were controlled with suitable analytical techniques. Generally, the gas productivities range from 0.5 to 1.2 m^3 per m^3 of digester and per day. But with pig liquid manure, gas productivity can be as high as 6.5 $m^3/m^3/j$ if an anaerobic filter is used.

6. METHANISATION OF INDUSTRIAL WASTEWATERS

Over the last 10 years, the main development and comprehensive studies in anaerobic treatment have been done in industrial wastewaters. The main aim is the purification of effluents in order to be discharged into watercourses. In this area, the anaerobic treatment presents some decisive advantages compared with aerobic process : first, the production of sludge is very low. Energetic expenses are limited and in many cases, energy recovery is achieved, particularly with hot wastewaters (blanching or peeling of vegetables, distillery ...).

But standard rules for discharge in watercourses are not achieved with methanisation. Residual COD is always too important and nitrogen pollution is about entirely recovered in digested water as ammonia. Thus, anaerobic process must be considered as a pretreatment for the recovery of carbon pollution as gaseous products with high yields (90-95%). Industrial applications of anaerobic treatment require to (1) choose a technology suitable for the treatment of the wastewaters and there is not one universal process ; (2) maintain the equilibrium between acidogenesis and methanogenesis ; (3) seed and start up the process. The inoculum will be as rich in methanogens and limited in volume as possible.

6.1. Characteristics of wastewaters

Organic loads of industrial wastewaters are generally dissolved and efficient acidification occurs rapidly. Organic load, expressed as COD, ranges from some thousands to several hundred thousands mg per liter. The ratio COD/total organic carbon varies between 2.5 and 3.0 leading to a methane yield of 250-300 liter per kg COD removed. Suspended solids vary also with the industry. Mineral contents are generally low. The majority of dry matter is composed of easily biodegradable organics (organic acids, simple sugars, neutral compounds, hemicelluloses). The ratio carbon/nitrogen is generally high but still remains compatible with the nutritional requirements of the digesting anaerobic sludge. Routinely, treatment efficiency ranges from 80 to 95% of total COD. Lastly, some industries like sugar industry have a seasonal activity leading to a great variability of the wastewaters flow. Some wastewaters are weakly mineralized and contain low concentrations of nitrogen compounds (for example canning wastewaters). After methanisation, the alcalinity is very weak. This leads to a low buffering capacity of the fermentation medium and high risks of rapid acidification in the case of overloading.

Indeed, methane production from soluble organics is a biological process which is extremely sensitive to excess of substrate. Acidogenesis is realised by rapidly growing bacteria which are relatively tolerant to pH medium. So, they adapt rapidly their number and activities in response to a substrate overloading. Conversely, produced acids are used by slow-growing bacteria, extremely sensitive to variation of the pH medium. Under normal conditions of methanogenesis, their kinetic potentials are saturated. So, control parameters are essential to point out an overloading of the process before it leads to a pH drop. Some analytical determinations are routinely used and well adapted to an early diagnostic of fermentation troubles. They are residual volatile fatty acids, ratio $CH_4/(CH_4 + CO_2)$ in the biogas and residual soluble COD of the effluent.

6.2. Processes

All the industrial processes are continuously fed. In all cases, one wishes to maximise the performances of the digester (and consequently the gas production) by maintaining high concentrations of methanogenic sludge in the digester. This is achieved by various concepts including recycling, settling or attachment to inert supports.

The process will be designed according to the organic load and the suspended solids of the waste water. However, in some cases several technologies could compete. Other criteria must be considered, particularly the overall volume, investment and the ability to retain the sludge even with important instantaneous variations of hydraulic flow. In the cases of low suspended solids waste waters, the anaerobic filter is particularly well adapted.

First generation processes are still employed in industry and include anaerobic lagoon, completely mixed digester, anaerobic contact process. A particular technology has been developped by IRIS in sugar industry. It uses both a mixed sludge bed and an internal settler. These processes are well adapted to medium and high strength waste waters. All these processes are limited in volumic performances because the sludge does not present the good settling characteristics which are necessary mainly in recycling process. This leads to maximal load of 2-4 kg COD per m^3 per day and large volume units. But when there is no restriction on the use of land, reduced costs and technology of these processes will give them the advantage.

In second generation processes, efficient systems have been developped to retain the sludge. They are based either on adhesion, either on trapping or on formation of granular sludge. Practically the efficacy of the retention is due to combination of these effects. As a result, the concentration of bacterial biomass in the digester is high and allows a net increase of performances. Volumic load as high as 20-50 kg COD per m^3 per day can be achieved in some processes. In the other hand, dilute waste waters can be treated with very low hydraulic retention times. This efficiency leads to reduction of the volume of the installation and to extend the application of methanisation, particularly to dilute or cold waste water treatment.

The sludge bed concept has been extensively investigated in the Netherlands since 1971. A proper gas solid separator has been developped resulting in the Upflow Anaerobic Sludge Blanket reactor (UASB). In this process, the high settleability and specific activity of the anaerobic sludge are the main conditions to achieve high loading rates. In fact, the sludge bed is heterogenous : a granular sludge settles in the lower part of the digester and a flocculant sludge predominates in the upper part. Factors affecting flocculation and granulation are not well defined. However, divalent cations (Ca^{++}) seem to have a beneficial effect on settleability. Stability of the bed to large sudden variations of pH, loading or temperature is not well studied.

The fluidized bed is based on efficient attachment of bacterial biomass to supporting finely divided particles such as sand, glass, ceramics. Furthermore, mixing by high flow recycling of the liquid phase provides an optimal captation of the substrate by active bacteria. The control of this process is still difficult and clogging may occur particularly during periods without feeding. However, some industrial applications are appearing.

The anaerobic filter was initially developped as an upflow process with a packing in bulk and without recycling. The result was a "plug flow" system, i.e. substrates and biomass gradients in the digester. With the packing in bulk, suspended solids retention is mainly due to trapping of bacterial biomass between or inside the support.
New advances were done in efficiency with :

- upflow feeding and recycling of the treated effluent

- down flow feeding with or without recycling
- directional inert support (fixed film bed).

Main differences are due to hydraulic characteristics. Support materials vary in nature (stone, clay, plastics) and in structure (bulk or directional). The anaerobic filters can overcome wide variations of process parameters (temperature, pH, volumic load) because they retain efficiently the bacterial biomass even in unfavorable conditions. However, the introduction of support material with low coefficient of porosity diminishes volumic loading efficiencies. Futhermore, clogging may occur in the presence of suspended solids or by extended development of biomass. This risk can be avoid by the design (recycling, upflow/down flow, periodic insuflation of gas) and by tubular inert support. This last system has been applied even with wastes containing suspended solids such as pig manure.

6.3. Some examples of industrial applications

6.3.1. Anaerobic contact and IRIS processes

These first generation processes are largely used in industries such starch pectin extraction and sucrose, distillery, vegetable canning, starch.

In vegetable canning industry, one of the most famous plant is working in France since 1981. After two years operation the energy balance is largely positive (520 TOE per year). But during operation of this anaerobic contact process, settling was sometimes disturbed by shock loads and C02 evolution due to high temperature in the settler (31°C) during summer. No other major problems were observed. Maximum loading rate was 4 kg COD/m^3/day with a net COD removal equal to 85%. Gas productivity was 2.4 m^3/m^3/day with 55% methane.

6.3.2. UASB process

This process is largely used in Netherlands and RFA for the anaerobic treatment of waste water in sugar industry (beet sugar, candy, distillery). Its application extends to potato industry. The volumic loading rates range from 10 to 20 kg COD per m^3 per day. Gas productivity ranges from 3 to 6 m^3/m^3/day with 70 to 90% methane.

6.3.3. Fluidized bed digester

The first European industrial plant will be operating during the first months of 1984 in France. Two digesters of 80 m^3 will treat 50 m^3 of baker yeast wastewater per hour. The estimated biogas production is 1645 m^3 per day at 65% CH_4. Sand (0,1 - 0,3 mm diameter) achieves bacterial attachment. In laboratory experiments, volumic loads were superior to 60 kg COD per m^3 per day. The hydraulic detention time ranges from 1 to 3 hours. This industrial process is theorically the most efficient of all processes. But practical and economical evaluation of the fluidized bed remains necessary.

6.3.4. Anaerobic filter

This process can achieve high gas productivity (1.5 to 4 m^3 CH_4 per m^3 per day) with high loading rates (12-13 kg COD per m^3 per day). The

Bacardi Corporation (Porto Rico) have built the biggest industrial anaerobic filter in the world. In a 12,800 m^3 down flow filter, the PVC support presents an effective area of 120 ha in a practical volume of 9400 m^3. One year after starting up, its performances have been reported. COD reduction is around 70% with a loading rate of 10 kg COD per m^3 per day and hydraulic retention time of 9-10 days.

In Italy, a double-digester system has been built near Bologne. The working volume is 2 x 1250 m^3 with a polymeric material. The estimated performances range 8 to 17 kg COD per m^3 per day with a retention time of 1.2 days.

In France, a 1,100 m^3 digester was built in a beet sugar industry. After 4 months, COD removal was 90% and 79 to 86% of the total energy of the liquid refuse was recovered as high pressure steam vapor.

CONCLUSIONS

Second generation digesters opened a new era in methanisation of waste waters and refuses. Important decrease of fermentation volumes has been obtained concurrently with an increase of performances, mainly volumic loading rates and gas productivity. The actual know-how in methanisation leads to an efficient choice among the processes according to the economical, biochemical and even space environments. But some biological limitations still remain and will remain in the future, mainly the rate of acetate assimilation by methanogenic bacteria and the rate of hydrolysis of biopolymers. The first has been partially solved by an increase of active biomass in the digester. The most attractive development is actually done with the industrial fluidized-bed reactor. But practical evaluation remains necessary.

High solid waste treatment is still under evaluation. Basic knowledge in microbiology, mainly in thermophilic bacteria leads to a new design for fermentation of solids, i.e. a two-step process. The first step will hydrolyse the solid residue in the thermophilic range (60-65°C). In the second step, the produced acids are converted to methane in second generation reactors.

BIBLIOGRAPHY

G. ALBAGNAC (1982) Valorisation par conversion en combustibles gazeux : production biologique de méthane - ADITEC- LYON (FRANCE).

G. ALBAGNAC, D. VERRIER (1983) Méthanisation des effluents d'industries agro-alimentaires. Biomasse actualités 9, S2, 17-21.

A. BORIES, D. VERRIER (1983) Les perspectives d'évolution des technologies de traitement anaérobie : les fermenteurs de deuxième génération. 5ème Colloque "l'Eau, la Recherche, l'Environnement" - LILLE (FRANCE).

D.R. BOONE, M.P. BRYANT (1980) Propionate degrading bacterium, Syntrophobacter wolinii sp. nov. gen. nov. from methanogenic ecosystems. Appl. Env. Microbiol. 40 (3), 626-632.

P. BOULANGER, B. VESPRILLE, K.C. PETTE, R.J. ZOETEMEYER (1982) Digestion méthanique des effluents. Application du procédé biothane UASB en sucreries-distilleries. Ind. Aliment. Agric., 99, 7-8, 565-569.

J. COILLARD (1983) La production de biogaz à la ferme. Techniques mises en oeuvre par les industriels. Problèmes rencontrés. Biomasse actualités 9, S2, 26-33.

M. HEDUIT (1983) Le traitement des effluents d'élevage. Biomasse actualités 9, S2, 12-16.

A. KIENER, T. LEISINGER (1983) Oxygen sensitivity of methanogenic bacteria. System. Appl. Microbiol. 4, 305-312.

P. LE RUYET, H.C. DUBOURGUIER, G. ALBAGNAC (1984) Thermophilic fermentation of cellulose and xylan by methanogenic enrichment cultures : preliminary characterization of main species. Syst. Appl. Microbiol., 5, 247-253.

P. LE RUYET, H.C. DUBOURGUIER, G. ALBAGNAC (1984) Homoacetogenic fermentation of cellulose by a coculture of C. thermocellum and A. kivui. Appl. Environ. Microbiol., 48, 4.

G. LETTINGA, A.F.M. VAN VELSEN, W. DEZEEUW, S.W. HOBMA (1979). The application of anaerobic digestion to industrial pollution treatment. In 1st int. Symp. on anaerobic digestion. CARDIFF (UK).

M.J. Mc INERNEY, M.P. BRYANT, R.B. HESPELL, J.W. COSTERTON (1981) Syntrophomonas wolfei gen. nov. sp. nov., an anaerobic syntrophic fatty acid oxidizing bacterium. Appl. Env. Microbiol. 41 (3), 626-632.

D.O. MOUNTFORT, M.P. BRYANT (1982) Isolation and characterization of an anaerobic syntrophic benzoate degrading bacterium from sewage sludge. Arch. Microbiol. 133, 249-256.

E.J. NYNS, H.P. NAVEAU, M. de MUYNCK (1983) Biogas plants in the European Community and Switzerland. Status-Bottlenecks - Future prospects in 3rd Int. Symp. on Anaerobic Digestion - BOSTON (U.S.A.).

P. OGER, J. CHANDESRIS (1982) Digestion anaerobie des vinasses de distilleries de betteraves avec valorisation du gaz produit. Ind. Aliment. Agric. 99, 7-8, 571-572.

K.C. PETTE, A.I. VERSPRILLE (1981) Application of the UASB concept for wastewater treatment. In "Anaerobic Digestion" (1981) D.E. HUGHES et al. Ed. Elsevier Biomedical Press B.V.

D. PETROFF (1983) Fermentation méthanique de polysubstrats. Aspects cinétiques et microbiologiques. Thèse de Docteur-Ingénieur Université LYON I.

J.T. PFEFFER (1979) Domestic refuse as a feed for digesters. In "1st Int. Symp. on Anaerobic Digestion". CARDIFF.

B. SCHINK, M. STIEB (1983) Fermentative degradation of polyethylene glycol by a strictly anaerobic gram negative, non spore forming bacterium, Pelobacter venetianus sp. nov. Appl. Env. Microbiol. 45 (6), 1905-1913.

M. TISSIER, M. DABAJI (1983) Fermentation méthanique : le lit fluidisé grandeur nature. L'usine nouvelle, 37, 119-121.

J.P. TOUZEL, G. ALBAGNAC (1983) Isolation and characterization of Methanococcus mazei strain MC3. FEMS Microbiol. Lett. 16, 241-245.

J.P. TOUZEL, D. PETROFF, G. ALBAGNAC (1984) Isolation and characterization of a new thermophilic methanogen strain CHTI 55. Syst. Appl. Bact. (in press).

D. VERRIER (1983) La méthanisation bilan et perspectives. Colloque "les Economies d'Energie et les énergies de substitution en agriculture et dans les industries agro-alimentaires" CAMBRAI (FRANCE).

D. VERRIER (1984) Le filtre anaérobie et ses applications pour le traitement et la valorisation énergétique des effluents. La Tribune du CEBEDEAU (in press).

D. VERRIER, R. MOLETTA, G. ALBAGNAC (1983) Anaerobic digestion of vegetable canning waste waters by the anaerobic contact process : operational experience. 3rd Int. Symp. Anaerobic Digestion - BOSTON (U.S.A.).

A.J.B. ZEHNDER, K. INGVORSEN, J. MARTI (1981) Microbiology of methane bacteria. In "Anaerobic Digestion" (1981) D.E. HUGHES et al. Ed. Elsevier Biomedical Press B.V.

GASIFICATION OF BIOMASS BY THERMAL PROCESSES

M. LEMASLE
CREUSOT-LOIRE
Division Energie
B.P. 31 Cedex
71208 Le Creusot
France

ABSTRACT. Biomass, the consumption of which was neglected in developped countries when the price of hydrocarbons was particularly low becomes now a new important ressource in the energetic balance of nations, particularly for developping countries where biomass is available in great quantities for some of them (Brasil).

Different methods are now proposed or studied to use biomass :

- direct combustion in kilns or boilers
- gasification
- liquefaction

In fact, biomass is affected by a major disadvantage which is to be a solid, difficult to handle and store, and then of a rather delicate use. Then, biomass (as coal) can be hardly substituted to liquid or gaseous fuels. It's the reason for which most processes concerning biomass now are developped in order to convert the solid fuel into a gaseous fuel.

Among the different methods which are currently developped, oxygen gasification seems to be the most advanced (Pilots built on that principle in USA, CANADA, SWEDEN, GERMANY, FRANCE).

USA : Pilot IGT (oxygen gasification in fluidized bed - 500 kg/h - 20 bar).
Pilot SERI (oxygen gasification in fixed bed under pressure)
CANADA : ENFOR Project
SWEDEN : Pilot Mino at Studsvik (oxygen and steam gasification in fluidized bed under pressure)
GERMANY : Pilot Lurgi (circulating fluidized bed. High temperature gasification with O_2 + steam)
FRANCE : Pilot Creusot-Loire (fluidized bed oxygen gasification of wood under pressure)

Biomass gasification with oxygen yields a medium BTU gas (2 000-2 500 kcal. Nm^3) which can be used in chemical synthesis (like methanol). Such a gas can also be converted into methane, easily substituted to natural gas.

Before concentrating on that aspect of biomass gasification to produce a medium BTU gas (for chemical synthesis) it is important not

G. E. Beghi (ed.), Synthetic Fuels, 235–258.

to avoid mentioning some other important uses of biomass :
- pyrolysis to produce wood char an pyrolysis gas (about 40 % of the Brasilian siderurgy based on wood char).
- air gasification of wood or biomass wastes to produce a poor gas which can be substituted to liquid or gaseous rich hydrocarbons. (a lot of applications in the world, for example in the paper mill industry).

But the purpose of that paper will be to deal with gasification of biomass for syngas production. Several aspects of that subject will be treated in the overall diagram of a wood to methanol plant, which are not restricted to the gasification section (drying, milling, influence of pressure). Before entering into the subject, some generalities have to be introduced, in order to well understand the problem.

As an introduction, we should say that the very special properties of biomass, that is high fraction of volatiles and reactive char, make these fuels excellent as raw materials for gas production.

1. PRELIMINARY CONDITIONS FOR BIOMASS TO METHANOL PROCESSES

The conversion of biomass to methanol is a process requiring several distinct steps as well as auxiliary equipment. A generalized flow scheme for such a process is shown on figure 1. Depending on the process and the fuel some of these steps will be less important or excluded, for example the oxygen plant in a non oxygen process.

When evaluating different gasification processes for the biomass to methanol route, the requirements on the syngas on one hand the fuel characteristics on the other have to be clarified. The choice of gasification process is most important since it defines the biomass pretreatment requirements and the gas handling and cleaning system.

1.1. Quality demands on syngas

Due to low conversion in a single pass through the methanol reactor, extensive recirculation of gases take place in all methanol processes. This together with the sensitivity of the catalyst means that the synthesis gas for methanol production have to meet the following requirements :
- Low content of inert gases (mainly $CH_4 + N_2 < 5$ %)
- A hydrogen to carbon oxides ratio close to stoechiometry
- Less than 0,1 ppm total sulphur.

The two latter requirements can be met by gas handling such as shifting and acid gas removal. The extent of shifting and the amount of carbon dioxide removed is dependent upon the choice of gasification technique. The amount of sulphur to be removed depends upon the fuel. The amount of inerts in the syngas is dependent upon the gasification process and the nitrogen content of the fuel.

1.2. Characterisation of biomass

Biomass generally contain C, H, O, N, S and ashes plus moisture.

For a dry-ash-free (daf) substance, the molecular formula can be

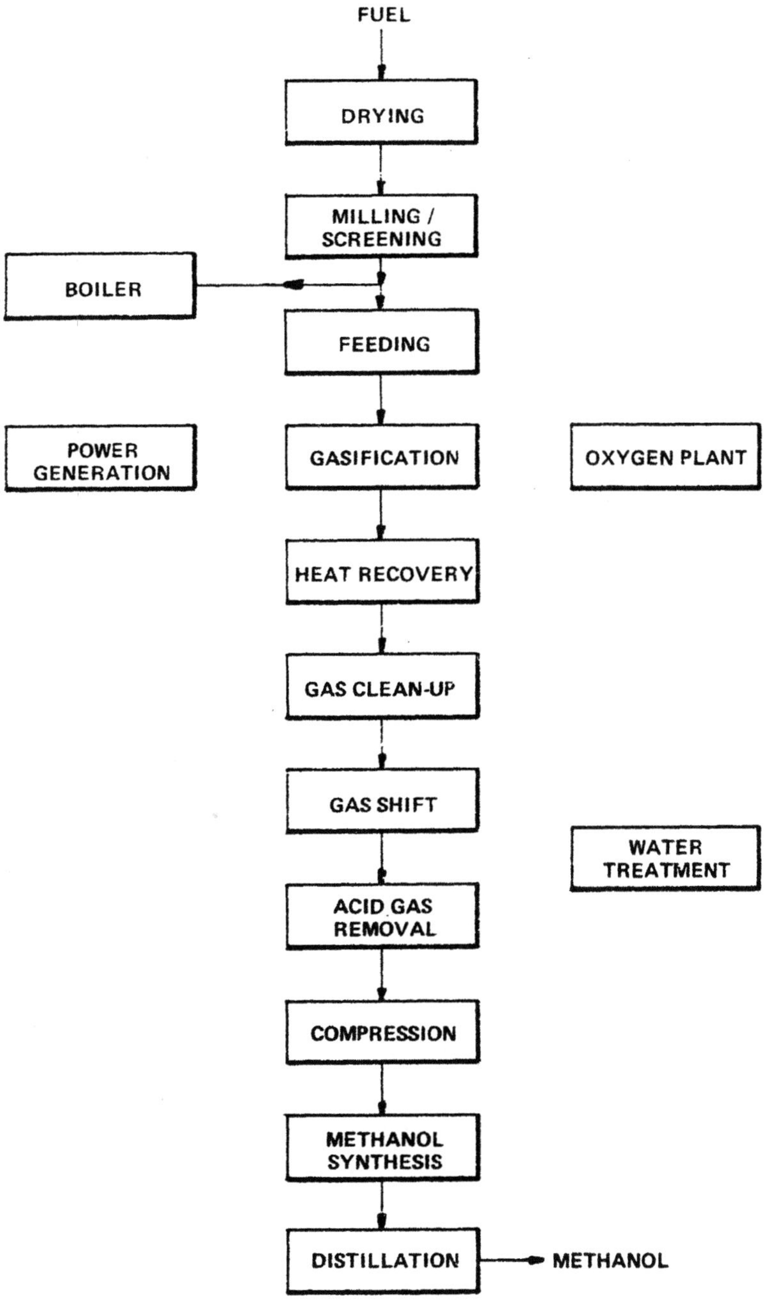

FIG. 1 – GENERALIZED FLOW SCHEME FOR A BIOMASS TO METHANOL PLANT

summarized as follows : $C_6 H_9 O_4$ where C represents 50 %, H = 5 % and O = 45 % on a weight basis.

N and S are generally in very low quantities.

Ash content depends on types of biomass. For wood, that content is most often bellow 1 % as for other biomass (straw, bagasse) it can reach 10 %. The ash melting point of biomass varies to a great extent and can be very low due to high concentrations of alkalines.

Wood and biomass have low bulk density and extremely bad flow properties and can contain up to 50 % moisture in the as-delivered state. Other biomasses such as straw can be delivered at a much lower moisture content (10-20 %).

The heating value compared to coal is low per mass unit and even lower if counted per volume.

Biomass has generally a high content of oxygen and volatile products, which confer it very high reactivy properties if compared for example to coal.

The following table gives a comparison between the main characteristics of biomass and coal :

CHARACTERISTICS OF MATERIAL	WOOD	COAL
% volatile matter	84	30
H.H.V. kcal/kg	4750	6650
Elemental analysis		
Weight percent		
C	48.8	73.4
H	6.0	4.8
O	44.2	10.5
N	< 0.5	1.4
S	0.014	0.22
Weight percent ash	0.5	9.8

1.3. Conversion chemistry

Conversion Chemistry

Gasification of biomass with steam and oxygen takes place in different steps. The initial reaction, pyrolysis, is very fast. The char fraction after pyrolysis is for biomass approximately 10 % of raw substance. The char combustion with oxygen is of course extremely rapid and the amount of char burnt depends primarily upon contacting scheme (char - oxygen or gas-oxygen).

The rate limiting solid-gas reaction is in most cases the char-steam reaction. Reaction rate depends primarily upon temperature. In steam/oxygen gasification of biomass only approximately 5 % of maf. substance is converted in this slow reaction. That means a very rapid total reaction rate.

The gas produced primarily from biomass gasification is rich in tar, methane and ethylene. The secondary gas phase reactions control the final gas composition.

Thermal cracking of tar leads to substantial amounts of "carbon black". The main part of the tar is rather easily converted e.g. by char acting as a catalyst. Low concentrations of methane seem to be possible only at very high temperatures (> 1 000°C).

2. INFLUENCE OF FUEL CHARACTERISTICS ON GASIFICATION TECHNOLOGY

2.1. Milling

Fluidized bed and entrained bed gasifiers both require milling of the fuel. Some biomasses such as wood require of lot of energy for milling.

Fluidized beds can use particle sizes up to about 50 mm but entrained beds have to use very fine powder probably in the range of 0.1 mm.

Fixed beds allow the use of big particles but can not tolerate the use of fines.

2.2. Drying

As stated earlier, wood can be received with 50 % moisture content. As gasification is done at high temperature, feeding of moist fuel to the gasifier increases fuel consumption. As the heat contained in the evaporated moisture can only be partially recovered and at a lower level of temperature, this leads to lower efficiency than drying before gasification.

On the other hand, fuels like wood and peat are difficult to dry completely due to their cellular structure and a good compromise is generally found in the range of 10 to 20 % moisture for the fuel entering the gasification section.

2.3. Feeding

The main problems with biomass as gasification raw material are feeding and handling due to the low bulk density and the extremely bad flow properties. All hoppers have to use stirres or other means for keeping a smooth flow.

The low bulk density of dry biomass (200–300 kg/m^3) means a large volume of all hoppers and thus especially at high pressure a high consumption of inert gas for the pressurization of the feeder system. Forced feeding with extruder type feeders might be possible but have up till now only been used for wet wood chips in the pulp and paper industry.

3. GASIFICATION REQUIREMENTS : Process point of view

3.1. Presentation of the problem

The problem in making a syngas from biomass is abruptly to produce carbon monoxide and hydrogen from a symbolic molecule $C_6 H_9 O_4$. The nical reaction that can be written "on the paper" is thus

$$C_6 H_9 O_4 \ + \ (2\ O) \qquad 6\ CO \ + \ 4{,}5\ H_2$$

But globally, the gasification is endothermic and it is also essary to provide heat to the gasifier in order to maintain an othermic gasification.

So, the syngas production from biomass can be roughly summarized the following requirements :

o provide at least two atoms of oxygen
o provide heat to maintain the thermal balance to equilibrium.

. Possible solutions

bally, wood gasification is endothermic. Then, it is necessary to vide heat to the reactor in which gasification reactions occur. eral solutions are possible to provide heat to the system as indied on figure 2. Heat can be provided from the outside towards the ctor walls. Or heat can be provided by circulating a hot solid ween the reactor and a "regenerating" reactor. Or heat can be proed directly in the reactor itself by producing exothermic reactions it.

The last method is the most used because the simplest. It conts to inject air or oxygen (into the gasifier) which leads to a tial combustion of biomass providing the heat which is necessary for endothermic gasification reactions. In that system, both heat and ms of oxygen are provided to the system simultaneously.

In some processes, to avoid the use of oxygen, a solid is used e lime which gives to the process the heat required by the exormic reaction $CaO + CO_2 \qquad CaCO_3$. In that case, it is necessary to ject steam into the gasifier to provide the atoms of oxygen necessary make $CO + H_2$. It is also necessary to have a second reactor in which) is regenerated at high temperature by residual carbone combustion air.

In an other process, also to avoid use of oxygen, calcium sulfate used which, by reduction with carbon, provides Oxygen and then heat uired for gasification. In that case, a second reactor (oxidizer) is so necessary to re-convert CaS into Ca SO_4 by combustion of char with r.

Gasification of biomass is a heat consuming process as the reacts must be heated and the chemical reactions are endothermic. The at can be supplied either by burning a part of the fuel internally in e gasifier or by using heat carriers or heat exchanger and external rning. These two ways are often referred to as autothermal and alloermal. In the autothermal case oxygen and not air is used when metha-l synthesis gas is to be produced.

When discussing processes in general it is convenient to consider th technical factors and factors which have a direct influence on the ermal efficiency of the process.

From a technical point of view the oxygen processes are simpler. the one hand and oxygen process requires an oxygen plant, oxygen mpressor and the amount of CO_2 which has to be removed is increased. the other hand, a great and perhaps decisive advantage is that it is

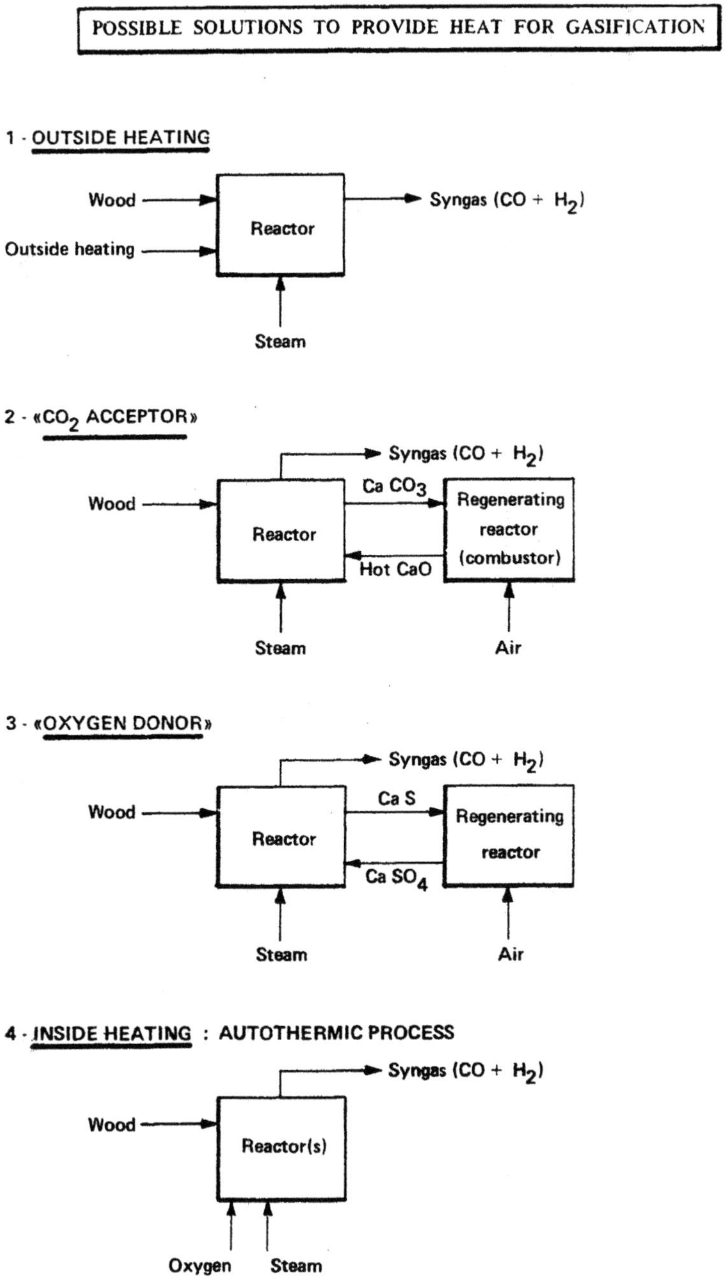
POSSIBLE SOLUTIONS TO PROVIDE HEAT FOR GASIFICATION
1 - OUTSIDE HEATING
Wood
Outside heating
Reactor
Syngas (CO + H2)
Steam
2 - «CO2 ACCEPTOR»
Syngas (CO + H2)
Wood
Reactor
Ca CO3
Hot CaO
Regenerating reactor (combustor)
Steam
Air
3 - «OXYGEN DONOR»
Syngas (CO + H2)
Wood
Reactor
Ca S
Ca SO4
Regenerating reactor
Steam
Air
4 - INSIDE HEATING : AUTOTHERMIC PROCESS
Syngas (CO + H2)
Wood
Reactor(s)
Oxygen
Steam

Figure 2

DIFFERENT PRINCIPLES OF REACTORS

1 - Fixed bed : Counter-current

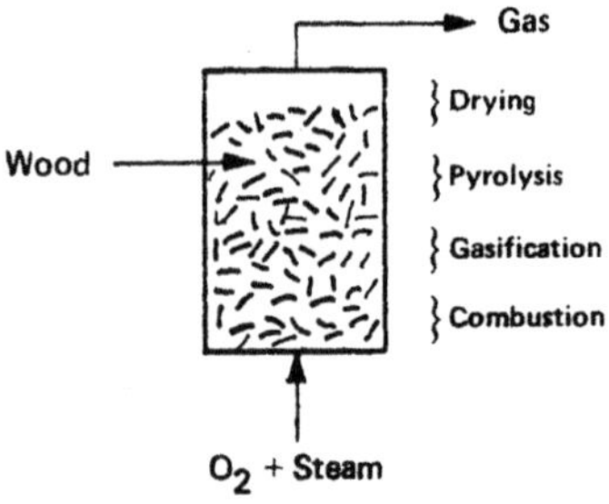

- Big particles allowed
- Fine particles excluded
- Emission of tars + oils

2 - Fluidized bed :

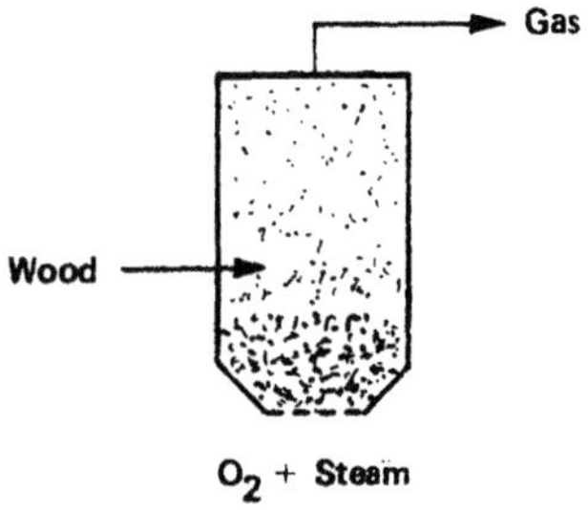

- 0 - 50 mm particles allowed
- > 50 mm particles excluded
- Uniform temperature ⇒ easier control
- Higher heat and mass transfer coeff.
- Limitation of oil and tars
- Working bellow sintering conditions

3 - Entrained bed :

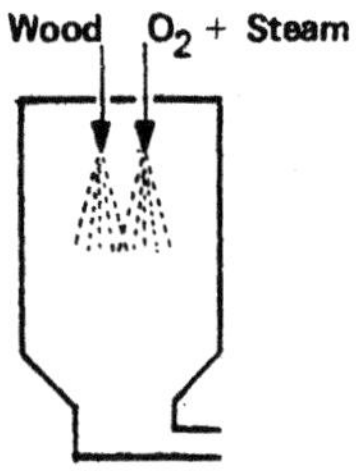

- Max particle size allowed ≃ 0,1 mm ⇒ high milling cost
- Very high temperature > 1 300°C ⇒ No tar, No CH_4
- Working in slagging conditions

Figure 3

easy to reach temperatures above 950°C with oxygen. It is necessary to bring the methane content down to a level which corresponds to at least this high temperature. Non-oxygen processes require a separate combustion chamber and sophisticated heat carriers or heat exchangers.

So, in the following presentation, the subject will be limited to the oxygen gasification in an autothermic process, which seems to be the most attractive and presently the most widely studied.

4. GASIFIER REQUIREMENTS (Technology point of view)

As it has been previously described, gasifiers can be classified (on a process point of view) by the type of providing heat + oxygen to the system.

But on a technologic point of view, they also can be classified in the three following well known different types, as shown on figure 3 :

- fixed bed gasifier
- fluidized - or circulating fluidized - bed gasifier
- entrained bed gasifier.

In the fixed bed (also called moving bed), solids remain jointed and it's necessary to select accurately granulometries in order to eliminate fine particules to allow a smooth distribution of oxygen in the overall reactor, avoiding plugging. In case of biomass wastes, such a selection is a disadvantage. Also, fixed bed don't have an homogeneous temperature profile due to the division of the reactor in several zones (drying - pyrolysis - gasification - combustion). Because of counter-current between solids and gases, this leads particularly for biomass to an extensive production of hydrocarbons and tars.

In the fluidized or circulating fluidized beds biomass particles have to be previously milled but can be tolerated in a rather wide range of granulometry. Temperature is easily controlled but has to be maintained below the ash fusion temperature. For some biomasses, that temperature can be as low as 800°C.

In the entrained bed, very high temperature (> 1 300°C) are reached, leading to slagging operating conditions. At such a temperature, hydrocarbons and tars become negligible. But such a gasifier needs a very fine biomass (probably less than 0.1 mm) and then a very high consumption for milling. Taking in account the rather low heating value of biomass, it is not proven that the entrained bed is the most economic solution in case of biomass gasification.

The low ash sintering point of biomass and the stability of methane lead to the conclusion that single step solid-gas reactors operating below ash melting point (e.g. fluidized bed) can not produce a suitable syngas. One exception might be a single step catalytic fluidized bed at low pressure, but that introduces additional problems such a catalyst recovery. The possible choice is high temperature slagging gasifiers or two-step processes with catalytic or thermal cracking of tar and methane in the secondary step.

It is thus concluded that in order to reach a high capacity, the temperature in the reactor should be maintained at the highest possible level with respect to the sintering of the ash. Especially in the case

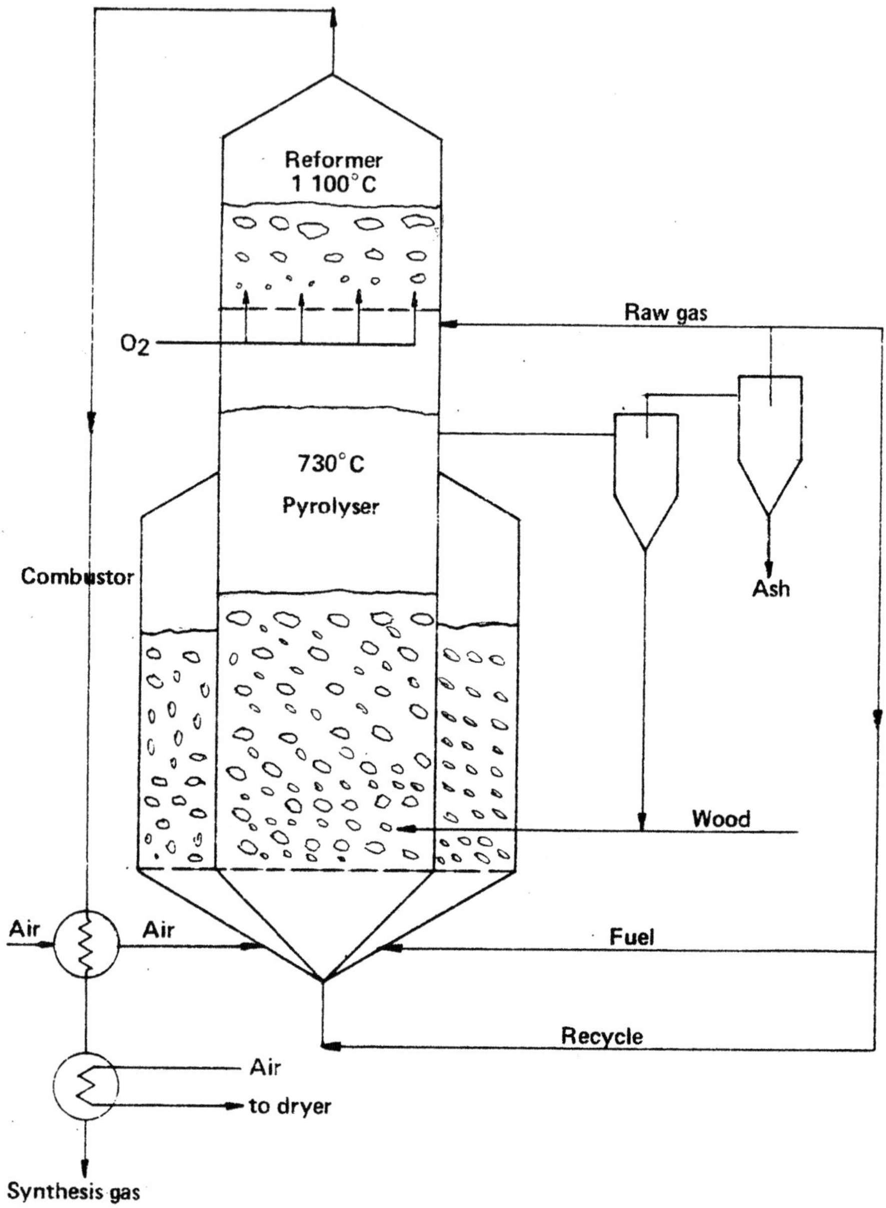

FIG. 4 – PYROLYSIS WITH INDIRECT HEAT TRANSFER.
THE AGIP NUCLEARE / ITALENERGIE PROCESS

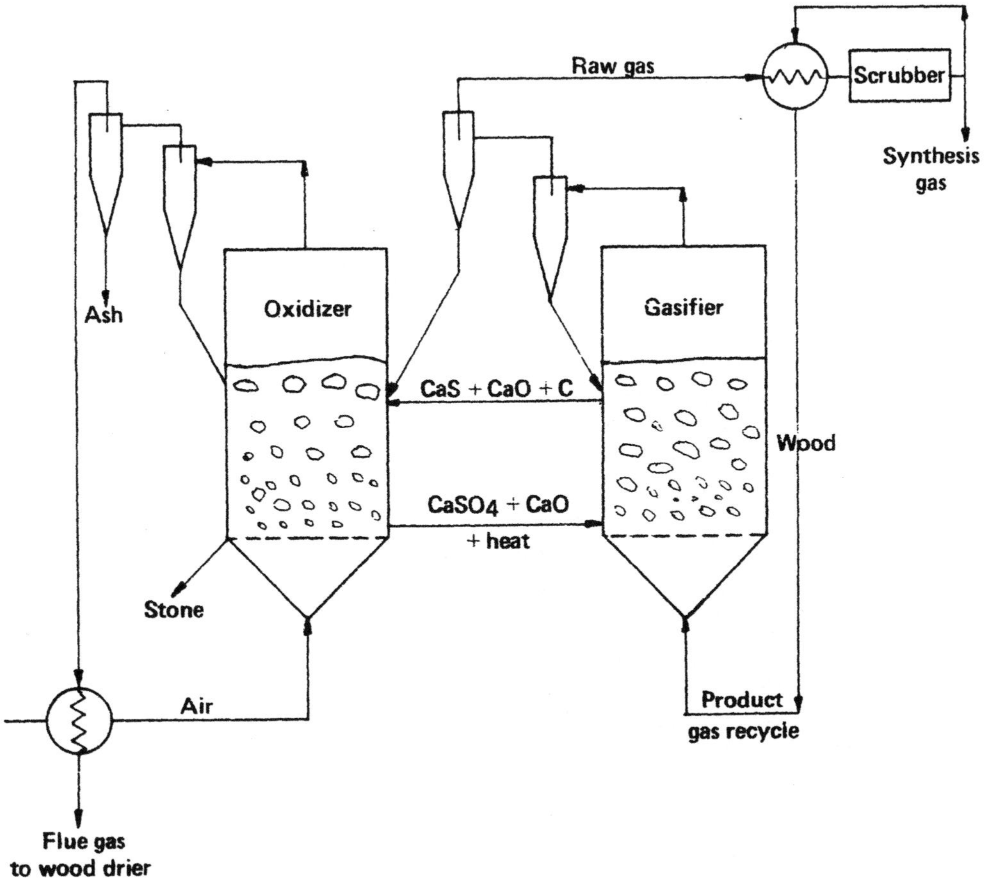

FIG. 5 – THE OXYGEN DONOR PROCESS
JOHN-BROWN WELLMAN (U.K.)

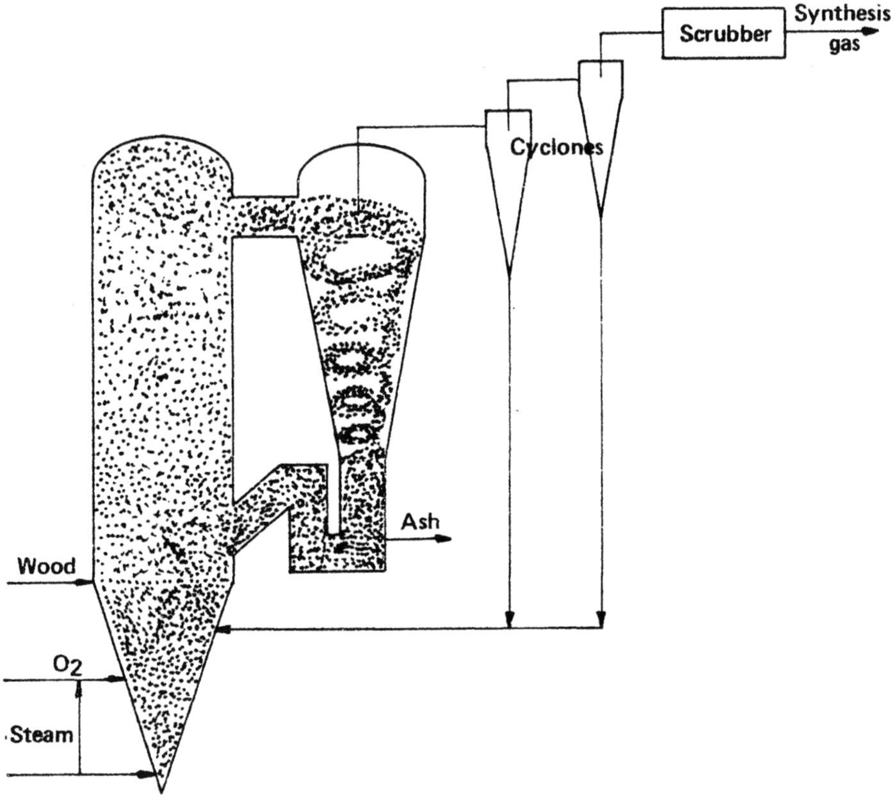

FIG. 6 – OXYGEN GASIFICATION IN A CIRCULATING FLUIDIZED BED, THE LURGI PROCESS.

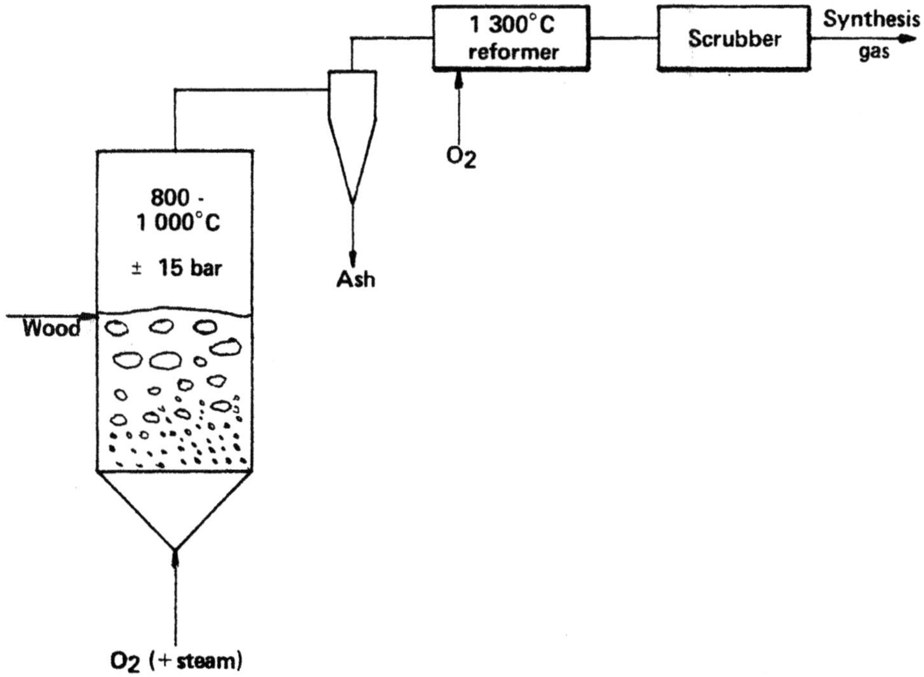

FIG. 7 – OXYGEN GASIFICATION IN A SINGLE FLUID BED, THE CREUSOT-LOIRE PROCESS.

where oxygen is used this calls for a good temperature control to avoid hot or cold spots.

As it has just been described there are, schematically seen, three modes of contacting the solid fuel with the reactant gas in a gasification process ; moving bed, fluidized bed, and entrained bed. When comparing the requirements of a good biomass gasifier with the characteristics of reactors based on the different contacting modes it is concluded that the fluidized bed or circulating fluidised bed is best. The fluidized bed seems best to fulfill the requirement to allow a wide range of fuels, biomass fines as well as wood chips. The rapid heating is most easily realized when feeding small particles directly into the fluidized. The gas phase has a smooth concentration profile and the intensive mixing of the bed dilutes the primarily formed tar rapidly as the gas film around the particles becomes thin. As to the temperature control of the reactor the fluid bed is superior.

5. INFLUENCE OF PRESSURE

The methanol synthesis is performed at a pressure of 50-70 bar. Therefore it is suitable also to carry out the gasification at a rather high pressure, avoiding excessive compression of the synthesis gas. There are, however, other reasons to use pressurized gasification. Most chemical reaction rates and gas treatment steps such as absorption are gained by a high pressure. Also the performance of a fluidized bed is improved and the equipment size is reduced. On the other hand there are some aspects supporting a low pressure. Thus a pressurized process requires all reactant streams including the solid fuel feed to be pressurized, which can be a difficult problem. The optimal pressure should thus be at approximately 15-30 bar.

6. EXAMPLES OF THERMOCHEMICAL PROCESSES FOR BIOMASS GASIFICATION

The processes which are presented below don't constitute an exhaustive list of all the pilot currently existing in the world. They can give an idea about the state of development of research on the subject for different of them.

In Europe, the European Community selected four processes on the subject of methanol from wood (see figures 4, 5, 6 and 7).

- Italenergy - AGIP NUCLEARE (Italy)
- John Brown/Wellman (Great Britain) -Fluidized bed + regenerator non oxygen process
- Lurgi (F.R.G.) - Circulating Fluidized bed oxygen process
- CREUSOT-LOIRE (FRANCE) - Fluidized bed + oxygen reformer oxygen process

In the world, some processes have reached the step of construction of pressurized semi-industrial pilots (see figures 8, 9, 10 et 11)

- I.G.T. (U.S.A) - Fluidized bed - oxygen process
- STUDSVIK + R.I.T. (SWEDEN) - Fluidized bed + catalytic reformer oxygen process

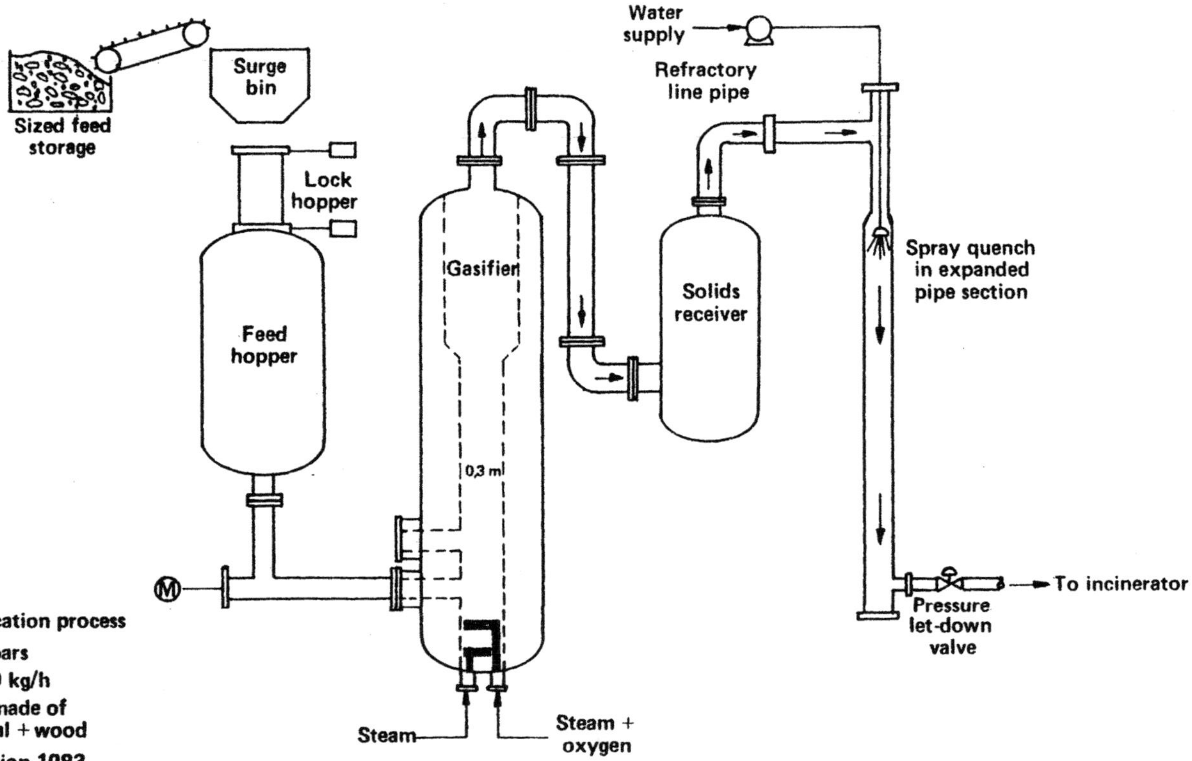

One step gasification process

Pressure = 20 bars

Capacity ≃ 500 kg/h

Fluidized bed made of
an inert material + wood

Begin of operation 1983

FIG. 8 – SCHEMATIC OF PDU SYSTEM
OF I.G.T. (U.S.A.)

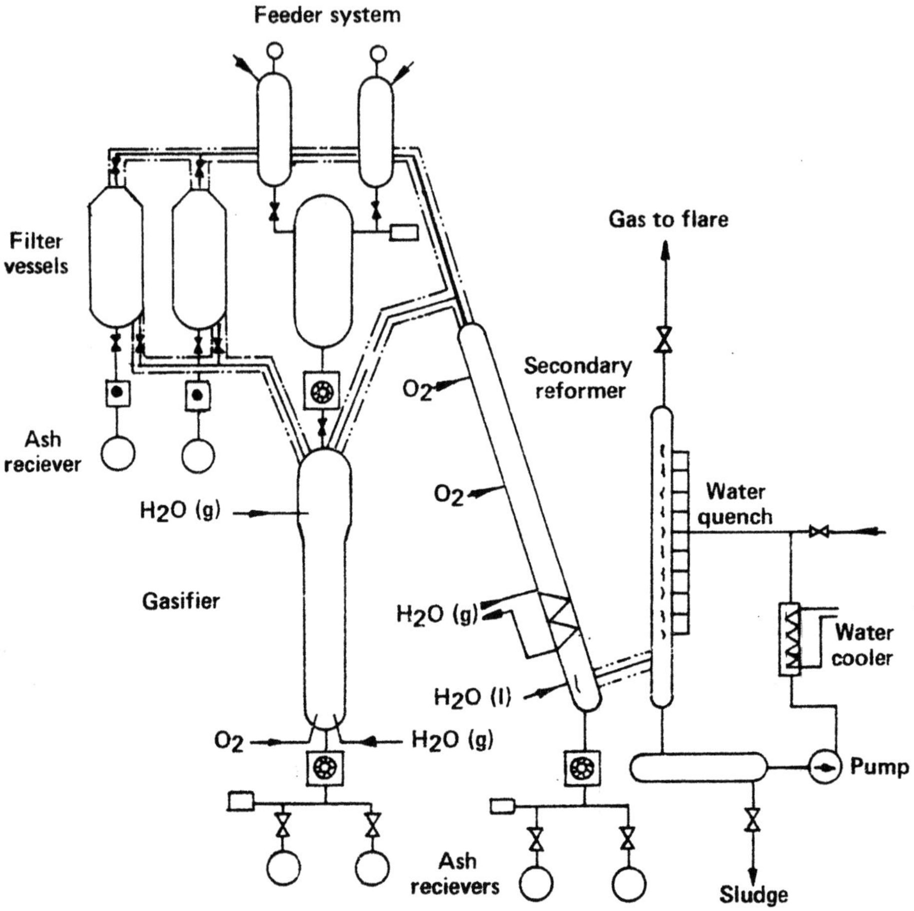

PDU STUDSVIK (SWEDEN)

Two steps gasification process

Pressure up to 20 bars
Capacity up to 500 kg/h
Fluidized bed made of carbon (700-800°C)
Secondary catalytic reformer with O_2 (1 100°C)

Begin of operation : Early 1983.

FIG. 9 – MINO PDU. FLOW SCHEME.

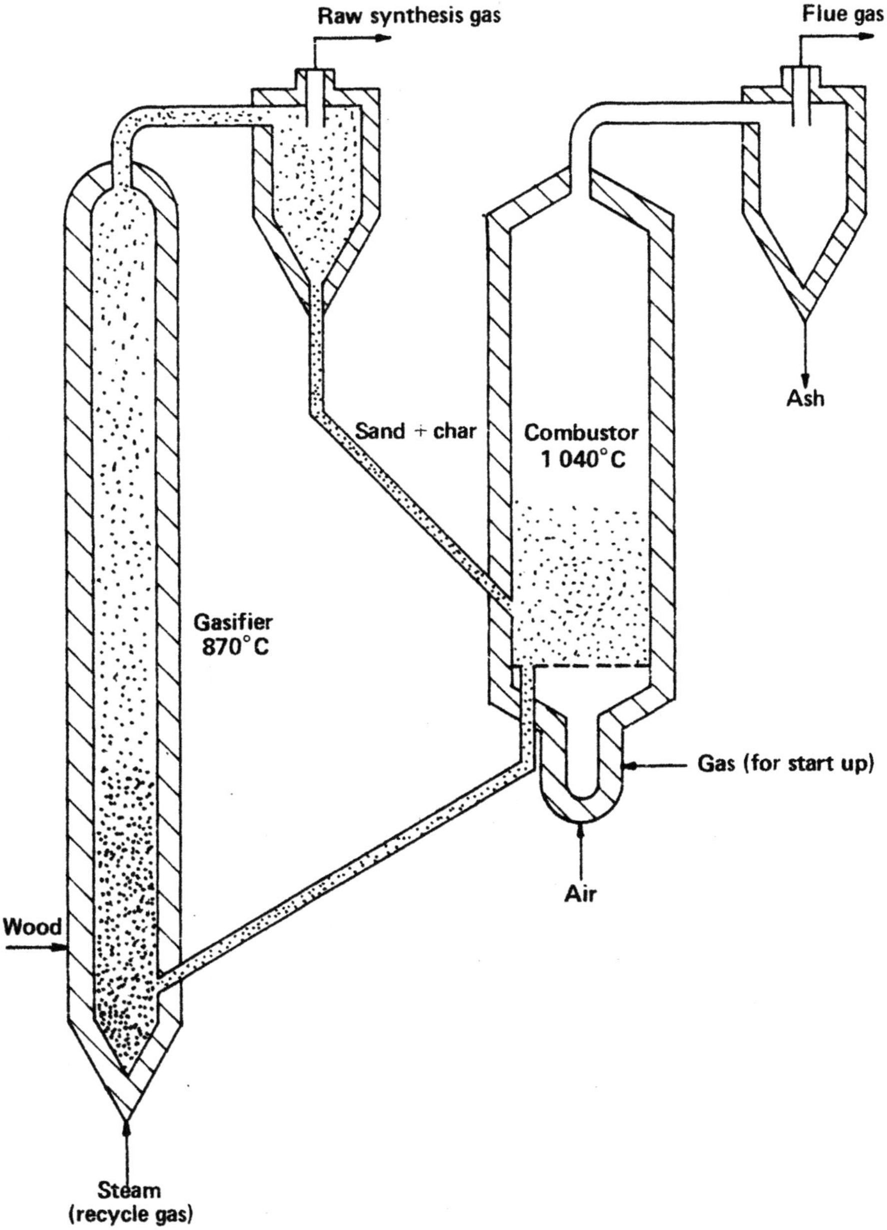

FIG. 10 – MULTI SOLIDS DOUBLE FLUID BED GASIFIER (BATTELLE) U.S.A.

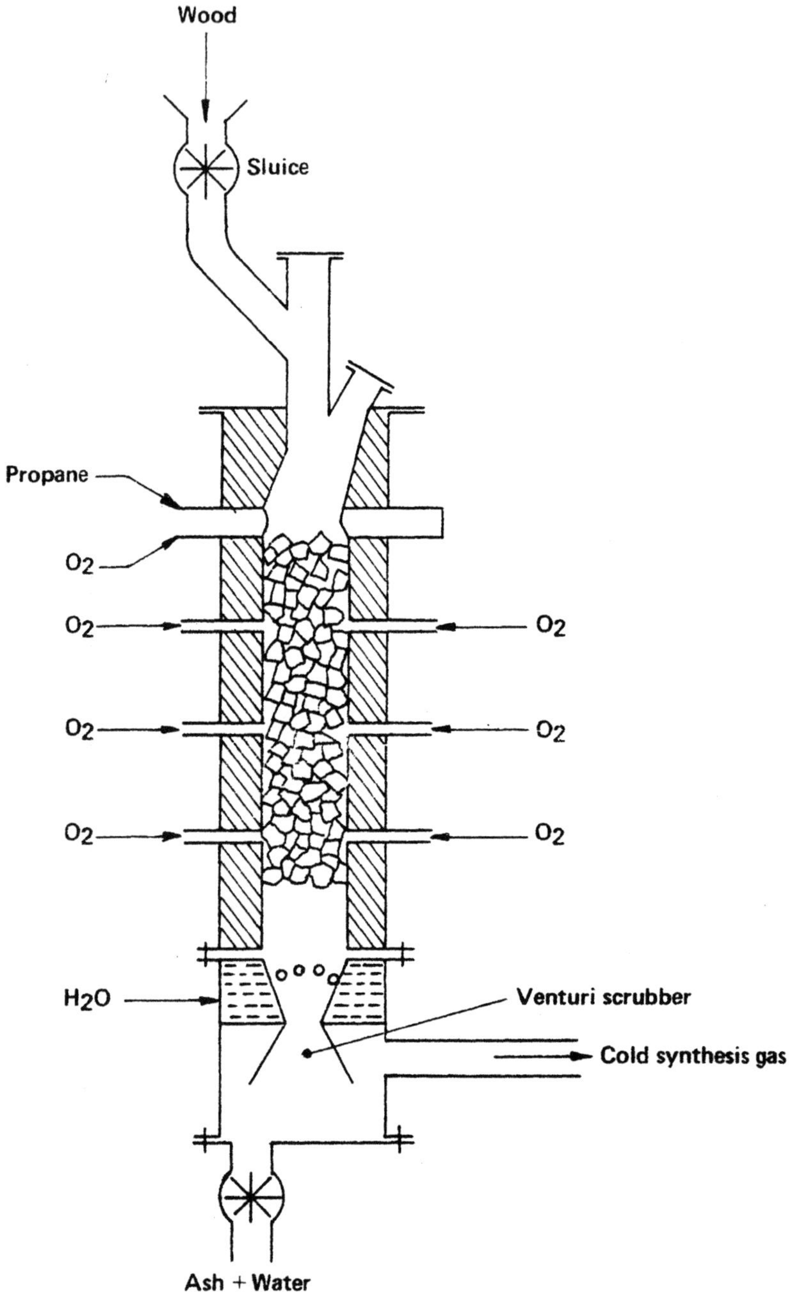

FIG. 11 – SERI'S DOWNDRAFT PRESSURIZED GASIFIER FOR OXYGEN-WOOD GASIFICATION (U.S.A.)

- NOUVELAIR (CANADA) - Fluidized bed oxygen process (little informations)
- BATTELLE (U.S.A.) - Multi solid double fluidized bed gasifier
- SERI (U.S.A.) - Downdraft pressurized fixed bed gasifier.

7. CONCLUSION

About ten medium joule gas demonstration plants at capacities of 10 to 60 tons/day of wood are operating or under construction in the E.C, the U.S, CANADA, SWEDEN ... Both oxygen, air and steam gasification are tested in various reactors and with several methods of heat supply. Nearly all processes are meant for methanol production. Preliminary relative cost estimates for the different processes agree within 20 % which is not significant in light of the still existing uncertainties in both capital and operating costs and in light of the present state of development for some of these processes.

List of references : The following list of references proposed by Dr VAN SWAAII and Dr BEENACKERS of the Twente University of Technology, the Netherlands, gives a good and complete overview on the thermochemical processes of biomass gasification.

REFERENCES

(1) Foley, G. and Barnard, G., "Biomass Gasification in developed countries, Earthscan report, London (1983).

(2) Bircher, K.G., Energy from Biomass and Wastes VI, IGT, (1982), 707.

(3) Moreno, F.E. and Goss, J.R., in : Energy from Biomass and Wastes VII, IGT, (1983) prepints paper 24.

(4) Meta Systems Inc. "An Examination of the Substitution of Woody Biomass Based Fuels for Oil in the Industrial Sectors of Costa Rica", Report prepared for the Instituto Technologica de Costa Rica (ITCR) and the Citizens Energy Corporation Program (CEC), Cambridge, Mass. USA (1982).

(5) Information from ADENAP, 1386 Holt Avenue, Los Altos, CB 94022.

(6) Kaupp, A., Energy from Biomass, 2nd E.C. conference (Strub, Chartier and Schleser eds) Appl. Sc. Pub., London (1983) 715.

(7) Stassen, H.E.M., "Gasification by Partial Combustion, Project in Tanzania", Twente University of Technology report (1980)

(8) Kaupp, A., The gasification of rice hulls for the generation of electricity and shaft power on a small (5-30 hp) scale, in : Proc. Producer Gas Conf., Sri Lanka, Nov. 1982, The Beyer Inst., Stockholm (1982).

(9) Van den Aarsen, F.J., Beenackers, A.A.C.M. and van Swaaij, W.P.M., Performance of a rice husk fueled fluidized bed pilot plant gasifier, in (9a) : Proc. Producer Gas conf., Sri Lanka, Nov. 1982, The Beyer Inst., Stockholm (1982).

(10) Groeneveld, M.J., Thesis, Twente University of Technology (1980).

(11) Groeneveld, M.J., Gellings, P.E. and Hos, J.J., Energy from Biomass and Wastes VII, IGT, (1983), preprints paper 15.

(12) "Moteur Duvant" BP 599 - 59308 Valenciennes, France.

(13) Susanto, H., Beenackers, A.A.C.M. and van Swaaij, W.P.M., in : Energy from Biomass, 2nd E.C. Conf. (Strub, Chartier and Schleser eds.) Appl. Sc. Pub., London (1983) 1112.

(14) Kaupp, A., in : Strub, A., Chartier, P. and Schleser, G. (Eds.), Energy from Biomass, 2nd EC Conf., Appl. Sci. Publ., London (1983) 715.

(15) Molle, J.F. in : Palz, W., Chartier, P. and Hall, D.O. (Eds.), Energy from Biomass, 1st EC Conf., Appl. Sci. Publ., London (1981) 575.

(16) Reed, T.B., Graboski, M. and Markson, M., paper presented at the 13th Thermochemical Conversion Contractors Meeting, Arlington, VA, October 27-29, 1981, issued as SERI/Pr-234-1455.

(17) Bliek, A. and van Swaaij W.P.M., to be published.

(18) Palz, W. and Grassi, G. (Eds.), Solar Energy R & D in the European Community, Series E, 2, Reidel (1982).

(19) Dubois, P., in (19), p. 76.

(20) Chartier, P. and Palz, W., Solar Energy R & D in the European Community Vol. 1, Reidel, Dordrecht (1982).

(21) Wilson, H.T. in (21), p. 203.

(22) Miller, H., Erdöl und Kohle, 28 (2), (1975).

(23) Hummelsiep H. in (27), p. 198.

(24) Fisher, T.F., Chem. Eng. Progr., Oct. 1976, p. 75.

(25) Arbo, J.C. and Dynecology, Inc., in : Energy from Biomass and Wastes IV, (1980).

(26) Gehrman, J. (Ed), Thermochemische Gaserzeugung aus Biomasse, KFA, Jülich, Conf. 46, November 1981.

(27) Chrysostome, G. and Lemasle, J.M. in (19), p. 28.

(28) Lindman, N., Energy from Biomass and Wastes V, IGT, 1981, 571.

(29) Nitschke, E., in (27), 231.

(30) Pers. Com. with Gravel, Byosyn, Montreal, Quebec.

(31) Schiefelbein, G.F., Sealock, L.J. and Ergun, S. in : ACS Symp. Ser. 130 (1980) paper 2.

(32) Class, D.L. in : Energt from Biomass and Wastes VII, IGT Conf. Orlando (1983), paper 1.

(33) Chrysostome, G. and Lemasle, J.M. in (19), p. 28.

(34) Chrysostome, G. in (21) p. 198.

(35) Strub, A., Chartier, P. and Schleser, G. (Eds.), Energy from Biomass, 2nd E.C. Conf., Appl. Sci. Publ. (1982).

(36) Chrysostome, G. and Lemasle, J.M. in (36), p. 402.

(37) Franken, G., Adlhock, W. and Koch, W., Chem. Ing. Tech. 52 (1980) 324.

(38) Mawer, D.J., Coal Technology, 3 (1980), 161.

(39) Brochures and papers from Rheinische Braunkohle, e.g. : H. Teggers, K.A. Theis and A. Tonnesmann, Latest status of the Rheinbraun High Temperature Winkler process.

(40) Rheinische Braunkohle A.G., German Patent DE 2949533A, Appl. 8-12-1979.

(41) Rheinische Braunkohle A.G., German Patent DE 3033115A1, Appl. 3-9-1980.

(42) Ekstrom, C., Lindman, N. and Pettersson, R., in : Abstracts, Fundamentals of Thermochemical Biomass Conversion, SERI/CP-2341719, Golden, Colorado, (1982), 73.

(43) Blomkvist, G., Liinaki, L., Lindman, N. and Sjöberg S.O. in (36), p. 896.

(44) Lindner, C. and Reimert, R. in : W. Palz and G. Grassi (Eds), Energy from Biomass, Vol 2, Reidel, Dordrecht, (1982), 115.

(45) Lindner, C., in J. Gehrman (Ed), Thermochemische Gaserzeugung aus Biomasse, KFA, Jülich, Conf 46, November 1981.

(46) Foster Wheeler Power Products Ltd. in (19), p. 89.

(47) Libbey, J.P., Evergreen Energy Corp., pers. corresp.

(48) Bailie, R.C., Energy from Biomass and Wastes V, IGT (1981) 549.

(49) Bowling, K. McG. and Waters, P.L., Mech. Chem. Engng. Trans., Nov. 1968 234.

(50) Feldmann, H.F., Energy from Biomass and Wastes V, IGT (1981), 529.

(51) Lancet, M.S. and Curran, G.P., US Patent 139915, Cocono, June 15, 1982.

(52) Rensfeldt, E., et al., Energy from Biomass and Wastes, IGT (1978), 465.

(53) Janesch, R. Werner, F. and De Santi, D.M., in (19), p. 101.

(54) Rowell, R.M. and Hokanson, A.E., Progress in Biomass Conversion, Academic Press, 1 (1979) 117.

(55) US Patent 7900761, 9th January 1979 ; Eur patent 13590.

(56) Bickle, R.S., Edwards, A.J. and Moss, G., in : W. Palz and G. Grassi, Energy from Biomass, Vol 2, Reidel, Dordrecht (1982), 43.

(57) Brecheret, V., Proc. Fifth Int. Alc. Fuel Tech. Symp. Vol I, J. McIdoe, Auckland, (1982), 1-107.

(58) Brecheret, V., Zagato, A.J.A., in : ACS Symp. Ser. 159, Monohydridic Alcohols, (1981), 32.

(59) Oliveira, E.S., 1980 Proc. IV Int. Symp. Alc Fuels, Vol. I, Inst. Pesquisas Technological, Sao Paulo, 1980 A 38, p. 199.

(60) Antal, M.J. Jr., Royere C. and Vialason, A., in : J.L. Jones & S.B. Radding, Thermal Conversion of Solid Wastes and Biomass, ACS Symp. Ser. 130, (1980) 237.

(61) Marshall, J.E., UK, ISES Conf. (C20) at the Royal Soc. : Biomass for Energy, July 1979, p. 52.

(62) Rock, K.L., Energy from Biomass and Wastes VI, IGT (1982), paper 31.

(63) Brandon, C.S., Duhl, R.W., Miller, D.R. and Thakker, B.R. in : Proc. Fifth Int. Alc. Fuel Tech. Symp 1, John McIndoe, Dunedin (NZ), (1982) 3-107.

(64) Cahn, R.P. et al, US patent 3993467, 23 nov. 1976.

(65) Gluckman, M.J. and Louks, B.M., 8th Energy Tech. Conf. March 1981, Washington DC, Government Inst. Inc., Aug. 1981, p. 920.

(66) Ader, G., Bridgwater, A.V. and Hatt, B.W., in : W. Palz, P. Chartier and D.O. Hall (Eds), Energy from Biomass, Appl. Sci. Publ., London, (1981), 598.

(67) Beenackers, A.A.C.M. and van Swaaij, W.P.M., in : W. Palz and G. Grassi Energy from Biomass, Vol 2, Reidel, Dordrecht, (1982), 1.

(68) Wan, E.I., Simmons, J.A. and Price, J.D., in : Energy from Biomass and Wastes VI, IGT, 1982, paper 37.

(69) Wan, E.I. and Price, J.D., in : Proc. 14th. Biomass Thermochem. Conv. Contractor's Meeting, Arlington, Virginia, June 23-24, 1982, Pacific Northwest Laboratory, Richland, Washington, October 1982, p. 230.

(70) Nitschke, E. and Keller, J., Chemie Technik, 9 (1980) No. 3, 121.

(71) SERI/TR-33-239. A Survey of biomass gasification, Vol III, Current Technology and Research, SERI, Colorado (1980).

2) Van Swaaij, W.P.M., in : Palz, W., Chartier, P. and Hall, D.O., ds), Energy from Biomass, 1st E.C. Conf., Appl. Sci. Pub., London 981) 485

3) Reed, T.B., A Survey of Biomass Gasification, Vol I : Synopsis and :ecutive summary, SERI/TR-33-239-VI, Colorado (1979).

4) Hokanson, A.E. et al., Chemicals from Wood, NTIS, PB-262489 (1975).

5) Bradford Holoway et al., in : Energy from Biomass and Wastes VII, :T (1983) preprints paper 40.

'6) Beenackers, A.A.C.M. and Van Swaaij, W.P.M., in : Grassi, G. and ılz, W. (Eds), Energy from Biomass, Vol 3, Reidel, Dordrecht (1982) 201.

'7) Wigmans, T., Ph.D. Thesis, Amsterdam (1982).

'8) Proc. 7th Int. Conf. on Catalysis, Tokyo (1980).

'9) Mudge, L.K., Baker, E.G., Mitchell, D.H., Brown, M.D. and Robertus, .J., in : Preprints Int. Conf. Energy from biomass and Wastes VII, ;T (1983), paper 13.

30) Mitchell, L.K., Mudge, R.J., Robertus, R.J., Weber, S.L. and ealock, L.J., Chem. Eng. Prog., Sept. (1980), 53.

31) Graham, R. and Fung, D.P.C., A review of Biomass Gasification, astern Forest Products Laboratory, Report OPX221E, Ottawa (1979).

BIOLOGICAL PROCESSES FOR ALCOHOL PRODUCTION FROM BIOMASS

Francesco ALFANI and Maria CANTARELLA
Istituto di Principi di Ingegneria Chimica,
Università di Napoli,
P.le Tecchio, 80125 Napoli
ITALIA

ABSTRACT. The motivations for an extended use of biotechnology in the area of energy from biomass are presented. An analysis of the state of art and works in progress concerning the enzymatic hydrolysis of lignocellulosic residues is specially made.

The physical and chemical properties of biomass and the mechanisms of biodegradation of biomass components are discussed. The mode of action of cellulase preparations and the kinetic and structural limitations which prevent enzymatic attacks are reported.

Finally attention is focused on the reactor configurations which are suitable for cellulose hydrolysis and on the economical bottlenecks which still limit the diffusion on the market of plants for enzymatic saccharification of cellulose.

1. INTRODUCTION

A great interest has been recently devoted to replace chemical with biological reaction steps. This tendency is already asserted in pharmaceutical and food industries but several new biological processes are becoming closed to the market.

Also "bioenergetic", an area which covers different aspects of biological conversion of lignocellulosic materials into fuels, is rapidly in progress. Several laboratory and pilot plants are under construction and in operation, the largest one in Canada for the enzymatic hydrolysis of 2 ton/day of dry biomass, and it is general opinion that in the next five years the technology will be ready for the design of full scale plants, even though not competitive on economic basis.

At first, the advantages offered by Biotechnology are briefly considered. In comparison with chemical catalysts, biocatalysts greatly enhance reaction rate under mild operational conditions, and assure high specificity, specially regiospecificity and stereospecificity. However several limitations, such as costs, instability, phase incompatibility, product and substrate inhibition, often still limit the use of

G. E. Beghi (ed.), Synthetic Fuels, 259–267.

biocatalysts on large industrial scale. Nowadays roughly 2000 enzymes are known, 200 enzymes are commercially available, 16 enzymes are industrially employed (1).

With regard to Bioenergetic, microorganisms with specific activity in the conversion of lignocellulosic compounds have been isolated. However, since these materials are heterogeneous in chemical composition and physical structure, the conversion of the different components is carried out with different effectiveness.

2. BIOMASS

Different meanings have been attributed to this word. In this context "biomass" represents the whole of agricultural wastes, woody materials and cellulosic residues.

Biomass production on the earth is abundant, on the average, the net production of the photosynthetic plants is 150.000 million tons of dry biomass per year. One hundred billion tons of this biomass occurs on the land area, 65% of which is represented by forest, 15% is generated in grasslands and cultivated lands, the rest is from different sources. In the oceans biomass production is 50 billion tons. It was estimated that roughly 360 million tons, equivalent to 1.25% of the total land biomass productivity is eventually used for human food and about 13.5 billion tons is lost as waste from harvesting or processing operations. These wastes, which can be considered as sources of cellulose, can be classified into three categories : i) primary cellulosic, harvested for their cellulose content, ii) agricultural wastes (including animal manure) generated as by-products harvesting or processing of agricultural plants, iii) municipal cellullose wastes. All these materials appear suitable for consideration in hydrolysis processes (2).

A crucial problem, associated with the availability of the wastes, is the cost of collection and transportation which is prohibitive for the establishment of large plants, thus low or medium size and simple technology processes are most important.

Another limitation to an extended use of raw material is the necessity of chemical or physical pretreatments to make the material accessible to biological conversion. Mechanical pretreatment is generally very energy demanding since its efficiency is low, on the other hand chemical procedures are accompanied with environmental and corrosion problems.

Other processes have been also suggested. At present time the major limitation is the cost of the suggested pretreatment (3). Nowadays the enzymatic hydrolysis of cellulose is still limited by expensive cost of enzymes but considerable progress has been reported in enzyme production which should lower drastically the price.

In order to fully understand the mechanism of degradation the chemical composition of biomass must be known. In a broad sense biomass is a miscellaneous material of three major components, cellulose, hemicellulose and lignin, the relative amount of which depends on the origine of the biomass, and pectic compounds and extractives in minor percentages. The composition of various type of biomass is reported in (4).

Cellulose is a homogeneous polymer of anhydro-β-glucopyranose with β-glycoside linkage and a tendency to have a well organized structure. This is characterized by sequences of high crystalline regions interrrupted by amorphous phases. The degree of polymerization (DP) ranges between 7.000 and 15.000 monomer units, depending on the source. In the agricultural wastes the DP is much lower, less than 1.000. Further indications and representations of the cellulose fiber showing transverse and longitudinal structure components and microfibrils can be found in (2).

Hemicellulose is a heterogeneous polymer of D-xylose, D-glucose, some D-mannose, L-arabinose, acetic acid and small amounts of furfural and is characterized by an amorphous structure and low degree of polymerization, 100-200 monomer units. Therefore, hemicellulose is quite easily hydrolyzed.

Finally lignin, which is the main component in woody biomass, has a chemical structure not yet totally known but it can be classified as a polyphenolic molecule which is very difficult to breakdown. Lignin is found to be interpenetrated by hemicellulose and also in close association with cellulose since it surrounds cellulose microfibrils as a "lignin seal" which increases the integrity of the cellulose molecules and the resistance to chemical or biological attack.

In addition to the cellulose, emicellulose and lignin; the three major components of the cell wall, the biomass contains extraneous materials including extractives and non extractives. The first ones are waxes, fats, essential oils, tannins, resins, fatty acids, terpenes, alkaloids, starches, gums and various cytoplasmatic constituents. They represent up to 0.4-8.3% of dry weight in wood and more in agricultural wastes. The second ones, the non extractives consist in the non organic components such as silica, alkali, carbonates, oxalates. In straw, they may be up to 10% of the dry weight.

3. DEGRADATION OF BIOMASS COMPONENTS

The degradation of cellulose is the process of breaking down the anhydroglucose polymer into smaller molecules. This process can yield oligosaccharides, di- or tri- saccharides such as cellobiose and cellotriose, the monomer glucose, or its degradation products (small alcohol, aldehydes, acids and ketones) or finally carbon dioxide and water. The distribution of the final product depends on the operational conditions of the degradation process. Processes which involve biological routes of degradation, employ different sources of cellulases as enzymatic catalysts while the non biological processes follow physical or chemical routes to hydrolyse cellulose but in this latter case the formation of degradation products occurs in larger extent.

Several difficulties arise during the bioconversion of biomass into fuels. The hydrolysis of hemicellulose is accomplished quite easily but the subsequent fermentation of the pentose stream is still under investigation and only in the recent years good yield of conversion in ethanol have been reached in pilot plants. Nevertheless, in order to develop an economic process for biological production it is crucial to reach high conversion of cellulose to glucose together with the fermentation of xylose to ethanol. Recently it has been reported in the literature that species of Clostridium thermohydrosulfuricum and some yeasts, such as Candida sp and Pachysolen tannophilis, are able to efficiently ferment xylose to ethanol (5).

On the other hand the fermentation of glucose, the main product of cellulose hydrolysis, is a well established process and is performed with very good yield, but since the previous breakdown of cellulose is hard to carry out the effectiveness of the whole process is low. The major problems are the crystalline structure of the substrate and its close association with lignin, which prevents the attack of hydrolytic enzymes. Thereby physical or chemical pretreatments or a combination of both are necessary to break the lignin seal and to expose cellulose to enzymatic action.

In conclusions, since the hydrolysis of hemicellulose can be performed under mild conditions the main problem for biomass bioconversion is the low efficiency of the enzymatic attack of cellulose.

This situation has stimulated the search and the study of numerous fungal and bacterial cellulases which have been isolated from different microorganisms and used to hydrolyze cellulose; the most known are Trichoderma spp, in particular T.reesei, and Aspergillus niger.

Taking into account the technical and economical limitations, several process schemes have been proposed in the literature which represent possible solution for overcoming the previous discussed problems. These are summarized in Fig.1. According to the first process scheme during the initial pretreatment hemicellulose is removed and cellulose is partially degradated. Two separate hydrolysis steps are successively performed and the products can be fermented either independently or together if a yeast is available. This process layout offers the possibility to optimize the different stages but it is expensive since several auxiliary equipments and a larger number of reactors are required.

The second flow sheet is based on a single reaction step, which is presented as alternative to acid hydrolysis of biomass; therefore the costs of plant are limited but it is difficult at present time to foresee the possibility that high biomass conversion can be reached. This design, such as the third one, would be promising if biocatalysts with high activity towards lignin biodegradation were available, but unfortunately it is not nowadays the case.

Finally, in the third scheme, hydrolysis and fermentation are performed in the same reactor in order to reduce product inhibition which mainly acts on the rate of cellulose degradation.

1)- Two - step Hydrolysis followed by fermentation

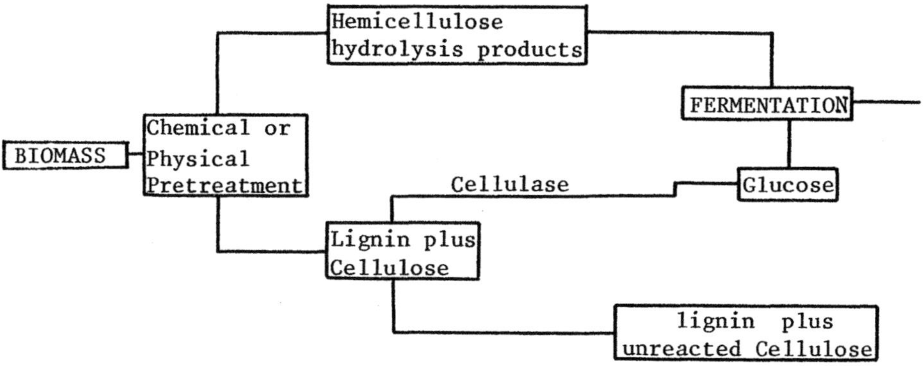

2)- One - step Hydrolysis plus Fermentation

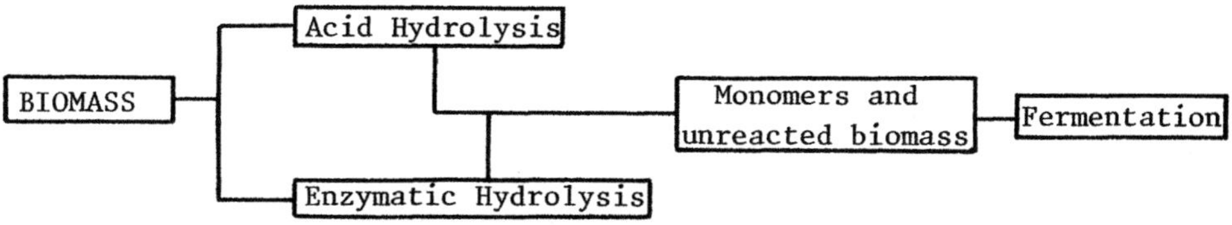

3)- Simultaneous process

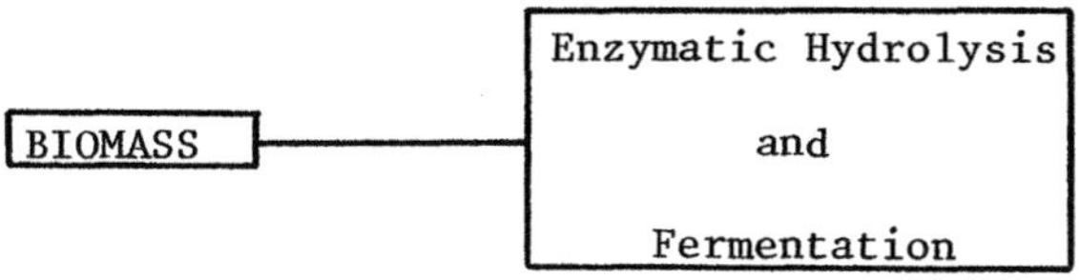

Fig. 1 - Suggested process schemes for Cellulose hydrolysis

4. MODE OF ACTION OF CELLULASE COMPLEXES

Recent researches and studies have shown the role of initial and intermediate substrates and the mode of action of single cellulase components.

The cellulase system of any organism can be classified as a mixture of three major components:
Endoglucanase or C_x 1,4-β-D-glucan glucanohydrolase (EC 3.2.1.3).
Exoglucanase or C_1 1,4-β-D-glucan cellobiohydrolase (EC 3.2.1.91) and cellobiase or β-glucosidase (EC 3.2.1.21).

Cellulose conversion can be schematically represented as in Fig.2, where the catalytic behaviour of each enzymic component is indicated by the arrows.

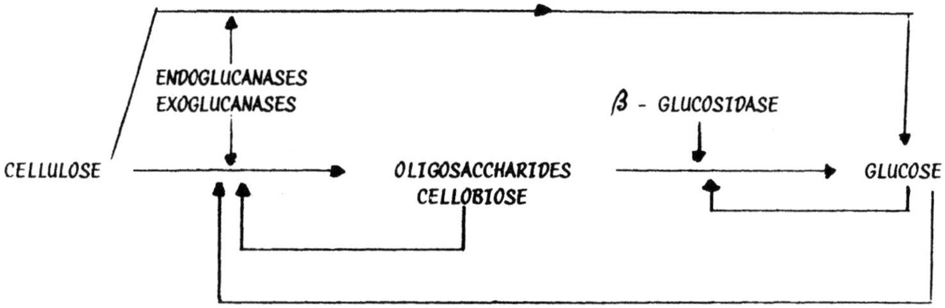

Fig. 2 - Overall reaction scheme of cellulose hydrolysis

The mechanism of molecular degradation is a much debated question in the literature, however it is recently prevalent the opinion that endoglucanase (C_x) (MW : 18.000 - 60.000) randomly attacks the amorphous regions of cellulose and removes glucose, cellobiose and cellotriose from the polymer chain liberating new end chains. The exoglucanase (C_1) (MW : 50.000 - 70.000) removes cellobiose units from the non reducing end chains and finally β-glucosidase (MW : 76.000) catalyses the conversion of cellobiose and cellodextrins to glucose.

The optimum temperature for these enzymes is 45°C; at higher temperature reaction rate increases but the biocatalyst is rapidly deactivated. It is convenient to work with buffer solution the pH of which varies depending on the cellulase sources. For many commercial preparations the range is between 4.8 and 5.0.

It has been proved that cellobiose and glucose are inhibitors of the cellulase complex. Cellobiose inhibits the exoglucanases and cellobiase while glucose mainly inhibits cellobiase. The mechanism of inhibition is an other debated question, however it seems possible to believe that the exo and the endo enzyme components are competitively inhibited by their products while β-glucosidase is non competitively inhibited by glucose.

The mode of action of cellulases on pure cellulose (free of hemicellulose and lignin) was detailed in the literature and indicates the synergetic attack of endo- and exo-glucanases on the polyglucose chain.

On the other hand the kinetics for the native cellulose is still matter of investigation since many factors mask the reaction pathway. The hydrolysis of cellulose is an heterogeneous reaction, which consists of several series and parallel steps, involving a soluble phase - enzyme and final products - and a water insoluble phase - substrates and many intermediate products. Moreover the presence of segragated zones and diffusional limitations increases the difficulties to discriminate between kinetics and transport resistances and to derive the relative rate equations. In fact the analysis of all reaction products is difficult to perform, since the intermediate products are not commercially available and therefore it is quite difficult to verify mass balance of each step. However, the production of glucose even in the presence of high concentration of δ-gluconolactone, specific inhibitor of the β-glucosidase component, and the possibility to correlate with an unique rate equation cellulase degradation appears at least indicative of the glucose production catalyzed by C_1 attack and the synergistic action of C_1 and C_x. Several models have been proposed to explain the possible mechanism of adsorption and desorption of C_1 and C_x components on the substrate, (β-glucosidase is not adsorbed) and their possible motions along the chains. Both the extended and folding chain models of the cellulose have been considered. For further details it is suggested to refer to (5)

5. KINETIC AND STRUCTURAL LIMITATIONS OF ENZYMATIC ATTACK

Many fungi and bacteria have been reported to be able to degradate cellulose but the products of growth are microbial cells and methabolic products such as carbon dioxide and methane. Only a few fungi including T.reesei and a few bacteria such as Clostridiumthermocellum thermonospora are able to degradate extensively insoluble cellulose to soluble sugars in vitro.

In enzymatic hydrolysis of cellulosic residues, stable cell-free enzyme preparations are required, which possess well balanced activity and good operational thermal stability for the C_1, C_x and β-glucosidase components.

However these different microorganisms generally have an enzymatic complex which is not satisfactory balanced in the three components. Cellulases from T.viride, which are able to attack crystalline cellulose with high efficiency are usually rich in the C_1 and C_x components but are deficient in β-glucosidase which is on the other hand the predominant component in cellulases from A.niger.

Further on as far as the operational stability is concerned, the β-glucosidase component generally exhibits a very poor thermal stability while C_1 and C_x are more stable.

These limitations, namely unsatisfactory balanced complexes and poor thermal stability can be overcome either supplementing cellulase from a microorganism such as T.viride with a more stable soluble β-glucosidase from an other source or immobilizing the β-glucosidase component. During the immobilization procedure, part of the initial activity is curtailed but half-life increase is significant and this together with the possibility to reuse the catalyst lead to an economic improvement. Several methods of immobilization have been tested in cellulose hydrolysis reactor (5).

6. REACTOR CONFIGURATIONS FOR CELLULOSE HYDROLYSIS

Nowadays several types of reactor configurations, on laboratory scale, have been studied for the enzymatic hydrolysis of cellulose: fluidized bed reactor, batch reactor, stirred tank reactor , plug-flow reactor.

Some of them imply the utilization of immobilized cellulases on different carriers such as collagen and solid supports. In order to improve the thermal stability and the productivity of β-glucosidase different immobilization techniques on cellulose activated with cyanogen Bromide, with chitosan through glutaraldehyde reaction and with Amberlite DP-1 have been suggested.

However, while the use of immobilized β-glucosidase is worthwhile, the use of immobilized enzymes, cellulase, on high molecular weight substrates, cellulose, still present difficulties.

At present, two reactor configurations are considered the most attractive, at least for fundamental studies of enzymatic saccharification, and are the bacth and membrane ones.

UF-membrane reactor is helpful since its configuration allows to achieve.

- separation of high molecular weight substrates and enzymes from low molecular weight products,
- possibility to control and reduce product inhibition,
- possibility to concentrate dilute sugar solutions prior the fermentation step,
- possibility to reuse the catalyst.

In addition to those possibilities, which can be considered technical solutions to adopt for full-scale plants, the UF-membrane reactor is also helpful tool for studying complex enzymatic reactions like those involved in cellulose saccharification: namely it offers the possibility to determine separately the behaviour of the enzymatic components, their thermal stability and their resistance to product and reactant inhibition. An analysis of the advantages to use UF-membrane reactors in connection with cellulose hydrolysis is reported in (6).

On the other hand batch reactor is suggested for being the simplest converter in homogeneous reaction phase with simple stirring device.

. ECONOMICAL BOTTLENECKS IN ENZYMATIC CELLULOSE SACCHARIFICATION

In conclusion, the economy of enzymatic hydrolysis of biomass is highly dependent on the overcoming of the following bottlenecks:

I) to locate cheap, readily available, large biomass sources in order to minimize collection and transport costs,

II) to set up cheap and efficient pretreatment methods which enhancing substrate reactivity will permit to attain 10-12% glucose concentration even with current enzyme preparations. At present state of art steam explosion appears to be the most promising pretreatment.

III) to select cheap and potent sources of cellulolytic enzymes for more efficient conversion of cellulose and hemicellulose content of biomass so that limited pretreatments will be necessary. Strains with high C_1 activity and β-glucosidase resistant to glucose inhibition would be welcome.

IV) to totally utilize the biomass in order to give cellulose hydrolysis processes an economical feasibility. The basic concept is to delignify the biomass in order to recover a valuable lignin product either for polymer or power use and to utilize the hemicellulose for alcohol fermentation.

V) to realize a combined saccharification and fermentation technique with saccharifying and alcohol producing organisms (ex: Cl.termocellum) in order to limit product inhibition. A biological system that would hydrolyze both cellulose and hemicellulose to monomeric compounds and then directly convert these sugars to ethanol would be ideal.

VI) to establish low-energy demand processes for the concentration of dilute products, glucose prior the fermentation step up to 10% and for removing the water from the fermented alcohol stream in order to produce high purity alcohol.

These conclusions are in accord with those reached by (7,8).

8. REFERENCES

1. A. Klibanov Proceedings of Enzyme Engineering VII, N.Y. Academy of Science, in press.
2. C.E. Dunlap and L.C. Chiang, in Utilization and recycle of agricultural wastes and residues, pag. 19-60, M.L. Shuler ed., CRC Press Inc., Boca Raton, Florida, (1980).
3. M.R. Ladisch, Process Biochemistry, pag. 21-25, (Jan 1979).
4. I.S. Goldstein, in Organic Chemicals from Biomass, pag 10-18, I.S. Goldstein ed., CRC Press Inc., Boca Raton, Florida, (1981).
5. M.R. Ladisch, K.W. Lin, M. Voloch and G.T. Tsao, Enzyme Microb. Technol., 5, 81, (1983)
6. F. Alfani, M. Cantarella, V. Scardi, Journal of Membrane Science, 16, 407, (1983)
7. A.E. Humphrey in Hydrolysis of Cellulose: 'Mechanism of enzymatic and acid catalysis', R.D. Brown jr. and L. Jurasek Eds., Advances in Chemistry Series 181, (1979).
8. H.E. Gretlein Biotech. Bioeng. 20, 503, (1978).

THERMAL AND CATALYTIC PROCESSES FOR DIRECT CONVERSION OF BIOMASS TO LIQUID HYDROCARBONS

Francesco ALFANI and Maria CANTARELLA
Istituto di Principi di Ingegneria Chimica,
Università di Napoli,
P.le Tecchio, 80125 Napoli,
ITALIA

ABSTRACT. The factors which have determined in the recent years a renovated interest towards thermocatalytic conversion of biomass to fuels and chemicals are presented. The economical and technical bottlenecks, which still limit a rapid diffusion in the market and have compelled to further process developments, are indicated.

The mechanisms of biomass conversion in pyrolytic reactions are discussed and references are given of the most important reactor designs which have been suggested by the constructors.

Moreover, the results of new studies and advanced researches concerning the catalytic pyrolysis and biomass liquefaction are discussed.

1. INTRODUCTION

Thermal and catalytic conversion of biomass are considerably attractive whenever dry raw materials with an humidity less than 50% and a carbon to nitrogen (C/N) ratio greater than 30 are available. Several agricultural and forestry wastes present these peculiarities and in the last years, during the recent energy crises, their conversion to liquid hydrocarbons was once again investigated.

The possibility to obtain chemicals and fuels from the biomass, mainly from wood, was known also to the ancient populations. Destructive distillation was practiced by Chinese while pyrolysis of wood was introduced by the Egiptians, Greek and Romans. In the 1800s and during this century the share of Chemicals and Fuels produced with biomass was increasingly surpassed by that of coal and crude oil.

Different factors have determined this lost of interest. Using surplus of other raw materials, such as naphta and the liquid content of natural gas, chemical intermediates could be produced cheaply. The use of crude oil and natural gas derivatives is not limited by local environmental conditions and further on petroleum price was for several years more stable in comparison with that of products from agricultural crops and forestry.

Moreover, since the biomass is spread over wide areas, the

G. E. Beghi (ed.), Synthetic Fuels, 269–281.

expenses of collecting and transport weigh heavily on the cost of supply and greatly increase with the distance from the plant. Therefore it is difficult to assemble and store the biomass that might be needed for a large plant in order to reduce capital investment which in the chemical industry depends on the reactor volume raised to 0.7 power.

Many of these factors are changing today and several biomass products have droppen in price relative to crude oil, even though they do not compete on an energy content basis. Taking into account only the conversion technologies of the biomass which are sufficiently developed, full scale plant have been constructed, but their capacity to compete with fossil fuels from the cost point of view is generally unsatisfactory.

With regard to different uses of biomass as energy source, the oldest and simplest one, direct combustion, is still the most convenient. On the other hand heating and electricity demand cannot be economically satisfied utilizing more sofisticated processes for conversion of biomass into energetic products (1).These considerations have compelled to : i) further on improve thermal and catalytic processes already known, ii) develop new ones, mainly biological conversion, iii) assure an integrated utilization of the different components of the biomass, iv) to restrict the interest not only to fuel product but to consider also the upgrading of all byproducts, as chemical sources.

In this paper the discussion is mainly devoted to present the developments of thermal and catalytic processes for direct conversion of biomass to liquid hydrocarbons. There are several advantages for using liquid fuels. These in fact can be more easily stored than gases, their mechanical transport in the plant is facilitated and their combustion is simplified with respect to solids. Therefore several attempts have been made to increase the percentage of the liquid fraction in miscellaneous products and develop new routes for conversion of biomass into liquid fuels.

Liquid mixture of hydrocarbons can be obtained following different thermal and catalytic processes, pyrolysis, gasification and subsequent synthesis of the products, catalytic liquefaction and acid hydrolysis. The gasification of biomass was exhaustively reported in other chapters whereas the conversion of gaseous mixtures to a carbon based fuel will not be commented here since this is a well establised chemical process. On the other hand acid hydrolysis cannot be considered a direct process for biomass conversion into liquids, since the hemicellulose and cellulose components of the biomass are initially degraded to sugars, which are subsequently fermented to alcohol. Finally catalytic liquefaction of biomass is some time away from full scale plant, and therefore attention will be mainly paid to discuss the pyrolysis process.

2. PYROLYSIS

Pyrolysis is the incomplete thermal degradation of a carbon based material carried out in the absence of air. However recently industrial processes have been developed in which traces of additives and air are present. In a broad sense pyrolysis describes any thermal degradation process in which the products of interest are char and liquids. Apart from these latters in pyrolysis processes a gaseous mixture is also produced. The relative percentage of the three fractions strongly depends on reaction temperature. In TABLE 1 average composition are reported.

TABLE 1

Phase percentage as function of reaction temperature

Temperature, °C	Solid	Liquid	Gas
400	40	30	30
500	26	4	70
900	6	--	94

At first, the convenience of biomass and wood pyrolysis are briefly discussed before analysing the process kinetic and the factors which effect the yield of bio-oil. Pyrolysis is generally convenient when the biomass is available at some distance from the consumer as compared to other conversion techniques. In the presence of either concentrated raw materials or residues spread over a wide area, the cost of transport can be reduced producing charcoal and oil in a large or medium scale plant. Then in a second stage bio-oil can be conveniently used to produce thermal energy or fractionated to obtain chemicals (2).

The pyrolytic conversion of biomass can be understood firstly analyzing the mechanisms by which the three major components, cellulose, hemicellulose and lignin, are converted. Cellulose is a natural polymer of glucose characterized by a sequence of amorphous and crystalline regions. The degradation of these two different structures follows different routes. At temperature below 300 °C the crystalline regions remain almost intact (3); and acids, aldehyde, and glucose are generated by the breakdown of the amorphous regions. The rate of degradation markedly increases in the presence of oxygen, water, acid or alkali. At higher temperature, above 300°C, the main reaction product is 1,6-Anhidroglucose (levoglucosan) and its yield depends on the operational conditions. In fact this chemical compound tends to undergo secondary reactions and therefore its fast recovery from reaction medium is important.

In comparison with cellulose, hemicellulose is more sensitive to the temperature. Its pyrolytic degradation occurs between 200 and 260 °C and gives rise to the formation of acetic acid, formaldehyde and furfural. The product mixture is rich in gas, poor in tar and contains almost the same amount of aqueous distillate formed during cellulose degradation. The relative large concentration of volatile compounds compared to that of furan derivatives, which could be expected from the breakdown of five carbon sugars associated with the hemicellulose, is very probably a consequence of the high reactivity of furfural under pyrolytic conditions.

The mechanism according to which the third component of the biomass, lignin , is degradated, is more sophisticated. This material has a complex chemical nature and a variety of products is generated during its thermal disruption. The main product, 55%, is a solid residue similar to char, while in less extent mixtures of phenolic compounds and gaseous molecules, such as CO, CH_4 and C_2H_4 are also produced.

In the biomass, cellulose, hemicellulose and lignin are simultaneously present and the product distribution reflects the composition of the native material. Recently, the results of a thermo-gravimetric and differential thermal analysis showed that the pyrolysis of biomass depends on the reaction atmosphere and that decomposition of the native material occurs in a temperature interval. The phenomenon starts in the case of wood at 225 °C, becomes maximum at 325 °C and ends at 345 °C. Small amount of oxygen or steam are important in the environment to combust a limited amount of biomass and to supply the energy requirement needed for the driving endothermic reactions (4).

In the recent years great effort was paid for the development of pyrolysis processes which can be mainly distinguished by reactor type. Its choice, in fact determines the conditions of pyrolysis, such as residence time, heating rate, temperature profile, gas atmosphere, and therefore the composition of pyrolytic products.

Gravitating bed reactors have been suggested by ITALENERGIE, TECH-AIR and BATELLE, rotating converters by KIENER,cross-flow reactors by FOSTER-WHEELER, fluidized bed reactors by TAG and GARRET, rack-furnaces by HERRESHOFOFEN and zig-zag converters by BALSTER (5).

TABLE 2

Phase percentage for different biomass vs. pyrolysis temperature,550 °C

Biomass	Water loss	Solid	Liquid	Gas
Oak chips	12.0	25.4	48.5	26.1
Pine	10.3	23.7	50.5	25.8
Tobacco	11.7	39.3	27.5	33.2
Almond shell	10.8	26.6	61.0	12.4
Tomato residue	11.6	21.7	47.7	30.6
Broom coppice	11.6	24.0	47.6	28.4

The operational temperature ranges between 400 °C and 600 °C and the yield of biomass converted in the final products decreases increasing the temperature. The gas fraction is generally low and the product distribution is function of the type of biomass as shown by the data reported in TABLE 2 (6). Water loss was measured after 3 hours drying at 125 °C. The relative amounts of the components largely vary in the biomass and since their thermal degradation follows different pathways, an unregular distribution of chemical products is expected.

In Fig. 1 and 2 a schematic flow sheet and a general view of the pyrolysis process designed by ITALENERGIE for the conversion of wood and agricultural wastes are drawn (6).

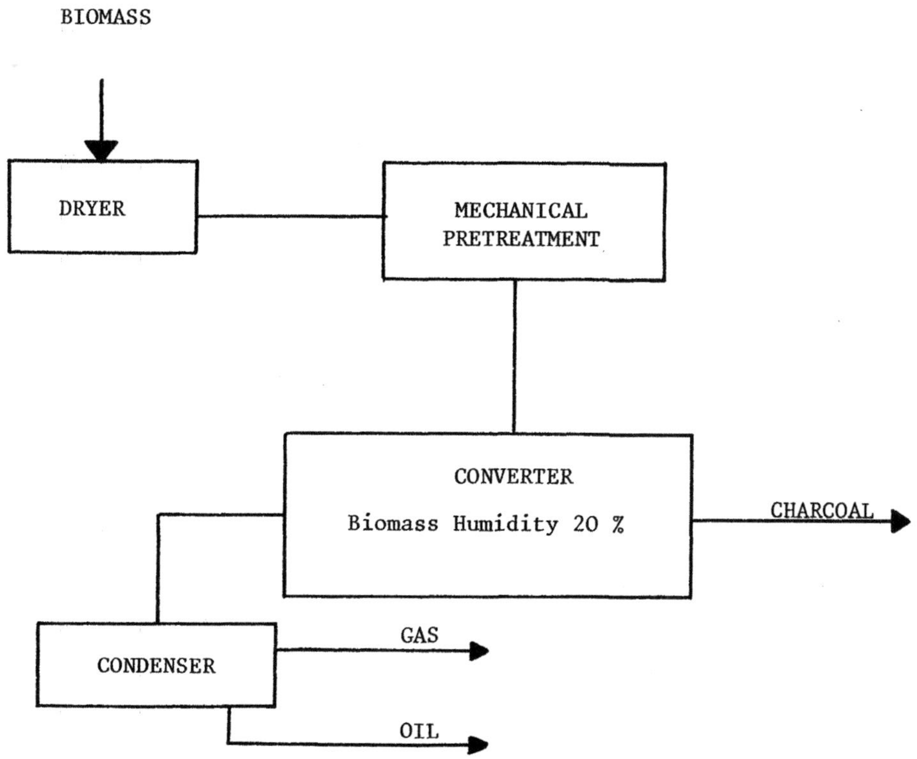

Fig. 1 - Schematic flow sheet of a pyrolysis plant

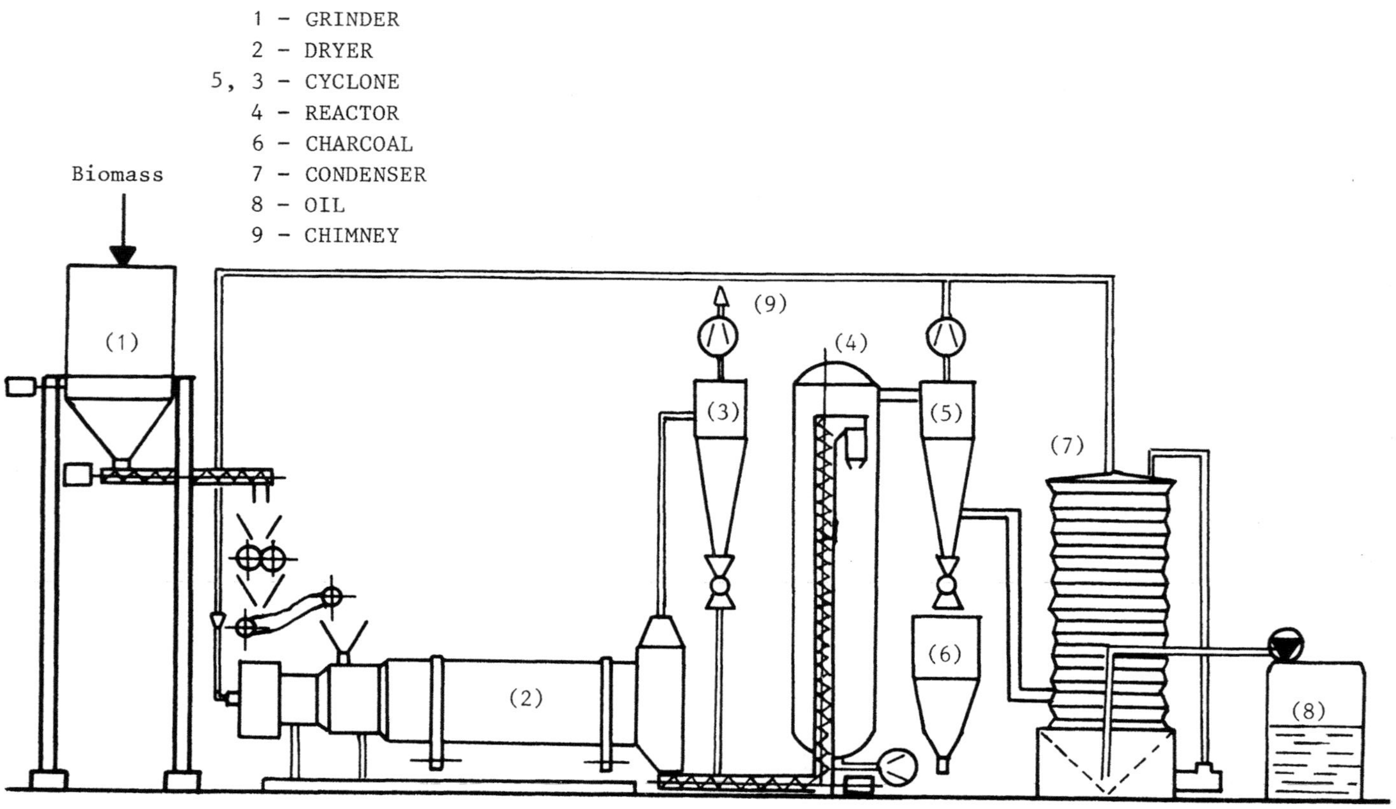

Fig. 2 - Pyrolysis plant designed by ITALENERGIE

Converterstation, System TECH AIR

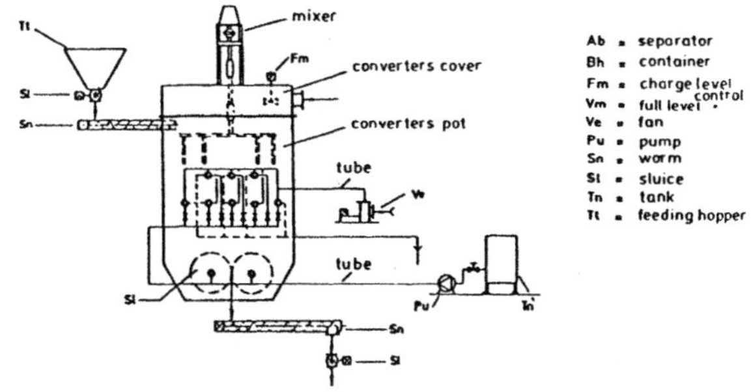

Pyrolysis in Pressure-Autoclaves (PPT-Process-Pyrolysis)
(Heyner)

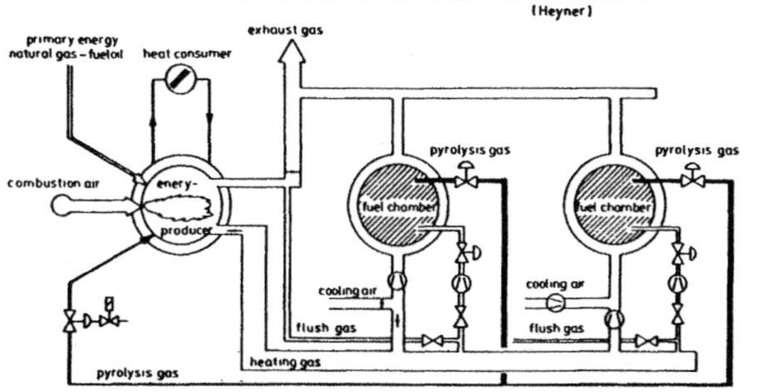

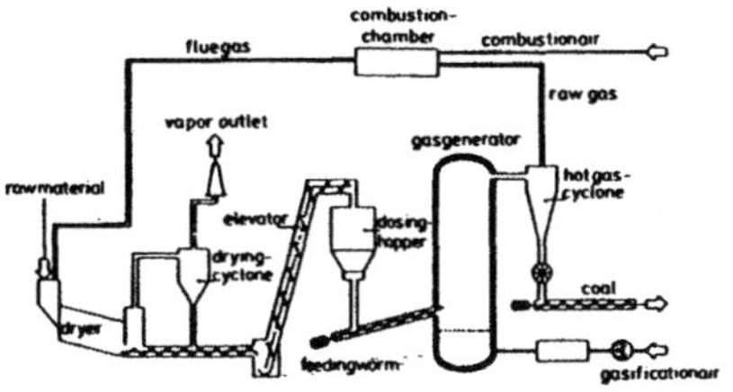

Fluidized bed plant for carbonisation of biomass
(TAG, ERCO)

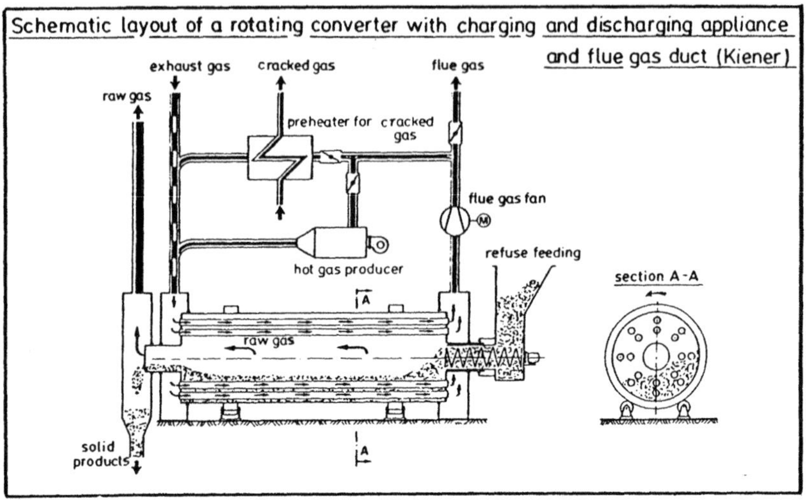

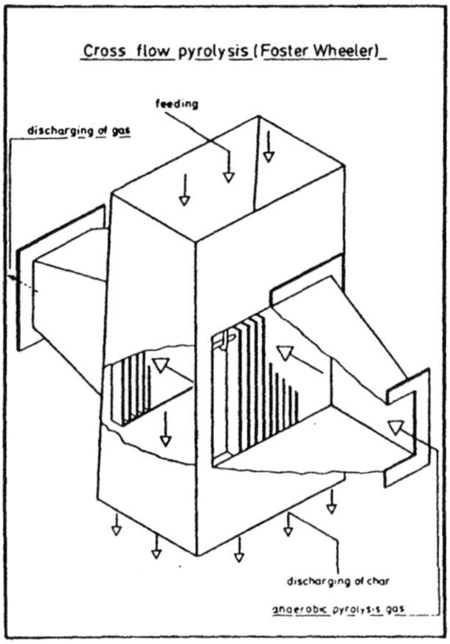

Schemes of different systems of pypolysis

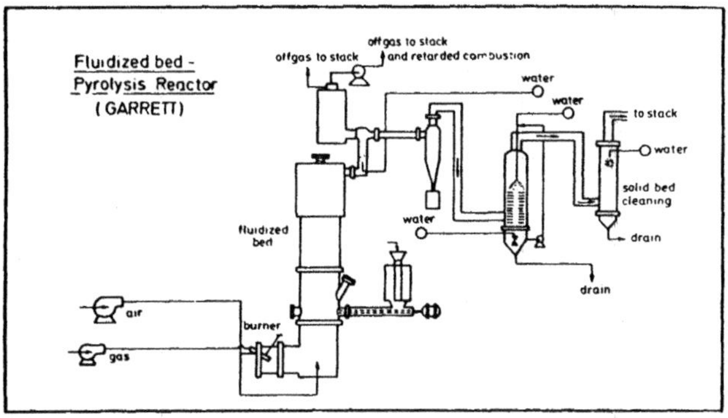
Fluidized bed -
Pyrolysis Reactor
(GARRETT)
offgas to stack
and retarded combustion
offgas to stack
water
water
to stack
water
solid bed
cleaning
water
fluidized
bed
drain
drain
air
burner
gas

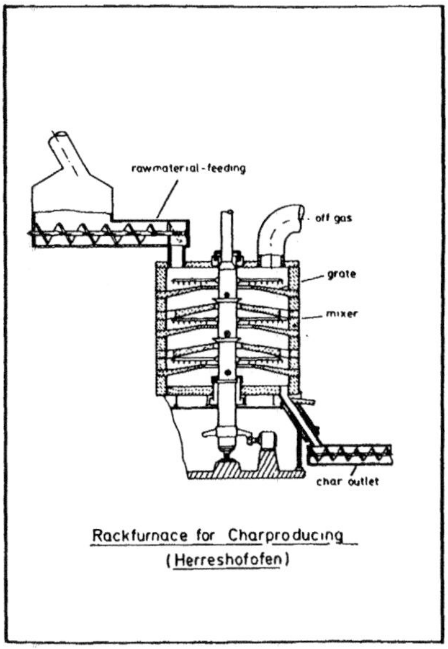
rawmaterial-feeding
off gas
grate
mixer
char outlet
Rackfurnace for Charproducing
(Herreshofofen)

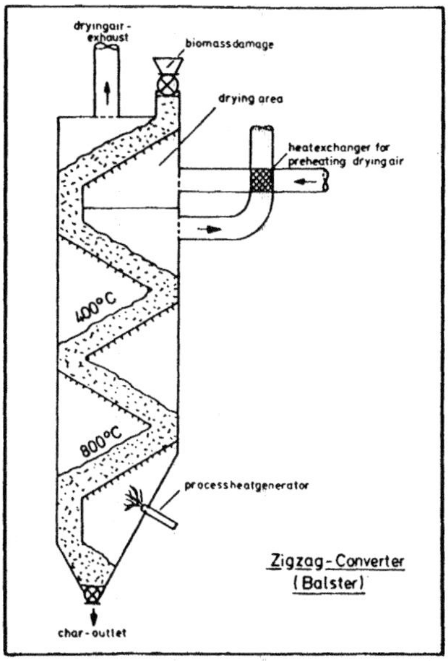
dryingair-
exhaust
biomassdamage
drying area
heatexchanger for
preheating dryingair
400°C
800°C
processheatgenerator
Zigzag-Converter
(Balster)
char-outlet

TABLE 3

Solid and liquid phase percentage; gas phase composition

Element	Charcoal	Oil	Gas	
C	83	51	14	CO
H	3	8	10	CO_2
O	11	40	9	H_2
N	0.3	0.9	18	N_2
S	--	0.01	43	H_2O
Ash	2.7	0.09	2	C_nH_m
			4	CH_4

In Table 3 the elementary analysis of solid and liquid phases together with the chemical composition of gas phase is reported.

3. CATALYTIC PYROLYSIS AND BIOMASS LIQUEFACTION

The use of various catalysts has been recently suggested in fast pyrolysis of biomass at short residence time (7) and at low pyrolysis temperature (8). When wood is impregnated with aqueous solutions of catalyst at room temperature, 0.01 mol of catalyst per 24 grams of sawdust, it is pyrolysed at higher rate and the percentage of gaseous compounds increases. The carbonates, K_2CO_3 and Na_2CO_3 are optimum catalysts, better than bicarbonates; clorures do not affect the yield of pyrolysis, while the presence of phosphate determines unfavorable conditions. Nichel acetate is also very effective towards biomass gasificatoin.

The catalytic action of alkali, specially of carbonates, results in the weakening of the polymeric structure of biomass components during impregnation as confirmed by the shift of the decomposition curve towards lower temperature. The char obtained from treated samples is more oxidized than from non-treated samples and the overall energetics of pyrolysis is altered by biomass impregnation producing an apparent exothermic effect which increases with salt content.

Catalytic liquefaction is the process of removing oxygen from oxygenated compounds to produce hydrocarbons. The reaction is a destructive hydrogenation of the biomass. It was demonstrated that cellulose could be completely converted in liquids and gases and that mainly phenol is generated from lignin conversion. The reactions were studied either in aqueous medium or in organic solvent, such as tetraline, and also in the crude oil produced in the process. Biomass liquefaction is carried out in the presence of hydrocarbon or carbon monoxide and among the different catalyst investigated, unsupported Nichel and Co/Mo seem the best in assisting the sequence of reactions.

Direct liquefaction of biomass is still under investigation on a laboratory level and researches are in progress to determine the best operational conditions, namely pressure and nature of reducing gases, temperature, reaction time, biomass to solvent ratio and to identify the mechanisms of conversion. As far as this last point is concerned many results seem to indicate that decarboxylation is the major pathway for oxygen removal and dehydration is the second most important route.

Generally liquefaction is performed at 300 °C - 400 °C, however higher tamperature operation results in an oil of somewhat lower oxygen content and lower molecular weight, a finding which suggests that the optimum temperature is in the 350 °C - 360 °C range. Carbon monoxide and hydrogen seem to play an equivalent role since elemental analysis, molecular weights and yields remain substancially unchanged. Residence time in the reactor ranges between 30 minutes and 1 hour and the concentration of wood solids in the slurry stream is 20 - 30% by weight. The efficiency of agitation appears to be one of the most important variables since char formation tends to occur preferentially near the external hot walls of the reactor and its formation is increased by an incomplete dispersion of the multiphase reaction mixture.

An extensive study of biomass liquefaction has been performed by Prof. Delmon and its coworkers at the Universitè Catholique de Louvain in Belgium in the framework of the last E.C. "Energy from Biomass" research and development project. The liquefaction tests were performed under relative low hydrogen pressure 60 - 120 atm and in 1 liter batch reactor. The interest was mainly focused on the catalytic liquefaction of the lignin fraction of the biomass and the characterization of light, phenolic, neutral, aqueous and heavy fractions. The separation of the product stream and the analysis of the effluents was satisfactory accomplished.

In USA direct liquefaction of biomass in the presence of alkali was object in 1971 of a preliminary report of H.R. Appell and coworkers at the Bureau of Mines (10) and on the basis of these experiences a process development unit for wood liquefaction was designed and constructed at Albany, Oregon and during four years of operation, 70 barrels of wood oil were produced. The reaction was performed in anthracene oil, and wood flour was fed to the reactor as a slurry. The performed experiments have shown that the mechanical pretreatment of the wood is very energy intensive, that plugging tends to occur at solid concentration greater than 10 % and that excessive heat is required to recycle the oil in the reactor.

A new low recycle process, water based alternative, has been recently developed at Berkeley by L. L. Schaleger et al. (11). Both CO and H_2 at relative high pressure, 300 atm, have been used, as reducing gas; all other conditions were in the range previously reported. The wood oil produced is similar in chemical and physical properties to that produced at Albany except for the lower yield in the water based process. However this discrepancy was attributed to the larger amount of water soluble product formed in this plant.

4. CONCLUSION

Wood and other biomass can be efficiently converted to a liquid fuel suitable for steam and power production by thermal treatment, either pyrolysis or gasification, under a pressure of water and reducing gas and in case also in the presence of a catalyst.

Several pyrolysis processes are commercially available which produce bio-oil with good yield. However the diffusion in the market of these plants is still limited. In order to implement their numbers a better classification of the physical and the chemical properties of the final products is required and the payback period of the investment must become more encouraging. Nowadays, the plants usually have small dimension and of course a limited productivity. Therefore the operational costs, mainly manpower, and the transportation costs greatly weigh, specially if the consumers are located far away from the plant.

Pyrolysis is at present the most attractive process from an economical point of view and the technology appears reliable enough to get over the demonstration period. On the other hand catalytic pyrolysis and direct biomass liquefaction are stimulating processes which need a further development.

It is possible to foresee that in the future an industrial scheme which integrates both selective hydrolysis of carbohydrates into hexoses and pentoses and catalytic hydrogenation of the lignin content would be the most attractive.

5. REFERENCES

1. P. Chartier, Proceedings of the 1st E.C. Conference Energy from Biomass , W. Palz, P. Chartier and D.O. Hall Eds, 22-33, Applied Science Publishers LTD, London, (1981)
2. A. Garkisch and K. Jaster, Proceedings of the 2nd E.C. Conference Energy from Biomass, A. Strub, P. Chartier and G. Schleser Eds, 410-417, Applied Science Publishers LTD, London, (1983)
3. E.J. Soltes and T. J. Elder, "Pyrolysis" in Organic Chemicals from Biomass, I.S. Golstein Ed., 63-100, CRC Press Inc., Boca Raton, Florida, (1981)
4. B. W. Hatt and H.H. Sheena, Proceedings of the 2nd E.C. Conference Energy from Biomass, A. Strub, P. Chartier and G. Schleser Eds, 919-924, Applied Science Publishers LTD, London, (1983)
5. E.M. Hofstetter, Energy from Biomass, Series E,5, 11-19, W. Palz and D. Pirrwitz Eds, D. Reidel Publishing Company, Dordrecht, Holland (1984)
6. F. Fonzi, ITALENERGIE, Private Communication, (1984)

7. A. Rolin, C. Richard, G. Martin and X. Deglise, Proceedings of the 2nd E.C. Conference Energy from Biomass, A. Strub, P. Chartier and G. Schleser Eds, 901-905, Applied Science Publishers LTD, London, (1983)
8. D.L. Pyle and C.A. Zaror, Proceedings of the 2nd E.C. Conference Energy from Biomass, A. Strub, P. Chartier and G. Schleser Eds, 950-954, Applied Science Publishers LTD, London, (1983)
9. A. Burton, D. De Zutter, G. Poncelet, P. Grange, B. Delmon, Energy from Biomass, Series E, 5, 279-288, W. Palz and D. Pirrwitz Eds, D. Reidel Publishing Company, Dordrecht, Holland (1984)
10. H.R. Appell, Y.C. Fu, S. Friedman, P.M. Yavorsky and I. Wender "Converting organic wastes to oil", Bureau of Mines Report of Investigation 7560, U.S. Department of Interior, Washington D.C., (1971)
11. L.L. Schaleger, C. Figueroa and H.G. Davis, Proceedings of the 4th Symposium on Biotechnology in Energy Production and Conservation, C.D. Scott Ed, 3-14, John Wiley and Sons, New York, (1982)

BIOMASS AVAILABILITY

Bernard Pellecuer
Chambre Régionale d'Agriculture du Centre
France

ABSTRACT. The resource of biomass cannot be reduced simply to a number of tep; this is because it is renewable, and also because one may be more interested in one form or another according to the plans of valorisation. The energy needs of a farm can be covered by biomass; the dry biomass surplus can partly cover the heating needs of the rural communities or be used to produce bio-fuel (unless the biomass is from alcohol producing plants). The importance of the resource depends as much on its mobilisation as on its destination. This must be considered in any macro- or micro-economic study, and also in any plans to insert biomass valorisation in a district.

INTRODUCTION

There are many ways of approaching this important subject of biomass availability in the various methods of energy valorisation that we know, and more particularly in the production of bio-fuel.

To estimate the value of biomass, it is essential to study the amount and the kind of biomass available, the technology involved, the constraints to its mobilisation and the possible competition to its use.

The ways of using biomass are varied. They may be quite simple, for example, in a short circuit where biomass is produced and used in the same place, like on a farm, which may, however, need to be slightly adapted to it. Its use may also be more complicated at an industrial level involving different professions and trades. In these two cases biomass will not be produced and converted in the same way. The producers of biomass (the farmer) will react differently. A farmer who produces and converts biomass for his own use will not accord great importance to the economic viability and perception of the work (it will be considered an additional activity but will not replace other activities). At an industrial level the farmer is concerned with the economic laws of supply and demand where each person involved will try to maintain or increase his profit margin. In either case, the con-

G. E. Beghi (ed.), Synthetic Fuels, 283–299.

straints of mobilising the biomass, the kinds of work sites and their organisation will be different. More important still, the precautions to be taken will not be the same.

The prime interest of a farmer is food production. Even if a farmer does become an efficient energy producer, this must not be to the detriment of his food production (or in the case of wood, to the detriment of the production of high quality wood). The notion of maintaining and preserving his estate (quality of the land, quality of the wood) is essential here.

All these points must be taken into account when considering the availability of biomass. The production of biomass must not upset the traditional productions, but be integrated with them. This may well be the limiting factor to its establishment at a local level.

However, do not let us forget the importance Europe accords to biomass. At the international Congress in Berlin (1982) organised by the E.E.C. we were reminded that in the year 2000, the potential production of biomass will exceed agricultural and forestry needs. It will be composed not only of agricultural waste that exists at the moment, but also of ligno-cellulosic energy crops. The main conclusions of this conference are the following:

1) According to Mr. Chartier, these ligno-cellulosic energy crops would come from 5.5 to 6 million hectars of land, which at present produce costly and useless agricultural surplus (costly because they are subsidised by organisations such as the F.E.O.G.A., useless because they are unsuitable for the third world). On this larger scale there would not, therefore, be any competition, but it would be necessary to change the production of the land. Financial aid could then be distributed in the form of subsidies to help such a change.

What would be these new crops? In France (but is it not the same for the whole of Europe) , 2/3 of imported energy is used for heating and 1/3 for fuel. It is, therefore, necessary to give priority to the development of ligno-cellulosic crops such as short rotation forestry. These have the advantage, not only of providing thermic needs, but also of being convertable to synthetic fuel (methanol); this is not the case of other traditional energy crops that all produce ethanol. Finally, according to Mr. Chartier, we should not develop a policy of ex-biomass fuel made only from agricultural surplus - even if the prices are strongly controlled the quantities available may vary -. It is essential that in this sector, production remains stable.

In this paper, we will only be concerned with the production. After a brief presentation of the different sorts of biomass, and their characteristics, we will then rapidly present the constraints associated with the production itself. We will give an example of a method for analysing the available biomass. We have adopted this method in a large French region: "La Région Centre".

1. THE NATURE OF THE RESOURCE

1.1 Forms of biomass

The crude forms of biomass are extremely varied. The biomass may, at first, be transformed physically to facilitate transportation, to make it dense and more "fluid", e.g. straw is chopped or transformed into pellets; wood is reduced to chips. It may be transformed biologically to produce ethanol (e.g. sugar beet, Jerusalem artichokes, grass, sugar sorthum, maize, wheat, lactoserum). It may be transformed thermochemically to produce methanol (e.g. ligno-cellulosic sub products). Animal dejecta, lacto-serum may be specifically transformed to obtain methane (methanic fermentation).

The value of a certain quantity of biomass may, therefore, vary greatly according to the way it will be transformed. Here we will simply enumerate the nature of theoretical energy productions.

Production in large agricultural areas:

- straw — 3 tons dry matter/hectar/year
- wood-energy — 4 tons dry matter/hectar/year (ordinary coppice)
  10-20 tons dry matter/hectar/year (short rotation coppice)
- Maier-cob — 4 tons at 60% humidity/hectar/year
- sugar beet — 50 tons/hectar/year
- Jerusalem artichokes
  20-50 tons/hectar
- grain — 50-60 tons/hectar
- Lactoserum
- biogas — (e.g. straw, maize stalks 160-270 m^3 CH4/ton of dry matter)
- grasses — 9 tons dry matter/hectar.

Thermic provision by dry biomass (cobs, straw, wood):

- 1 tep (equivalent ton of petroleum) $\simeq$ 3 tons dry matter.

Fuel fabrication:

- ethanol, e.g. sugar beet — 1 hecto-litre/ton, or
  47 hecto-litres/hectar
  - e.g. Jerusalem artichokes
    0.85 hecto-litre/ton, or
    34 hecto-litres/hectar
  - e.g. Sorghum — 40 hecto-litres/hectar
  - e.g. grain — 3.7 hecto-litres/ton (+ 0.3 t of dry matter of products 28% protein), or
    20.5 hecto-litres/hectar (1.75 tons dry matter 28% protein - if the grain is distilled)
  - e.g. Lactoserum

e.g. grasses 3.5 hecto-litres/ton dry matter
45.5 hecto-litres/hectar
(e.g. cellulosic residues
2-3 hecto-litres/ton dry matter)

- methanol e.g. ligno-cellulosic residues
5-7 hl/ton dry matter
e.g. grasses 4 hecto-litres/ton dry matter, or
52 hecto-litres/hectar
e.g. giogas 2.5 to 4.6 hl/ton of dry matter.

1.2 Characteristics of the production

At an equivalent level of tep (equivalent tons of petroleum) a production of biomass will never be equal to a traditional source of energy. This is because their characteristics are so different. If we wanted to substitute biomass for a traditional energy, it would be necessary to know precisely the qualities and the characteristics of the particular resource (here we will not go into the very specialised production of methane from animal dejecta which is very limited).

Biomass is a renewable production associated with solar energy. The cycle depends on traditional agricultural and silvicultaral practices. E.g. straw-subproduct of cereal production is produced yearly. Wood from coppices is produced after 20 years and clearance wood according to the demands of silviculture. Traditional energy crops are annual.

Biomass is, therefore, a continual production-renewable, but whose quantity may vary from one period to another. This is especially the case of the subproducts of annual crops; their quantity per hectar is dependent of the yield, the climate, diseases and rotation.

Biomass is spread out over a wide area. In one particular region a multitude of land owners decide whether it is worthwhile to transform it into energy. The production and decisions taken may be scattered over a wide area.

Biomass is a poor resource - diluted over a large area - low in energy density. It does not weigh much and is, therefore, voluminous. If it is not transformed at all, the transport is costly and its valorisation will only be sound financially within a determined distance.

This resource is often partly used elsewhere, so there is competition between its use for energy and a constant priority use, e.g. straw is used in priority for cattle; the quantity left over for energy from the whole variable amount, after the constant priority demand has been satisfied, is often impossible to foresee. This makes any planning of supplies very difficult. In the case of wood, there may be speculative tension arising from a possible competition for the wood to be transformed into paper. It is the resource that can only be used in certain conditions.

1.3 Constraints concerning the recuperation of dry biomass

The activities necessary in the growing and harvesting of annual crops like sugar beet are well known. Here we will only be concerned with what straw recuperation and wood energy entail (on modern maize sites we are not able to recuperate the cob when the harvest is performed by a corn sheller).

Straw energy (2). A survey has been made in a small cereal producing region concerning 400 farms. 37% farmers were against bringing in the straw (3). Apart from quite legitimate reasons like the fear of lowering the humus content of the soil, other the low price paid for straw, considering the high cost of collecting it (57%). Unlike mixed farms large cereal producers no longer have collecting machines. They would have to buy them specially for collecting the straw. This activity would then no longer be considered of small importance. Added to this, the time available for collecting the straw is 20-30 days at the most between the harvest and sowing, and in 4 out of 5 years it is only 20 days (excellent collecting machinery does exist, however). At present in the large cereal producing regions 3/4 of the farmers do not collect the straw. They prefer to burn it in the field (as much as 80% of the production sometimes) (4).

Wood energy (5). Time is not so severely limited here. Where wood is worked in a traditional way by hand-access to the site may be had at any time. On the other hand, if the work is entirely mechanised (with sophisticated and very heavy machinery) the time available is greatly reduced.

The greatest hindrance for wood energy lies in the vast quantity of wood owners. It is difficult to put together a management plan that will be efficient and financially viable. Private woods are generally very small and their owners often absent.

1.4 The kinds of market (6,7)

The biomass market (at an industrial level) is a new developing market. The difficulties it meets are due to the very nature of this resource.

- As it is difficult to foresee the exact amount of biomass available for any length of time, concerned with its transformation are not willing to involve themselves in long-term contracts that offer little security. The producers, on the other hand, are generally not sufficiently motivated financially.

- The cost of mobilising biomass may vary considerably from one situation to another. Paradoxically, costs for straw on a large cereal producing farm are greater than on a smaller farm. Certain markets for straw energy are, therefore, set up on the border of wooded pastureland, where straw, although less abundant, is cheaper and supplies seem more certain. These farms need a supplement to their income.

SOME EXAMPLES OF RESOURCES - THEIR CHARACTERISTICS

BASIC RESOURCE	POSSIBLE RESULTING PRODUCT	RIVAL ACTIVITIES (HYPOTHETIC)	OTHER CONSTRAINTS TO BE CONSIDERED	STATE OF TECHNOLOGY FOR PRODUCING OR CONVERTING THE RESOURCE
STRAW	Bails	Cattle (litter) fodder mushroom growing (chemistry)	Climatic constraints for collecting	Perfect
	Pellets	Fodder	At present only a marginal activity on a farm	Perfect
	Methanol	Chemistry		Experimental
	Ethanol	Chemistry		Research
MAIZE COBS	CObs	Chemistry	Method of collect to be adapted	
WOOD	Wood-rope chips	None None	Wide scattering of land	Perfect Perfect
SUGAR BEET	Ethanol	Sugar		Perfect
GRAIN	Ethanol	Flour		Perfect
ALL PLANTS	A.B.E. (Acétone butanol, éthanol)			Experimental
ANIMAL DEJECTA	Methane	None	Danger : gaz. in the neighbournood	Perfect and experimental

From : (3)

The great danger here is a deterioration of the organic matter in the land, that is generally already difficult to work.

- The biomass market is a market that is impossible to foresee. One of the best ways to stabilise the system would be to associate the farmers, themselves, with this market in an interprofessional cooperative (SICA). Here they would engage themselves by a contract and the profits obtained from the whole production would be shared out. The farmers would only mobilise their biomass if the price offered were sufficiently attractive, which would be comparable to what they might obtain from other agricultural activities.

2. GLOBAL ANALYSIS OF BIOMASS POTENTIAL: THE EXAMPLE OF THE "REGION CENTRE"

The "Région Centre" is the second region in France for its production of nuclear electricity. It also offers the greatest resource of biomass. This region is, in fact, the leading agricultural region in France. It consists of two large cereal exporting zones; of large wooded areas more or less pastureland; of mixed farms and specialised crops. From the biomass point of view, we must expect great diversity.

It has been possible to define the biomass potential (not counting energy crops) thanks to various studies made by the OREAC, in collaboration with the Chamber of Agriculture and the agricultural and forestry administrative departments.

2.1 The method used to analyse the potential

To calculate the regional potential it was necessary to bring together different sources of complementary information and then complete them after new research (9).

2.1.1 Map of land usage. This was not constituted from statistics, but from aerial photographs enabling us to distinguish the different zones of cultivation:
- cereal crops from which the essential resource is straw;
- woods;
- woodlands and heldged-in pastureland, from which the resource is wood from the hedges, manure and shurry;
- vineyards and orchards from which the resource is pruned wood;
- areas of mixed farming. The availability of the resource here is not so clear.

2.1.2 Map showing soils and organic matter. The type of soil determines whether straw may be exported. A map has been made showing precisely how much straw may be agronomically available. These results put together enable us to know the potential and position of the available biomass in each region.

2.1.3 Map showing the availability of wood. This distinguishes the various characteristics of the forests or full grown trees (or areas becoming so) and coppices allowing wood for fire to be taken away.

2.2 The potential (8)

The global potential of this region is 1 million tep (oil equivalent ton). This is equivalent to 20% of the total energy consumption of the region. It covers all the agricultural needs and partly the rural ones.

- energy from woods not utilised	300-400,000 tep/yr
- straw from woods not utilised	400,000 tep/yr
- animal dejecta	200,000 tep/yr
- twigs and pruned branches	50,000 tep/yr
- maize cobs	60,000 tep/yr

2.2.1 <u>Woods</u>. Forest covers 880,000 hectars - or 23% of the region. Coppices with standards represent 60-70% of the wooded area. This is the essential source of "small wood", only 20-30% of which is used at present. The wood which is not used represents 1.8 million m^3/year.

BIOMASS IN THE REGION CENTRE

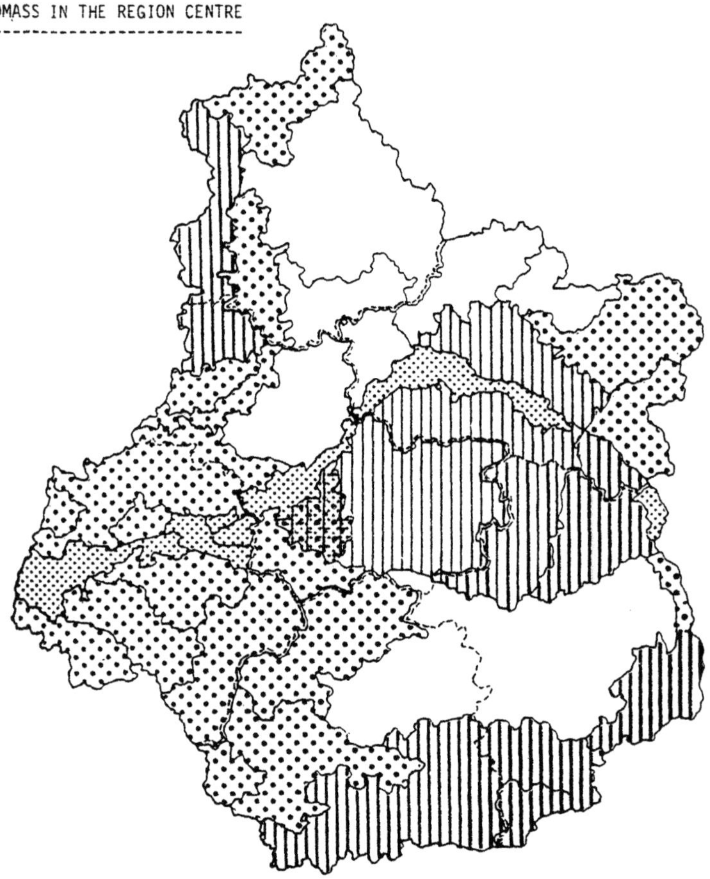

FORM OF CULTIVATION	RESOURCE
arable farming	Straw, corn cobs
mixed farming	Dejecta, wood (straw)
wood lands & hedged in pastures	Wood - déjecta (straw)
specialised crops	Twigs and pruned branches (compost consumption)
mainly wooded	Wood

(From : 8)

its use does not compare at all with traditional silviculture and may even help to improve the management of the coppices (this resource consists of coppices, cleared wood, twigs, etc.).
In addition to this wood, there are the hedges that represent more than 50,000 km (pruning = 5 m^3 wood/km/year) and also sawmill waste (50% wood coming to a saw mill).

2.2.2 Straw. Cereals are cultivated on 71% arable land. This explains the important place held by straw. The quantity of recuperable straw is the complement of the quantity already used.

- Rival uses: This gives an idea of the competition at a national level (quoted from 4).

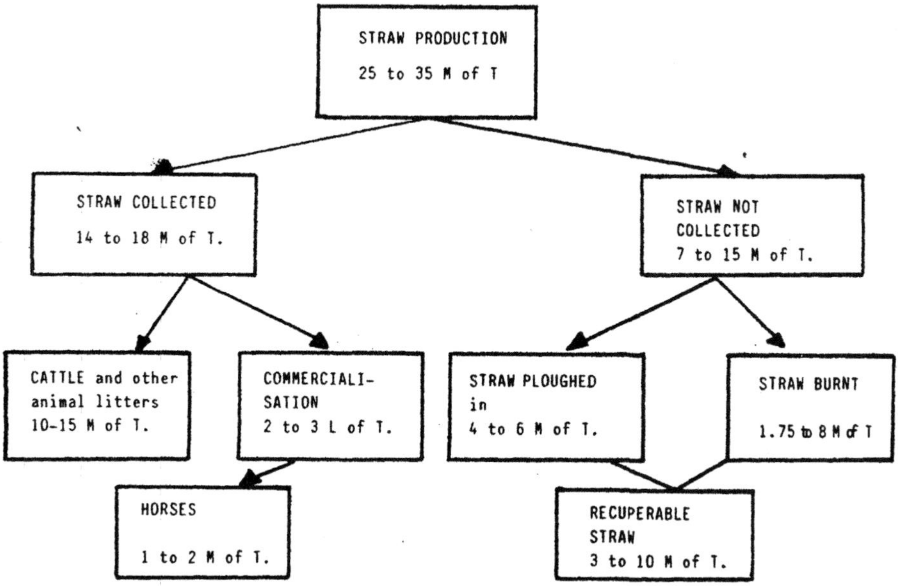

- Agronomic security: Here is an example of the proposed acceptable level of straw export for each agro-pedological case in the region: (from (9))

NATURE OF THE ZONE	SOIL TYPE	POSSIBLE EXPORT
Massive collection possible	Sols argilo-calcaires de Beauce Rendzine de Champagne berrichone	75 % 100 %
Medium possibilities	Argile à silex de Beauce Argilo-calcaire Limon de Beauce (à rendement élevé)	50 %
Moderates possibilities	Limons de Beauce "Bournais francs" "perruches" limons sur argile à silex limons non battants de polyculture élevage	30 %
Low possibilities to none	Limons battants Sables Terres fortes de Champagne Rendement	0 %

2.3 Example of a report on a small agricultural region

Here is an extract of recommendations concerning the mobilisation of biomass in a small cereal producing region (9):

Champagne Berrichonne

On the whole, the situation in Champagne is similar to that in Beauce, however, attention must be given to certain details:

- The percentage of exportable straw is higher than in Beauce as Rendzine can support large losses. However, the yields that are not so high and the lower percentage of straw cereals in the rotation, compendate this difference. A regional study shows straw to be available at 0.9 T/ha.
  During this inquiry, the potential, which was much higher, rose to 1.2 T/ha of S.A.U. (usable farm land).
- The farms are larger, but they consume less energy. Although their energy potential is similar to that found in Beauce, possibilities of energy autonomy and commercialisation of straw are greater.

Possibilities. On cereal producing farms in this region (which nearly all the farms are) the possibilities are similar to those in Beauce in order of decreasing economic importance:

1) domestic heating by straw, wood, cobs;
2) grain drying - this was possible in all the cases we met;
3) commercialisation of straw.

75% farms could achieve energy autonomy by using only agricultural subproducts.

3. ANALYSIS OF AVAILABLE AND COMMERCIALISABLE BIOMASS ON FARMS

The farm as a resource of biomass is the most viable unit to be observed. At this level, it is possible to study in great detail the real-potential in relation to the activity of the farm. It is also possible to analyse rival uses to its valorisation; i.e. non-energy uses, but also energy commercialisation of products leaving the farm.

It is obvious that the farmer prefers, first of all, in most cases, to satisfy his own needs, before commercialising his surplus. Consuming biomass on the farm is always worthwhile, except in the case of methanisation (11) (but here commercialisation is impossible).

Aftermaximum home-consumption for thermic uses, and according to the importance there to methanisation, the regional biomass surplus is between 460,000 and 580,000 tep (not counting alcohol producing plants). This represents 50% of the non-agricultural energy needs in rural communities.

To study the possibilities of biomass energy valorised on the farm, and the commercialisation of biomass, we have closely observed 30 farms in the "Région Centre". On these farms we have tested various heating techniques (12,13). An inquiry has also been made into energy production and potential energy consumption on 55 farms. As we know the thermic yield of each system, we have been able to find out the proportion necessary to cover domestic needs and, therefore, the commercialisable potential (not counting alcohol producing plants). The results are analysed according to the size and type of farm (technical-economic orientation = OTEX).

3.1 Average results by OTEX (12)

The results for four main types of farms in the region (65% of all the farms) and of one artificial type of farm, growing cereals and drying maize on the farm, are the following:

OTEX	AVERAGE SAU (*)	BIOMASS POTENTIAL TEP(*)/Yr					CONSUMPTION TEP/YEAR			
		STRAW	WOOD	COBS	WET BIOMASS	TOTAL	HEATING (DRYING)	TRACTION	ELECT.	TOTAL
11	70	21	1	1	-	23	3	7	1	11
115	135	47	-	6	-	53	7+11	17	1	35
12	120	42	1	8	-	43	5	16	1	22
810	66	5	1	2	6	14	4	6	1	11
819	84	3	2	8	13	18	4	6	1	11

11. cereals
115. cereals + maize drying
12. other general farming
810. general farming + grass eating cattle
819. grass eating cattle + general farming
(*) S.A.U.: useable farm land
(**) TEP : oil equivalent ton

Dry biomass which is the only resource on cereal farms (composed of 90-98% straw) represents only 27% of the total resource on cattle rearing farms (composed of 60% straw).

3.2 Results according to the size of farms

Cereal farms are measured according to the useable agricultural surface (SAU) and stock breeding farms according to the category of the economic size of the farm (CDXE). Here are two extreme cases:

OTEX general agriculture (11,12)

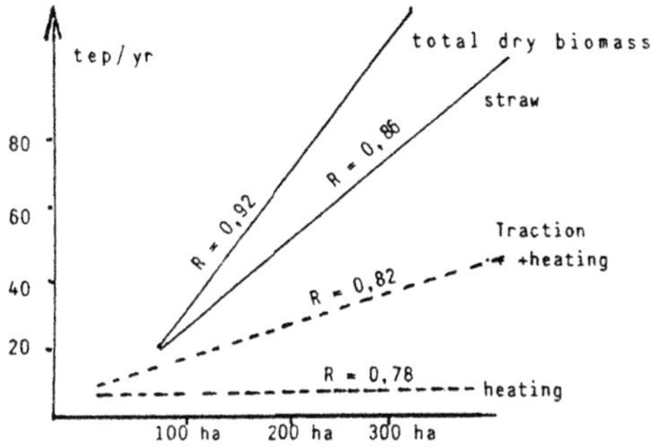

In this case, dry biomass always covers the thermic needs of the farm. After the dry biomass has been valorised for heating on the farm, there is 50-75% of the potential that can be commercialised or possibly be used, in extreme cases, for traction (gas-run engines).

OTEX grass eating animals (810 + 819)

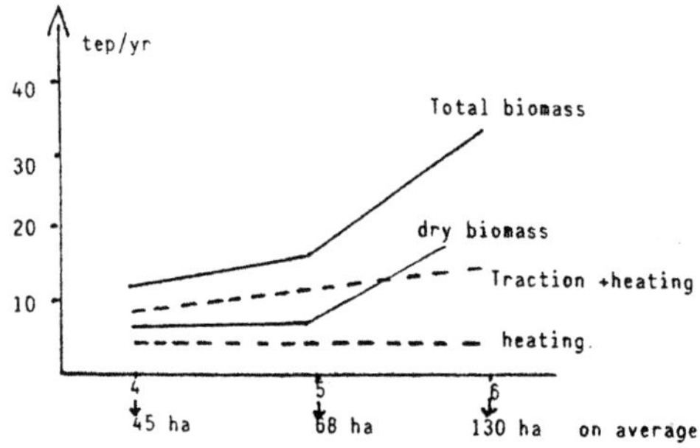

In 30% of these cases dry biomass is not available for energy (livestock) and in 12% it is insufficient. The only potential would be damp biomass (methane) which can cover all needs and is not in any circumstance commercialisable.

(ENQUETE REALISEE PAR L'OREAC)

EXEMPLE OF INFORMATION OBTAINED FROM THE INQUIRY ABOUT THE FARMS (from 9)

```
*************************************************************************
* CODE   :        410* DEPART.        41* REGION:PERCHE V.*  O.T.E:        140
*************************************************************************
*    SURFACE(HA)     *SURF.(OU LONG.)  *ANIMAL  :   U.G.B. * PAILLE :TONNES *
* S.A.U.:         96* VIGNE  :       0.00* BOVIN  :       0.00* TOTAL  :       210
* LABOUR:         96* VERGER:        0.00* OVIN   :       0.00* LITIERE         20
* CEREAL:         60* BOIS   :       5.00* PORCIN:        0.00* RECOLTE         20
* S.T.H.:          0* HAIES  :       0.00*AUTRES  :     35.00*
*************************************************************************


*************************************************************************
*BIOMASSE*TONNES  *        T.E.P.        *FUMIER  * TONNES *   M3     * T.E.P.
*        *        *RECOLTE *POTENT.  *CALCULE:       259       432     11.66
*PL.MOB.        190               62.70*DECLARE:        0         0
*RAFLES :     13.23       4.37      4.37*TYPE    FUMIER
*VIGNE:ARRACH.:           0.00      0.00************************************
*VIGNE:TAILLE  :          0.00      0.00*  HUMUS (KG)       :FACULT.  :OBLIGAT.
*VERGER  :                0.00      0.00*REST./CULT.ASSOL.        191       769
*HAIES                    0.00      0.00*REST.PAR FUMURE            0       405
*BOIS    :                0.00      3.50*  TOTAL                  191      1174
* TOTAL  :                4.37     70.56*
*************************************************************************
          PAILLE TOTALE / HA CEREALE:                  3.50
          P.MOB.(=TOT.-BETAIL) / HA SAU+BOIS  :        1.88
*************************************************************************
CONSOMM. *DOMEST. *         CHALEUR                *     MOTEURS       *
   EN    *  SAUF  *DOMEST. *ELEVAGE *SECHAGE *   DES    *A POSTE * TOTAL
 T.E.P.  *CHALEUR *         *SERRE,..*               *VEHICULE* FIXE    *

ELECTR.        3.11      0.00      0.00      0.00                 2.22      5.33
ID.EN KWH     14000         0         0         0                10000     24000
GAZ            0.00      0.00      7.74      0.00      0.00       0.00      7.74
CHARBON        0.00      0.00      0.00      0.00      0.00       0.00      0.00
BOIS           0.00      0.00      0.00      0.00      0.00       0.00      0.00
FUEL LOUR      0.00      0.00      0.00      0.00      0.00       0.00      0.00
F.O.D.         0.00      3.34      0.00      0.00      7.52       0.00     10.86
ESSENCE        0.00      0.00      0.00      0.00      0.00       0.00      0.00

 TOTAL  :      3.11      3.34      7.74      0.00      7.52       2.22     23.92
*************************************************************************


*************************************************************************
          *% HUMUS * % HUMUS SOUHAIT.*SURFACE *% HUM. A L'EQUIL.*PAILLE
   SOL    *MESURE  *INF:I .... SUP:S *         *100%EXP * 0%EXP. *EXPORT.
ARG.CALC.       2.5       2.3       2.8      0.25      2.87      3.33      100
ARG.CALC.       1.5       2.3       2.8      0.25      2.87      3.33        0
ARG.SILEX         1         2       2.5      0.50      3.06      3.56        0
*************************************************************************
PAILLES EXPORTABLES SANS  DANGER   :          25  %
                               SOIT:       0.47 T/HA SAU+BOIS

POTENTIEL VOIE SECHE                          CONSOMMATION
  SANS PAILLES                         7.86     CHALEUR             11.08
  PAILLES SANS DANGER                 15.68     ELECTR.SPEC.         5.33
  TOTAL SANS DANGER AGRON.            23.54     CARBURANTS           7.52
POTENTIEL VOIE HUMIDE
  FUMIER+LISIER                       11.66     TOTAL               23.92
  PRODUITS SECS                       32.07
  TOTAL                               43.73
```

4. STUDY OF THE POTENTIAL CONCERNING RURAL DEVELOPMENT - THE EXPERIMENT IN A RURAL DISTRICT (3)

The study of the regional potential, and the more detailed but essential study of the farms, show that concerted work is necessary in a small region to organise and set up the mobilisation and valorisation of biomass.

We were able to take advantage of a rural modernisation programme (Plan d'Aménagement rural d'Outarville) in a rural district in "Région Centre", to include the development of "energy" in the plans of local modernisation concerted work-groups were set up, information diffused and studies were made concerning its feasibility. Unlike the earlier methods of study, which considered essentially the most obvious mobilisation of biomass, i.e. thermic valorisation, this study was much wider. It began by examining the energy conversions possible considering the resources of the district and the present state of technology.

- straw, wood, cobs, animal dejecta, for thermic use;
- energy crops (sugar beet-grain) and various subproducts to produce ethanol, methanol or A.B.E. (acetone, butanol, ethanol) (14) (oily crops were not accepted as it would not be possible in this district to set up the infrastructure necessary for their transformation).

We present only the final conclusions of the enormous enquiry concerning all the farms (422 out of 490 farms). They take into consideration the constraints whether agronomic, of mobilisation or of marketing.

5. CONCLUSION

Biomass is integrated in the policies and plans of many countries, and interests many people. Oil companies are present in the debate. However, many of them seem desarmed faced with this new promising energy which is so different to the others. It is an agricultural energy, linked with farming activities and dispersed over the whole country. Some are waiting for farmers to organise themselves first, so that they can take over afterwards. The resource itself is, therefore, very important in the establishment of valorisation procedures.

This resource is widely scattered over an area, poor, not dense, and therefore, the cost of collection is high. It is found in different places and belongs to different owners, who take the decisions? By its definition it is decentralised so, what can be done without local initiative? It is a renewable energy. This is of great value, but from one year to the other the quantity produced is not constant.

It is an agricultural energy - and by its very nature is in many different forms - dry or damp; stockable or unstockable, useable in its crude state or having to be transformed; convertable for thermic needs, or fuel; on the farm or exported. It is, therefore, essential when

CONCLUSIONS OF THE INQUIRY CONCERNING THE DISTRICT OF [illegible]

BASIC RESOURCE T=Quantity available in 1981 D=available for energy uses U=used for energy	POSSIBLE CONVERSION	AVAILABLE ENERGY tep from D	ENERGY CONSOMMED IN THE PRODUCTION	NET ENERGY AVAILABLE	AVAILABLE AT A COMPETITIVE PRICE		POSSIBLE INCREASE OF THE RESOURCE
					Short term	Long Term	Irrigation, increased yield
STRAW T = 71 180 D = 42 660 U= 80 T - 29 tep	Bails of straw Pellets Methanol (5) Ethanol (4)	15 360 or 15 360 or 11 370 or 4 710	120 (3) 2 900 ? ?	15 240 or 12 460 ? ?	15 240 12 460 ? ?	15 240 12 460 11 370	increase in the proportion straw-grain. Better yield in the transformatio
MAIZE COBS T = 2689 T = D U = approx 50+=18tep	Cobs (for marking éthanol) Méthanol=CF Straw	940	3 (only transport)	937	937	937	Maize with several ears
WOOD T=D= 1020 m3 U=? + twigs & small wood	Wood rope Chips	132 and 62				1 200	Intensive SYLVICULTU Short Rotation Coppi
SUGAR BEET T = 267 780 T-D? U = 0	Ethanol	12 260 from T	730 (6)	11 930	0	According to amount cultivated	
GRAIN (2) T = 11 578 tons	Ethanol	23 000	10 350	12 650	Depends on the market of protein co-products	According to protein co-products	
Other plants	A B E	?	?	?	?	According to amount cultivated	
ANIMAL DEJECTA T = D = oo U = approx 50 tep	Méthane	without irrigation		316	according to use	150-250	Increase of herd

(2) All grains, source ONIC
(3) Ready of the farmyard
(4) 2.5 hl ethanol/T. dry matter
(5) Declared efficiency of this process 74 %
(6) Including energy valorisation of sub-products

considering biomass availability, to define what form you are concerned with, for which conversion, for which use and under which conditions.

We know that a farmer may be autonomous in energy, which would give him security in food production. In all cases thermic needs of a farm can be covered by biomass form the farm, and fuel needs could too, if 5-7% of the SAU (useable agricultural surface) was devoted to alcohol producing plants (sugar beet) converted in a short circuit.

Biomass surplus can be exported from the farm for commercialisation in its crude state, or transformed in industrial units using modern technology to produce heat or fuel.

To evaluate the resource it is necessary to work at a macro-economic level (to discover the situation, possible competition) and at a micro-economic level (analysis of valorisation potential, of constraints). It is necessary to study systems that would mobilise the maximum biomass with safety (covering contracts), also giving priority to preserving the value of the agricultural or wooded estate (studies of impact, soil tests, zoning). To do this, the usual networks of information diffusion in rural areas should be used, to set up the movement or encourage it, and not isolate the resource itself from the other procedures (marketing other professions involved).

All biomass resources must be complementary among themselves and complement conventional energy. In this way their use will benefit local economy (creation or maintenance of activities) and more generally the whole district.

REFERENCES

1. B. Pellecuer (1982) 'Valorisation de la biomasse sèche sur les exploitations agricoles - Exemple d'un réseau-pilote de référence et de démonstration en région Centre', Chambre Régionale d'Agriculture du Centre/Congrès Internationnal sur la Biomasse, Berlin.
2. B. Pellecuer (1982) 'La paille, source d'énergie en région Centre', Chambre Régionale d'Agriculture du Centre/Biomass Actualité No. Spécial "Paille", Juillet 1982.
3. P. Denoroy (1983) 'L'Energie dans un canton Beauceron - Outarville (Loiret), situation et perspective d'une maitrise énergétique locale', Chambre Régionale d'Agriculture du Centre, Septembre 1983.
4. Statistiques du Service Régional de la Statistique Agricole, Région Centre.
5. D. Auclair (1982) 'Le bois énergie', Forum des Energies/INRA.
6. B. Pellecuer (1982) 'Le marche de la biomasse', Colloque Energies Renouvelables en Milieu Rural.
7. B. Pellecuer (1981) 'Le bois énergie', Chambre Régionale d'Agriculture du Centre.
8. Potentiel Energétique de la Biomasse en Région Centre, Une Première Evaluation - OREAC, 1979.

9. L. Piermont, F. Brondel (1981) 'Valorisation énergétique de la biomasse en région Centre - Les filières possibles', OREAC.
10. M. Formery (1978) 'Valorisation des petits bois dans la région Centre', I.E.A./Chambre Régionale Agriculture Centre/ C.R.P.F.
11. J. Sourie (1982) 'Valorisation énergétique de la biomasse - Perspectives, obstacles à son développement', Colloque Energies Renouvelables en Milieu Rural.
12. B. Pellecuer (1982) 'Valorisation de la biomasse sèche sur les exploitations agricoles', Chambre Régionale d'Agriculture du Centre, Congrès Internationnal de Berlin.
13. A. Bouillet, F. Cayrol, F. Foucault, B. Pellecuer, C. de Silguy, J.C. Sourie, C. Thoiron (1982) 'Les chaudières domestiques à paille', AFME.
14. P. Raugel (1983) 'Ethanol et autres carburants d'origine agricole, Etude des filières et des produits en région Centre', Chambre Régionale d'Agriculture du Centre.

EXPERIENCE WITH AUTOMOBILE ENGINES, OPERATING ON BIOGAS IN COGENERATION SETS

H.-U. Huss
Daimler-Benz AG
Forschung und Entwicklung
Postfach 202
7000 Stuttgart 60
Germany

ABSTRACT. The potential of energy from biomass and organic waste materials is presented. The anaerobic digestion of organic materials in order to produce biogas is very promising. Biogas is well suited to run combustion engines. The modification of a standard diesel and gasoline car engine operating on biogas is described. Difficulties which arise from corrosive gas components can be solved. These modified engines integrated into a cogeneration set produce from 1 m³ biogas approximately 1 - 1.5 kWh of electricity and 3 - 4 kWh useful heat. In certain applications biogas production and use in cogeneration sets is economical. The difficulties of the storage of biogas onboard of vehicles are not yet solved. Thus it is better to use biogas in stationary plants, especially in cogeneration sets.

1. INTRODUCTION

Since the beginning of the 70ies the use of renewable energy sources, that means basically the use of solar energy, has once again been taken intensively into consideration. Two areas of research have been

- the direct use of solar radiation in collector systems, producing heat and
- solar cells, producing electricity directly.

In addition to that, the use of biomass appears to be very promising as well. Biomass is a chemical form of solar energy which is stored in plants.

The energy potential in plants on land areas is about 6 times greater than our current energy demands. However the photosynthesis process has only an average efficiency of about 0.3 %.

Figure 1 shows the annual energy demands splitted up according to the different primary energy sources. In comparison with the demand, biomass has a huge theoretical energy potential of about the equivalent of 40 billion tons oil. Today 400 million tons, mainly fuel wood, are used for energy.

The chart shows clearly our excessive dependence on oil. Consequently we have to use oil more effectively and we should try to

G. E. Beghi (ed.), Synthetic Fuels, 301–321.

substitute it by other energies whenever it is possible. For this purpose renewable energy sources can make a contribution.

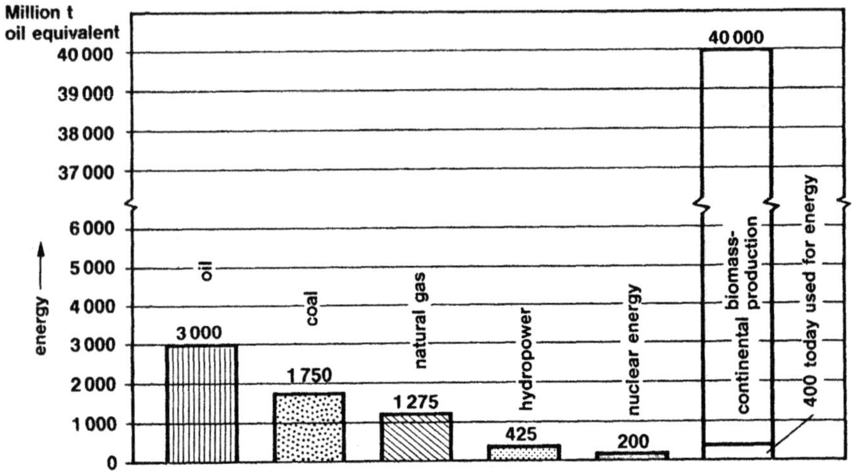

Figure 1. Annual energy demand and energy storage by photosynthesis (1981)

2. BIOGAS FROM BIOMASS

2.1 Processes for extracting energy

Plant materials or residuals from plant processing, which are primary energy sources, are in general very unhandy substances for energetic use. Aside from wood or woody materials, which can be burned directly, secondary energy fuels must be processed from biomass.

Figure 2 shows a survey of available processes. Liquid fuels, for example ethanol, can be applied allround, especially for transportation purposes.

Plants containing sugars, starches and even celluloses are used for fermentation, in which yeasts convert these materials into ethanol. A well known exponent is sugar cane. Currently about 4 billion liters of ethanol are produced in Brazil per year.

Liquid hydrocarbons and other energy intensive products (like vegetable oils and latex), which have to be refined in general, can be obtained by extraction or hydrogenation from a special kind of biomass (i.e. rape seed, euphorbias).

Gaseous fuels are better used in stationary power plants. Dry, cellulose containing plants can be used for gasification or pyrolysis, which leads to hydrogen and carbonmonoxide.

Organic matter with high water content can be converted by the anaerobic digestion process (oxygen-free environment), which produces a mixture of methane and carbondioxide. The production and use of this so called biogas is the subject of the following sections.

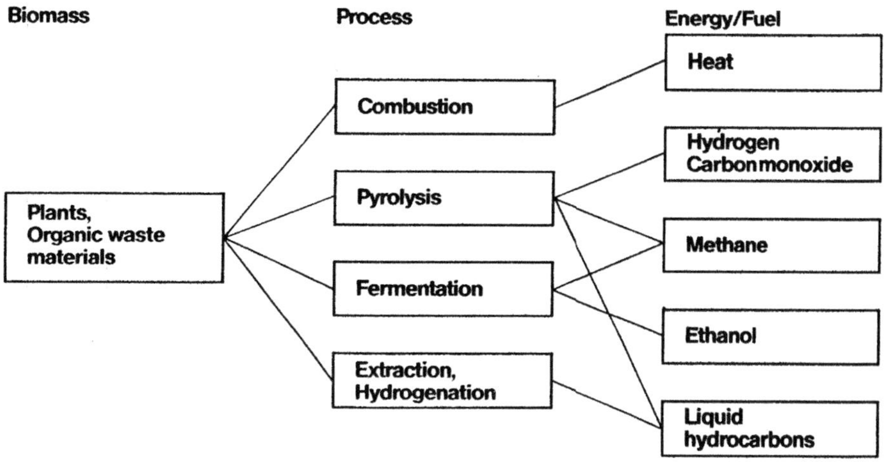

Figure 2. Processes for extracting energy from biomass

2.2 Potential of biogas production

The growing of plants for pure energetic use, so-called energy farming, can create a lot of problems. Plants need a huge land area and take up a large volume. In addition you have to spend a lot of energy for collection, storage, transportation and processing. Because of its huge land area demand, energy farming competes with food production.

For the short term, it would be better for industrialised countries to simply use their available organic waste materials for energy production. Whether they are used or not, waste materials have to be eliminated and they are available in sufficient concentrated amounts.

In figure 3 you can see the total generation of waste materials in Germany 1980 /1/. Other industrialised countries with large population density have the same ratios. For biogas production you can use the following three types of waste:

1. Municipal solid wastes
   These are domestic garbage and garbage-like industrial wastes. They contain about 50 % organic matter. 80 % are deposited in landfills, so that 10 million tons of organic wastes are available for biogas production per year. The gas production in the landfill starts after a certain time. This comes about as soon as the layer of deposited wastes are higher than a minimal thickness.
2. Sewage
   Sewage is a byproduct of waste water cleaning. In the municipal waste water cleaning plants, an annual amount of 36 million tons are produced. Most of it is treated today with anaerobic digestion, producing about 300 million m^3 of biogas.
3. Agricultural wastes
   The largest group is the agricultural waste. The yearly ani-

mal manure production is about 190 million tons. It is well suited for biogas production.

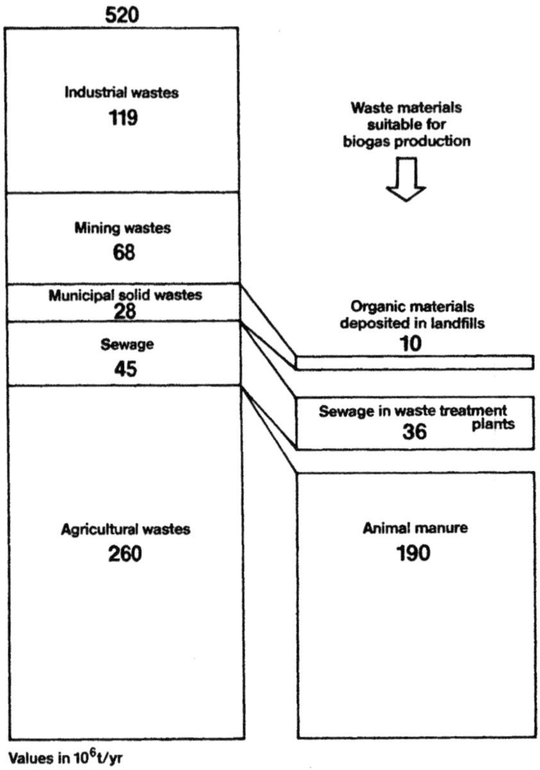

Figure 3. Total generation of waste materials in Germany 1980

Figure 4 gives some characteristic values of methane production from animal manure and an estimation of the theoretical energy potential in Germany as well as in the total European Community /2/.

In order to give an estimate, one can obtain biogas energy, which is equivalent to 1 liter of fuel oil per day, from the manure of 1 cow or 10 pigs or 100 chickens. These numbers of animals are called one livestock unit.

If all animal wastes are theoretically digested, you can get about 36 billion kWh biogas in Germany and about 180 billion kWh in the whole European Community. About 75 % of this energy can be produced by cattle manure.

Type of animal	Cattle	Swine	Poultry
Manure [kg/day]	50	4	0.2
Dry organic matter [kg/day]	5	0.4	0.035
Gas yield [l/kg dry matter]	300	400	500
Daily gas yield per livestock unit [m^3/day]	1–1.5	1–2	1.5–3

GERMANY		EC
$190 \cdot 10^6$ t/yr	Total production of manure	$900 \cdot 10^6$ t/yr
$6000 \cdot 10^6$ m^3/yr	Estimated gas production	$30000 \cdot 10^6$ m^3/yr

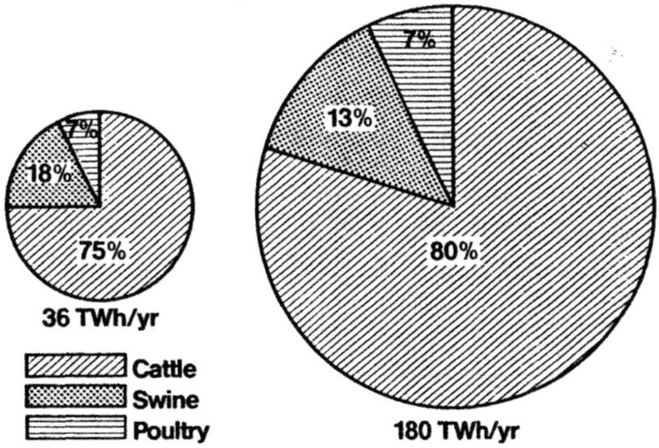

Figure 4. Specific gas yield and total energy potential of animal wastes in agriculture

2.3 Biogas process

2.3.1 Anaerobic Digestion. Anaerobic digestion is a very complicated biochemical decomposition process (figure 5) /3/. For simplification you can divide the total process into three parts:

hydrolysis, acid- and methane production. At each decomposition stage, there are respectively other kinds of bacterias active. Having different metabolism rates there must exist balanced conditions for an optimal gas production. In addition the digestion temperature and the amount and composition of the fed slurry has to be kept constant.

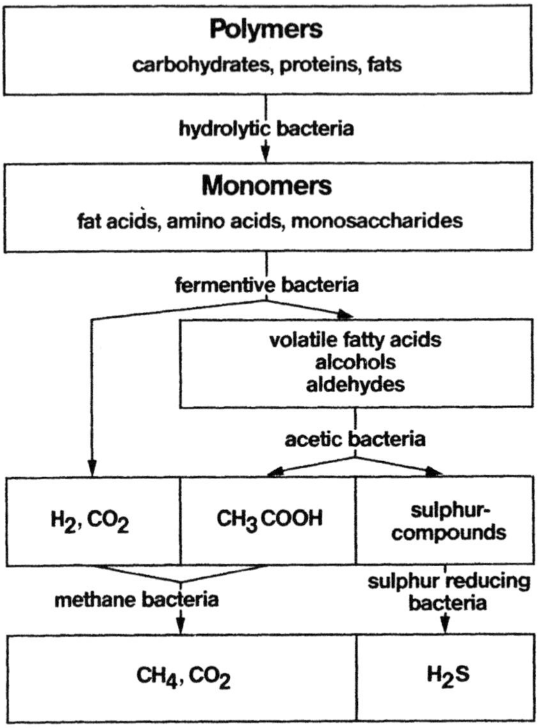

Figure 5. Anaerobic methane recovery from organic materials

2.3.2 Process engineering. From the point of engineering, biogas production is a relatively simple process. Figure 6 shows a scheme of an agricultural biogas plant. The most important part of the plant is an isolated digestor, which can be heated. In the picture you see a vertical digestor of a large biogas plant. A certain amount of diluted manure with about 10 % content of dry matter is fed into it per day. The same amount of digested slurry leaves the digestor and has to be stored before it can be distributed onto the fields. Usually a gasholder is necessary to adjust gas production to the gas consumption. In the picture a water sealed gas holder is illustrated which is a low cost method for gas storage. In order to have high net energy yields, the processing energy needs - that means heating an mechanical energy - should be as low as possible.

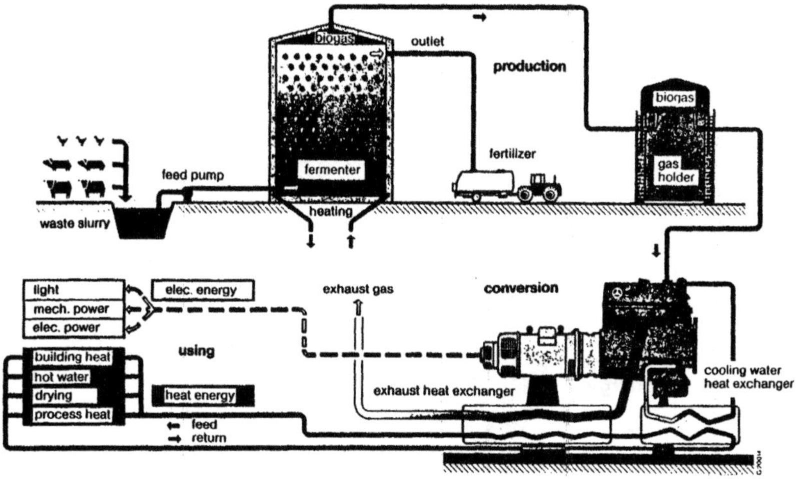

Figure 6. Production, conversion and use of biogas

2.4 Biogas energy in agriculture

2.4.1 Importance. Agricultural biogas plants serve mainly for the production of energy, which in turn allows one to save on classical fuels, for instance oil and electricity. Besides this the biogas process offers some other advantages:

- anaerobic digestion contributes to less pollution of the environment
- fertilizing property of processed animal manure is increased.

2.4.2 Problems concerning total biogas use. The lowest cost of investment is when biogas is used only for heat production. Figure 7 shows a schematic energy balance for this biogas use. In contrast to the heavy fluctuating of heating demand over the year, biogas production has almost a constant level. Because of low heating demand in the summer, surplus energy is created.

This reduces the profitability (economical efficiency) of biogas plants considerably. However, in summer time, a big demand for tractor fuel exists, and as a consequence, the farmers would like to use biogas to run their tractors. But there are many problems here, especially with the storing of gaseous fuels on board these vehicles. In section 5 this will be explained in more details.

Another way to use the surplus energy is to produce electricity from biogas. This makes sense, especially for all those farms, which need a lot of electrical energy throughout the year. This applies particularly to all poultry-farms, which have a high electrical baseload for lighting and ventilation of the chicken-houses.

However, the electrical demand on other farms is rather small. But there exists the possibility of feeding surplus electricity into the public mains.

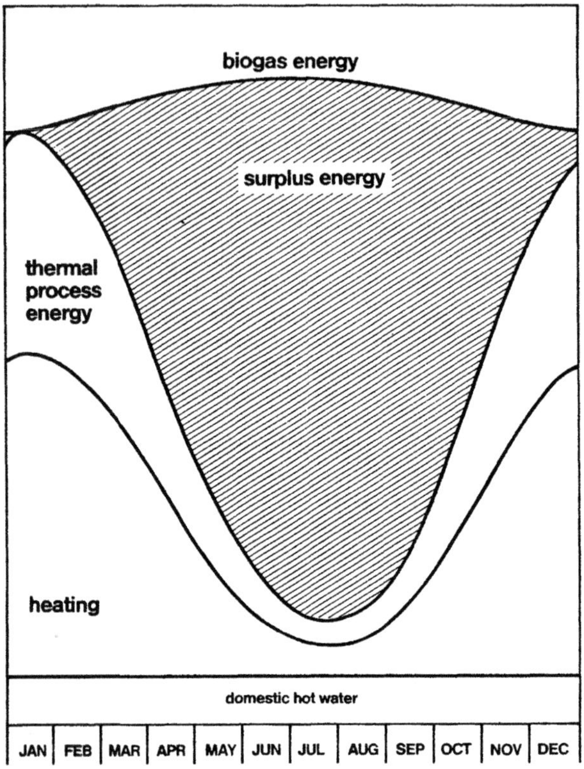

Figure 7. Energy balance for heat generated by burning of biogas

One point that should be mentioned for the electrical production from biogas are combustion engines which are set up to operate with biogas.

3. USE OF BIOGAS IN ENGINES

3.1 Properties of biogas

Biogas can in principle be used to run engines. In table 1 you can see a listing of some important data for biogas.

Biogas has a relatively high content of methane and therefore the biogas/air mixture has a relatively high heating value, which is similar to those of gas/air mixtures. Above all it has an extremely high knocking resistance. The specified methane number of 130 corresponds to an octane number of over 100.

Problems results in the high hydrogensulphide content which can amount up to 0.5 vol. %. This means, that sulfur content in biogas related to fuel energy can be more than 10 times higher than that of gasoline.

Methane	[Vol%]	50–70	
Carbon dioxide	[Vol%]	30–50	
Hydrogen sulphide	[Vol%]	<0,5	
Sulphur content	[g/m³] [g/kWh]	<7.5 <1.2	 [<0.08]*
Mean lower heating value	[kWh/m³]	6.0	[9 kWh/liter]*
Theoretical air demand	[m³/m³]	5.7	[10 m³/liter]*
Heating value of the mixture	[kWh/m³]	0.90	[0.95]*
Methane number		130	
Ignition limit	[Vol% fuel in mixture]	6–12	[0.6–8]*

* Data for gasoline

Table 1. Composition and Properties of Biogas

In principle there are possibilities to remove these components out of the biogas before it is used in engines. Some processes are based on physical and chemical absorption. Recently there have been tests using membranes for separating gas mixtures.

The ability for self ignition of biogas/air mixtures is low. Therefore we have to choose a spark-ignition engine (OTTO-Process) if we want to run an engine with biogas only.

A DIESEL-engine running with biogas has to be modified to operate in a dual-fuel mode. In this case a biogas/air mixture is introduced to the engine and ignited by a certain amount of diesel fuel injected into the cylinders.

Today there are only a few biogas-engines on the market. Normally they are converted from standard gasoline or diesel-engines.

3.2 General modification of standard automobile engines

The modification of a gasoline car engine so that it can run on gaseous fuels is relatively simple. The only change one has to do is to replace the carburetor by a gasmixer with throttle valve. On the market, special biogas-gasmixers are available (figure 8).

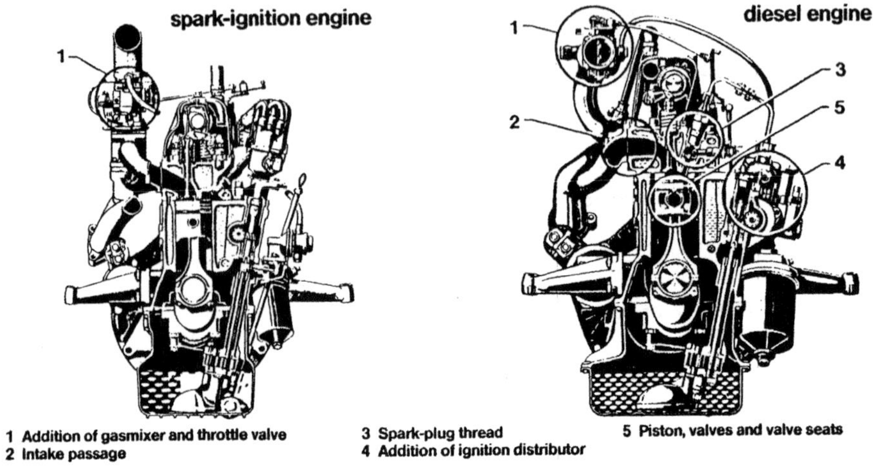

Figure 8. Modifications to basic otto- and diesel engines for operation on gaseous fuels

Much more efforts are necessary to modify a diesel engine running on biogas. On the intake manifold a gasmixer has also to be attached. If the basic engine is a direct injection diesel, we have to replace the inlet port, which provides swirls, by an inlet port with maximum charge.

Instead of the injection nozzle and the prechamber, shown on the picture, a spark-plug thread has to be made. The diesel injection pump is no longer necessary. The ignition distributor of the ignition system can be placed on its position. The compression ratio must be decreased to at least 14 : 1. Thus we have to change the piston size.

3.3 Special modifications of an M 115 spark ignition engine

In the research department of Daimler-Benz, we modified a Mercedes-Benz gasoline engine, type M 115, to run on biogas. The engine has a displacement of 2,3 liters and a compression ratio of 9 : 1. This engine is connected to an electric generator and is now producing electricity from biogas on a pilot biogas plant, set up at a poultry farm. The engine has been in operation since the end of 1981.

Table 2 contains all modifications applied to the standard gasoline engine. To keep the costs as low as possible, we begun only with a few basic modifications.

A gasmixer (IMPCO DG 200), which is essential for external mixture formation, is attached to the intake manifold. To keep maintenance costs low, we used a breakerless, transistorised ignition system and removed the V-belt which drives the generator and the water pump. Because the V-belt has been removed, we are now using a power pack for charging the starterbattery. Likewise an electrical water pump has been attached. Synchronous engine speed is controlled by an electronic governor (HEINZMANN E6), which is coupled to the

Basic modifications

Carburetor →	Gasmixer
Breaker-triggered inductive ignition system →	Breakerless transistorized coil ignition
Engine cooling water pump →	External 220-V-waterpump
Electric generator →	Power pack Governor (n_S = 3000 min^{-1}) Oil cooler

Modifications after operational experience

Standard component	Problem	Improvement
Intake valve guides (Copper alloy)	H_2S-corrosion (Failure after 500 h)	Valve guides made of grey cast iron
Exhaust valve seats	Wear (Failure after 5 000 h)	Valve seats made of different material
Exhaust system	SO_2-corrosion	Exhaust system partially made of stainless steel
Spark plugs	Electrode erosion	Cooler plugs (Operational life >2 000 h)

Table 2. Modifications to a standard spark-ignition engine (type M 115) for operation on biogas

throttle valve. An additional oil cooler is coupled to the engine because there is no airstream.

After the initial operational experience, we had to make further modifications. Many problems, which occured, have been related to the hydrogensulphide content in biogas. We had strong corrosion problems at the intake valve guides, made of copper alloy. With valve guides made of grey cast iron, we obtained a longer lifetime. Corrosion occured also in the exhaust system, especially in the part after the exhaust heat exchanger. Parts of the system have now been constructed of stainless steel.

After 5 000 hours operating time the exhaust valve seats failed. We replaced them by seats of different material. Because of strong electrode erosion of the standard spark plugs we are now using much cooler plugs. They have a lifetime of more than 2 000 hours.

3.4 Engine waste heat recovery

The engine waste heat, which is produced from cooling water and exhaust gas, has to be used for heating. In the exhaust system a heat exchanger extracts heat out of the exhaust gas. With two water circuits the heat is transported to a hot water storage container of 5 000 liters, from which the useful heat can be removed depending on the heat demand (figure 9).

We need a heat storage possibility because the engine is operating in accordance to the electricity demand of the farm.

In this case, the heat is produced in a certain proportion to the electrical load, and has to be stored to adjust heat production to the heat consumption. At night, or whenever the engine stands still, the heat demand has to be provided by discharging the heat storage.

The heat of the oil circuit is also extracted by a heat exchanger

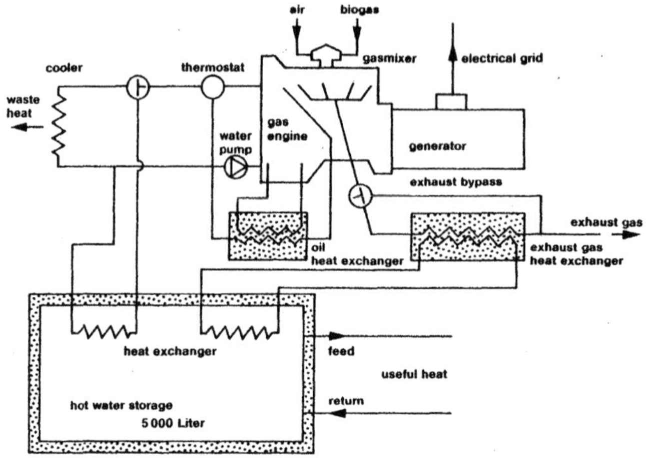

Figure 9. Engine waste heat recovery system

(oil cooler) and fed into the cooling water circuit. Therefore, at full engine load, the oil temperature can be kept constant under 110 °C.

If there is not enough heat demand, the temperature in the heat storage increases. At a temperature of 80 °C we bypass the exhaust gas heat exchanger. Also, we must lead away the cooling water heat over a cooler to the environment.

3.5 Influence of ignition time and equivalence ratio

Besides the hardware modification of the engine, we had to choose other values for equivalence ratio and ignition time. Figure 10 shows schematically the influence of inverse equivalence ratio to engine load, specific fuel consumption and exhaust gas emission.

Additionally rough specific maximum loads at 3 000 rpm. for gasoline, diesel and biogas fueled engines are specified.

Because of external mixture formation a biogas engine sucks 18 % less air - 11 % relates to methane and 7 % to CO_2 - than a diesel engine. At the same equivalence ratio we therefore had to put up with a load reduction. Because diesel engines are operated relatively lean to avoid smoke emission, we have a possibility to compensate load reduction by operating the biogas engine with a rich mixture. But in this case, we have disadvantages, caused by increased emissions and higher fuel consumption and thermal overstress which shortens engine life.

For long operational life the inverse equivalence ratio should be close to 1.3. In this case, we have a good fuel efficiency and somewhat decreased exhaust temperatures (figure 11). The load reduction of about 25 %, related to a diesel engine, can be accepted.

Figure 11 shows also the influence of ignition time on exhaust temperature. The exhaust temperature decreases considerably only

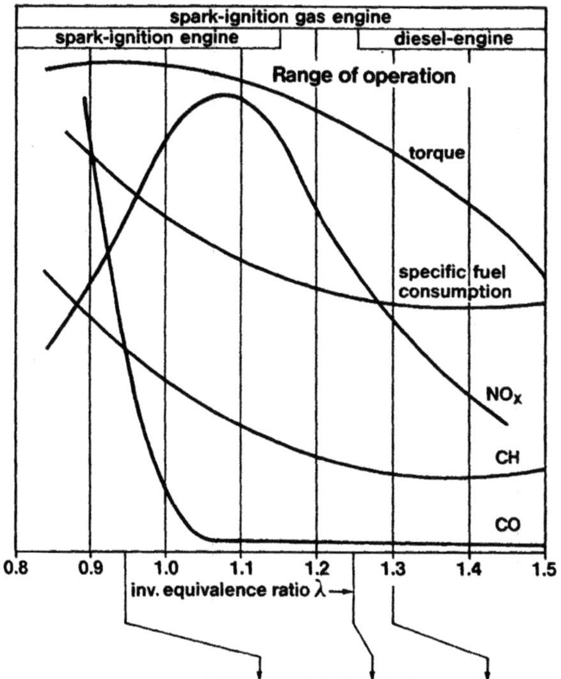

	gasoline	diesel	biogas
Power output per liter at 3000 min^{-1}	~ 25 kW	~ 20 kW	~ 15 kW

Figure 10. Influence of the equivalence ratio on engine operation

at very advanced ignition times, compared to that of gasoline engines. This is due to comparatively low burning velocities of methane/air mixtures. In addition, the CO_2 content in biogas, which has a high heat capacity, can lower the flame front too.

At an ignition angle of 50° before TDC, the engine has best efficiency. It is in the order of 30 % at a full load of 35 kW. Typically for spark ignited engines, there is the strong decrease of efficiency in the partload range (figure 12), partially due to breathing losses. The mixture, adjusted by the gasmixer under practical conditions, tends to become richer as the engine load increases.

The coupled synchronous generator has an efficiency of about 85 %. Therefore we have to multiply the values of figure 12 by 85 % in order to get the electrical efficiency. Maximal electrical

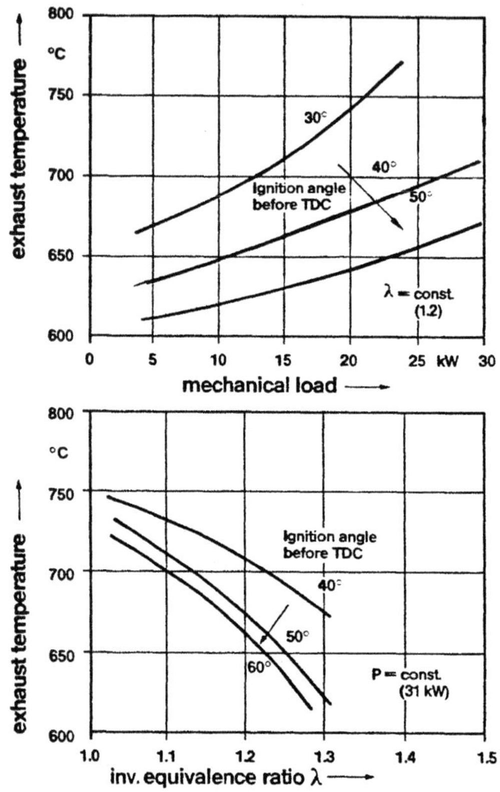

Figure 11. Influence of mechanical load and equivalence ratio on exhaust temperature

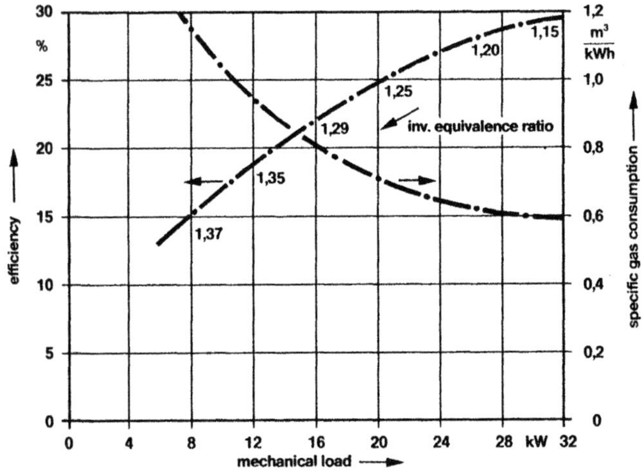

Figure 12. Efficiency and specific gas consumption of the M 115 engine operating on biogas

efficiency of about 25 % is reached at full load conditions. It falls down to only 14 % for 1/3rd full load. As rule of thumb, it can be stated that at half load, 1 kWh electrical energy can be produced from 1 m³ biogas fueled to the engine.

3.6 Problems concerning the gas quality

It happens that the methane content of biogas, especially biogas from landfills, can fluctuate considerably. If this is the case, and the mixture ratio is kept constant by the gas mixer, the result is the equivalence ratio will vary widely and will attain unallowed operation conditions. If the mixture is too rich (low λ), thermal overstress will occur. On the other hand, we will get missfire if the mixture is too lean (high λ). That means irregular operation of the engine, decreasing power output and increasing fuel consumption and hydrocarbon emission (figure 13).

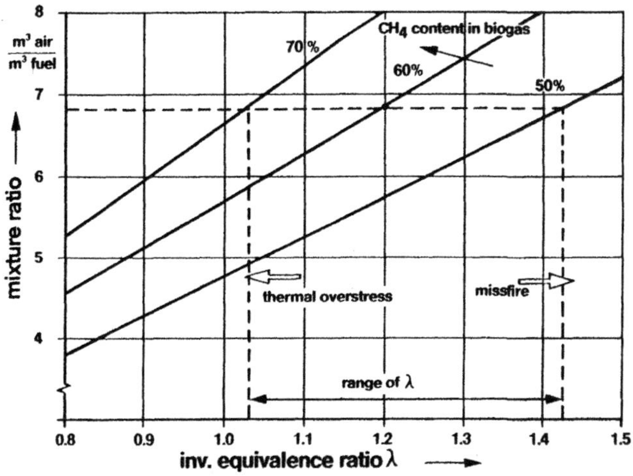

Figure 13. Variation of λ at constant mixture ratio for biogas of various CH_4 content

Therefore it would be advantageous if the gas mixer can be automatically controlled to keep the operation conditions constant. But for this, we need special sensors to pick up engine input and output data.

3.7 Improvement of operation

One of the first steps for mixture controlling is to measure the exhaust gas oxygen content /4/. Because methane has a constant chemical composition, there exists a definite relation between equivalence ratio and exhaust gas oxygen content (figure 14). For measuring this oxygen content, so-called λ-probes can be used. The classical λ-probes are not suitable. They are currently connected to

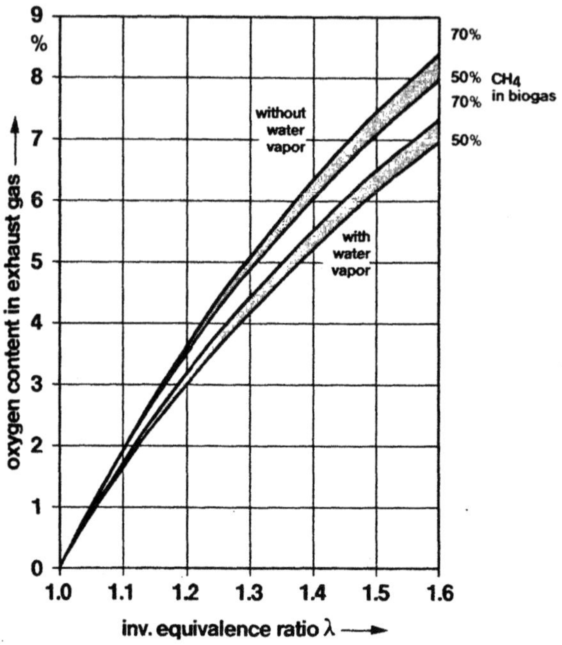

Figure 14. Relation between equivalence ratio and exhaust gas oxygen content in a biogas engine

cars to reduce the exhaust pollution and can detect only the transition from oxygen containing to oxygen free exhaust gas ($\lambda = 1$).

For the first experiments we used a prototype of a λ-probe which is based on a zirconiumoxide solid electrolyte, but operates like an oxygen ion pump and is able to detect oxygen rates up to over 20 % in a gas mixture. These probes have been developed to measure oxygen contents in exhaust gas in order to control the equivalence ratio of boilers fueled by natural gas and fuel oil /5/.

The current experiments with these probes are very promising and will be continued.

Beside this mixture controlling, we expect further improvements by the other measures mapped out on table 3. We are developping a modified ignition system without a high tension distributor in order to get even lower maintenance costs and better reliability.

Increase of compression ratio will improve efficiency and therefore lowers the specific fuel consumption. Because of high knocking resistance of biogas we expect that it will be possible to utilize compression ratios of the order of 14 : 1.

We hope that a watercooled exhaust manifold will decrease the thermal stress in the cylinder head region. Besides this, the thermal efficiency, which is now in the order of 55 %, will be further increased. By way of experimenting we are planning to accelerate the burning velocity by adding a certain amount of hydrogen to biogas.

If wanted, the required hydrogen can be produced by split-cracking out of biogas.

Near-term plans	**Increase of compression ratio**
	Ignition system without high tension distributor
	Watercooled exhaust manifold
	Control of mixture formation by measuring the oxygen content in the exhaust gas
Further experiments	**Acceleration of combustion by addition of hydrogen produced in a gas generator**

Table 3. Possibilities for improving the operation of stationary biogas engines

Another problem is the rapid decrease of engine oil lubricity caused by acid combustion products especially sulfurdioxide. The total base number (TBN), which is a degree for the buffering effect of the oil, decreases by 40 % within 250 hours of operation. After this time, the oil has to be changed. To increase this period, the hydrogensulphide has to be removed out of the biogas.

4. ELECTRICITY PRODUCTION FROM BIOGAS

4.1 Operation conditions of the cogeneration set

As mentioned before, our engine is used for producing electricity from biogas on a poultry farm near Stuttgart. From the manure of 12 000 chickens, about 250 m^3 biogas is produced per day. The methane content is nearly 60 % and the hydrogensulphide content is about 0.2 %. Poultry farms should have a standby power unit. Therfore the engine is coupled with synchronous generator which is able to supply the farm grid with electricity without being connected to the public power supply grid. But normally the cogeneration set operates in parallel connection to the grid. This operation mode guarantees a large reliability, and peak load demands of the farm exceeding the maximum output of the power unit can be delivered from the grid. The power unit can be controlled in a manner, so that it never exceeds the demand of farm's electricity needs.

The daily operation time is between 6.00 h a.m. and 10.00 h p.m., that means, 16 hours per day. During this time the total electricity demand of the farm is supplied, and thus the daily biogas production is totally used.

Meanvalue of daily electrical load is about 16 kW. That means about 50 % full load of the power unit. As mentioned before, we can produce 1 kWh electricity from 1 m^3 biogas in this operation mode. In the case of the highest efficiency (full load of the power unit) we only need 0.7 m^3 biogas to produce 1 kWh.

Over one year we have a total operation time of about 5.500 h which corresponds to a driving distance of about 270 000 km.

4.2 Energy balance

At half load operation, we have a thermal power output of about 43 kW at our disposal as shown in figure 15. This is enough for heating farm buildings and digestor even at low ambient temperatures. The total efficiency related to heat and electricity production is about 73 %.

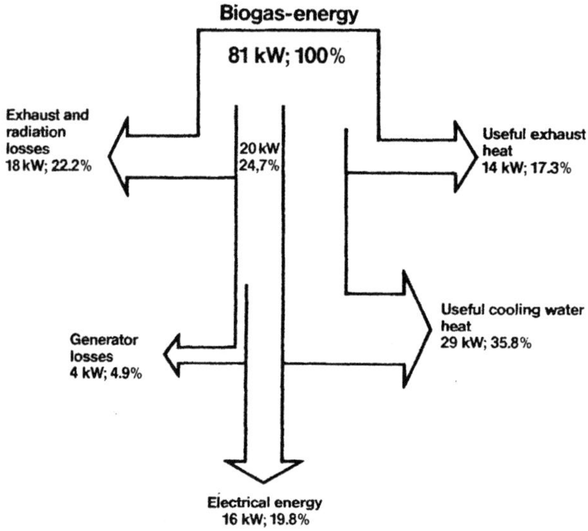

Figure 15. Energy balance at average operating conditions

Figure 16 shows an averaged daily energy balance from February 83. During the heating period 82/83 this month had the highest heating demand. Nevertheless, total heat consumption could be satisfied by engine waste heat. Total electricity consumption could be supplied at a rate of 85 %.

Another factor influencing the economic efficiency of the power plant is the amount of process energy demand, which plays an important role. In this case, the process energy demand is relatively low. Digestor heating needs only 15 % of total biogas energy. The electrical energy demand for stirring and pumping is under 10 % of the self produced electricity. Electricity supplied from the grid occurs at night, when the power unit is not in operation.

At the end of July 1984 we had a total operation time in the order of 12 000 hours. During this time we produced 200 000 kWh electricity and used engine waste heat equivalent to over 20 000 liters of fuel oil.

The plant is working near the breakeven point of economy.

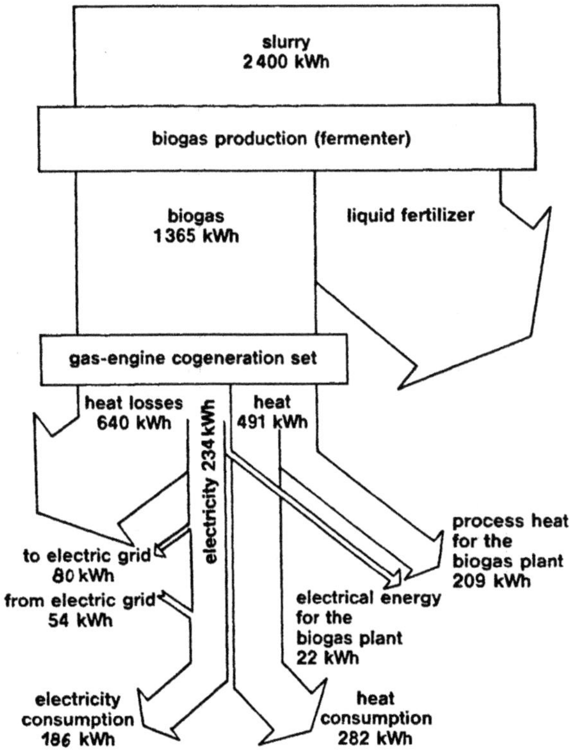

Figure 16. Averaged daily energy balance (February 83)

5. USE OF BIOGAS FOR TRANSPORTATION

Another possibility for using biogas surplus is to drive biogas vehicles. Tractors, which are currently driven by diesel fuel, are well suited for this purpose. However, the following three problem fields must be solved:

- cleaning of biogas
- storage of biogas on board such vehicles
- tractor engines operating with biogas.

To avoid corrosion problems we have to remove hydrogensulphide and water vapor out of the biogas. To increase energy storage density parts of the carbondioxide have to be removed, as well.

The key problem is the biogas storage on board the tractor. Basically the following methods can be used:

- Liquid natural gas (LNG) storage. We have to cool methane down to -161 °C to get LNG. In this case we have to use very expensive dewar containers on the tractor. In addition we will have high demand of investment and energy for liquefaction as well as problems with methane evaporation rates. This method seems to make sense only on a large scale.

- Another method is to compress methane to moderate pressures (up to 50 bar) in order to store it in molecular sieves or other adsorbents. In this field, there exist up to now only poor experimental results.
- What is currently used is the pressure storage method with up to 200 bar. In this case we have to invest in pressure vessels and a gas filling station for pressures up to 300 bar.

Despite safety problems and difficulties in getting an operation authorization, it seems that the pressure storage method can be realized today with minimal efforts. But this method has very low energy densities. Even by using filament-reinforced aluminium pressure vessels, which have about one half the weight of conventional metal cylinders, the energy densities are only in the order of about 1.2 kWh/kg or 1.1 kWh/liter.

As a consequence the biogas fueled vehicle would have great reductions in distance and loading capacity vis à vis a diesel driven vehicle. If we want to have the same distance as a diesel vehicle, the required biogas storage unit would have at least 8 times more volume and weight.

Another problem is that the biogas surplus will not usually be sufficient to run the tractor for the whole summer. Therefore the engine should to operate with two different kind of fuels. Diesel engines, operating on a dual-fuel mode, are well suited. Those engines can be operated with 100 % diesel fuel as well as only 20 to 40 % diesel fuel delivery to ignite the biogas/air mixture in the cylinders.

When everything is considered, it seems to make more sense today to use biogas in a stationary power plant than to drive vehicles. In this way oil fuels, used on the farm to produce heat, can be substituted . Theoretically they could be used for driving vehicles, without any problems, with the end effect of net savings in oil fuel costs.

6. REFERENCES

/1/ Handbuch "Umwelt und Energie"
Rudolf Haufe Verlag, Freiburg, 2/441

/2/ W. Palz; P. Chartier:
Energy from biomass in Europe
Applied Science Publishers Ltd., London

/3/ C. F. Siegfried; H. Sixt:
'Behandlung organisch verschmutzter Abwässer mit anaeroben Verfahren zur Minimierung des Energieaufwandes'
13. Essener Tagung vom 05.03. - 07.03.80

/4/ M. Leuchs
'Probleme beim Motorenbetrieb mit schwankender Gasqualität'
VDI-Berichte 459

/5/ E. Lüth; F.-J. Rohr:
'Untersuchung der Einsatzmöglichkeiten einer Lambda-Sonde für die Sauerstoffkonzentrationsmessung und Luftzahlregelung an Gas- und Ölheizkesseln'
34, HLH, 3

AUTOMOTIVE SYNFUELS: A POSSIBLE STRATEGY FOR THE FUTURE

P.P. Garibaldi
ASSORENI
Via Fabiani, 1
20097 San Donato Milanese (MI)
Italy

ABSTRACT. Transportation is an essential factor to be taken into account in evaluating the energy demand of industrialized countries. The present structure of the transport sector including fuel consumption and composition of the car park is reported for the main western european Countries as a baseline to set up a future energy strategy. The objective of this strategy is neither a reduction of the consumption by means of mileage reduction or by shifting from private to public transport nor a forced replacement of oil with other energy sources. This new proposal is aimed at the optimisation of the engine-fuel system from the energetic point of view in order to achieve the rationalization of the energy use in transports. A research study carried out by EEC is used as an introduction. In this work the determination of the best balance between the compression ratio of the engine and the fuel quality has been attempted in order to minimize the total energy consumption per kilometer. The results of this study are now obsolete due to the forecast impact of synfuels on the future energy scene. The different possibilities of utilization of synfuels for automotive vehicles are therefore investigated in the light of the experience gained during these last 10 years in order to reach the objectives of energy policy that are pratically the same for almost all the european Countries. Some solutions related to both the Otto cycle engine and the Diesel cycle engine are considered. A long term research and development program on a new approach to automotive transport is proposed.

INTRODUCTION

Automotive transportation accounts for a considerable share of the energy demand in the industrialized Countries.

G. E. Beghi (ed.), Synthetic Fuels, 323–348.

The present structure of the automotive population in the major european Countries is reported in terms of its composition and fuel consumption as the base line for setting up an energy strategy and examining its possible effects.

The fuel consumption in automotive applications in 1981 for some european Countries is reported in table I /1/ and the car park composition is given in table II /2, 3/.

TABLE I. TRANSPORTATION BRANCH-ENERGY CONSUMPTION (10^6 toe) (1981) /1/

	EUR-10	ITALY	FRANCE	WEST GERMANY	UNITED KINGDOM
Transportation branch energy consumption (a)	152.3	25.4	32.2	39.3	32.0
Gross inland consumption (b)	909.9	130.4	180.5	257.8	194.4
(a)/(b) %	16.7	19.5	17.8	15.2	16.5

TABLE II. THE CAR POPULATION IN MAJOR EUROPEAN COUNTRIES (10^3) (1980) /2,3/

	ITALY	FRANCE	GERMANY	U.K.
Otto engine cars	17,070	18,231	22,052	15,422
Diesel cars	430	899	1,185	15.5
Diesel cars %	2,5	4,7	5,1	0,1
Total cars	17,500	19,130	23,237	15,438
Industrial vehicles	1,547	1,730	2,650	1,913

In the transport sector as well as in other applications characterized by a high energy demand (power stations, residential heating, industry) a great effort was made in the last ten years for reducing the energy consumption in general and the oil consumption in particular.

The main objectives of the various Countries in terms of energy policy can be summarized in three general issues:

1. To rationalize the use of the energy sources.
2. To maximize the use of the domestic energy sources.
3. To diversify the imported energy sources.

There is one additional objective which applies to the automotive field and is posed by geografical constraints:

4. To enable the free circulation within the EEC in particular and within the european territory in general.

The synfuels can play in the future a determining role in achieving the above objectives.

A risky and difficult global strategy is imposed by the simultaneous actions needed, however the synergism derived by the achievement of those four objectives appears to offer a great benefit, well above the overall results obtainable separately.

The objectives are at first examined one by one and the possible use of the synfuels is evaluated in relation to the achievement of the objectives and the anticipated effects.

Next, a synfuels strategy is proposed that should enable the concurrent achievement of the four objectives.

This approach to the problem is obviously an attempt which requires a careful cross-check with all the interested parties.

An accurate technico-economical assessment of this strategy could be carried out in each Country and be coordinated by the European Community.

The research, development and demonstration programs suggested by the results obtained could then be initiated in order to prepare for the long term the appropriate technical and technological solutions. The advancement into the so called "commercial" phase would obviously take place when required by the overall social-political-economical conditions.

FIRST OBJECTIVE: Rationalized use of the energy reserves

Rationalization of the use of the energy sources in the transport sector means to optimize the ENGINE-FUEL system or in more general terms the CAR POPULATION-REFINERY system as to minimize the overall energy consumption per unit distance for both the Otto engine and the Diesel engine, these being the two type of engines that will last for a long while in road transport.

Otto engine

The approach to the energy rationalization was first reported at the European Motor Vehicle Symposium in Brussels in 1975 /4/.

On the basis of the energy scenario at that time and of a number of assumptions, the energy required for covering 10 Kms was calculated by considering the fuel consumption of the engine together with the energy required as refinery process heat for obtaining that amount of fuel.

The calculation was extended to several cases of the "ENGINE-FUEL" system having as variables the Research Octane Number (RON) of the fuel and the engine compression ratio appropriate for that particular RON.

The RON determines the energy consumption in the production of the fuel, while the compression ratio determines the fuel consumption of the engine.

The results obtained are reported in table III and illustrated in fig. 1.

The validity of this approach has been testified by the natural development of the system that has evolved as follows:

- RON at 97-98 for a high octane requirement car park of low specific fuel consumption in areas where the lead level is 0.4 g/l (i.e. Italy).
- RON at 92-93 with higher specific fuel consumption where the maximum lead level is 0.15 g/l (i.e. W. Germany, whose car population consumes regular-grade gasoline in large quantities).
- RON at 89-91 with even higher specific fuel consumption where unleaded gasoline is sold (i.e. US).

Although ten years have elapsed, this approach is still applicable. Obviously the boundary limits have changed. However the changes occurred are still favoring this approach and justify its further theoretical and experimental development.

The most important changes occurred meanwhile are:

- the possible total removal of lead from gasoline also in Europe
- the probable future availability of synfuels at high octane rating.

The possible total removal of lead from gasoline will have an important effect on the present refining structure.

The development of refining in the 60's was based on the antiknock effect achievable with the lead alkyls; indeed, the most favored refining processes were those leading to the iso-paraffins which are highly respondent to the lead alkyl. In example, the alkylate produced by reacting an olefin and a paraffin has a high susceptibility to lead even at low concentrations (fig. 2).

TABLE III. ACTUAL ENERGY CONSUMED IN COVERING 10 KM (*)

RON	Percentage decrease in engine consumption	Apparent engine consumption (Joule/10 km) (**)	Internal consumption at refinery (Joule/kg)		Actual consumption (Joule/10 km)	
			Pb (g/litre)		Pb (g/litre)	
			0.4	0.15	0.4	0.15
90	0.0	43524	5148	6319	48672	49843
91	1.3	42917	5491	6637	48337	49467
92	2.6	42310	5888	7048	48044	49174
93	3.9	41708	6290	7508	47751	48923
94	4.9	41231	6550	7914	47458	48755
95	5.9	40758	6943	8412	47290	48672
96	6.9	40289	7348	9094	47123	48755
97	7.9	39816	7889	9843	47081	48881
98	8.7	39431	8424	10534	47123	49048
99	9.4	39092	9048	-	47290	-
100	10.0	38795	9672	-	47500	-

(*) The calculations are based on a private motor vehicle which consumes 1 kg of gasoline in covering 10 km.

(**) The calculation of apparent engine consumption (column. 3) takes into account the variation of calorific values with ONs.

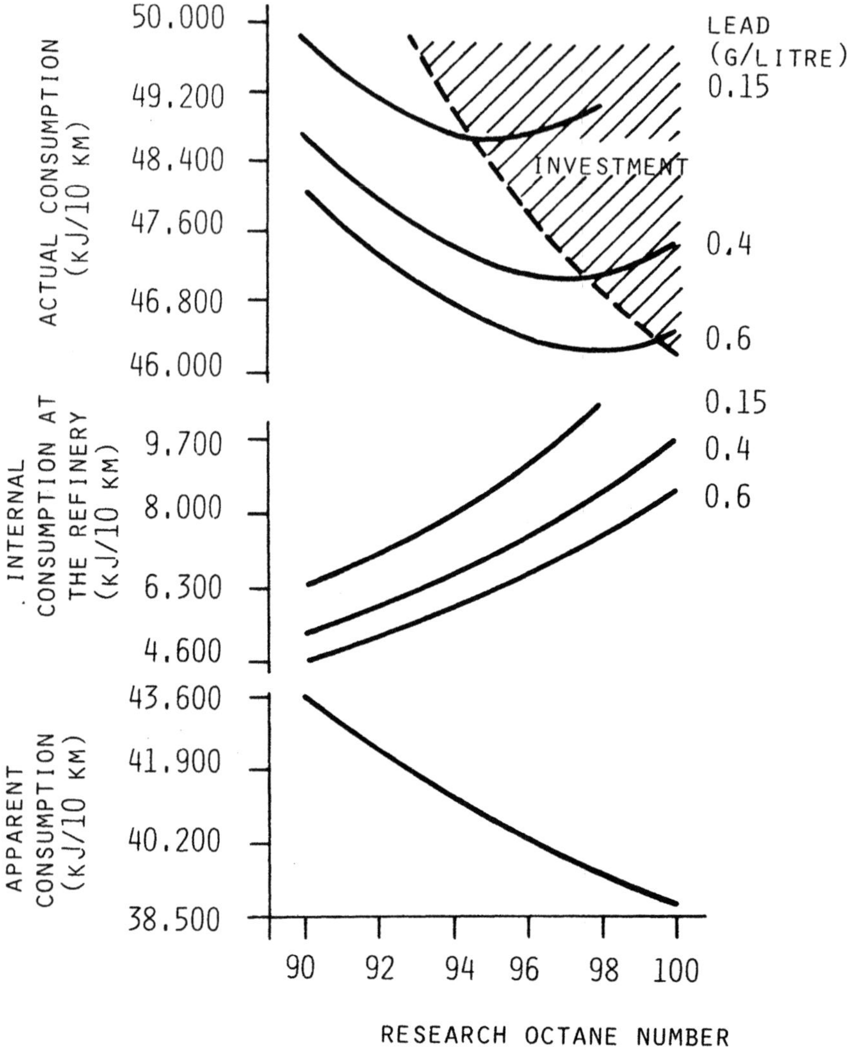

Figure 1. Variation in apparent and actual consumption (consumption in the engine and at the refinery) with O N.

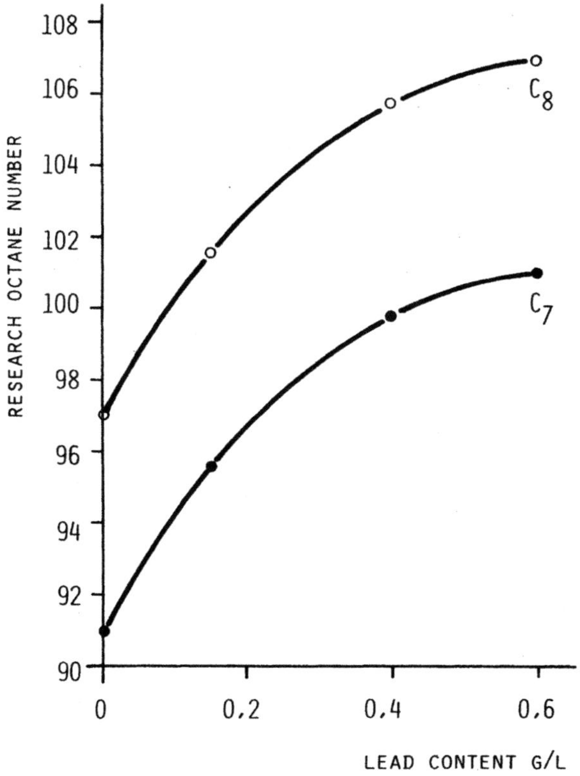

Figure 2. Susceptibility to lead of C_7-C_8 alkylates.

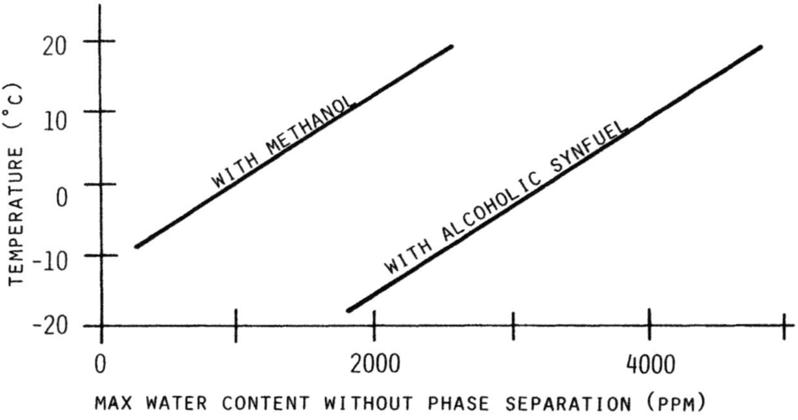

Figure 3. Comparison between methanol and alcoholic synfuel using blends composed of 80%v hydrocarbons and 20%v alcohol (base fuel composition: A = 29%, O = 6%, S = 65%).

As a result, the total removal of lead in a standard refinery would shift the minimum energy consumption of Fig. 1 to very low RON's only suitable for low compression ratio, poor efficiency engines resulting in an overall energy consumption per 10 Kms well in excess of the 49,400 Joule.

The production of high octane unleaded gasoline would in fact require a high processing energy consumption besides important investments for new equipment.

A part from the necessary investments, a calculation based on a refinery mathematical modelling has shown that the production of 1 kg of 98 RON unleaded gasoline with an energy content of 43,000 Joule would require a process energy of 13,600 Joule.

This indicates that the traditional petroleum system is limited in very narrow boundaries.

The possible availability of high octane synfuels as premium gasoline components could drastically change the system by enabling the production of high octane blends with a rather small internal refinery consumption for the production of the hydrocarbon base.

So far the energy consumption for the synfuels production has not been accounted for because the starting raw materials (coal, tar sands, oil shale, biomass) have an energy value much lower than that of oil.

Indeed, the production cost of synfuels should be considered.

The antiknock power of some types of synfuels is reported in table IV as blending ON values without any contribution of alkyl lead additives.

TABLE IV. INDICATIVE LEVELS OF ANTIKNOCK POWER OF SYNFUELS

	RON	MON
Methanol *	120-150	90-120
Ethanol *	110-135	90-115
MTBE *	115-135	98-110
Snamprogetti MAS *	117-132	90-100
Mobil gasoline (unleaded)	93	83
F-T gasoline + 0.3 gPb/l	94.6	83.8

* Blending Octane Number

For better clarifying this issue, the results of a study of the impact of methanol on several refining schemes carried out in Italy as part of the CNR-Finalized Energy Project N° 1 /5/ are reported.

The unleaded RON of the hydrocarbon streams obtained in a T-CR-CC-A refinery for the production of 98-99 RON gasoline as a function of lead and methanol content are reported (tab. V).

It must be pointed out that in some cases the blends do not meet the present MON specifications.

TABLE V. CLEAR RON OF BASE GASOLINES FROM T+CR+CC+A REFINERY SCHEME (1, 2)

Lead content	g/l	0.6	0.4		0.15		0	
Methanol (3)	%vol	zero	10	20	10	20	10	20
Base gasoline clear RON		91.2	88.8	81.3	91.5	84.5	96	92.1

(1) Final blend RON is 98-99 in all cases

(2) Final blend MON ranges between 85.3 and 88.5

(3) European Community is approving a directive that will regulate the use of oxygenated compounds as gasoline components. In such directive the maximum content of oxygenated compounds in gasoline is established at 10% volume, including methanol that must not exceed 3%.
Therefore, being the data reported in the table relative to 10 and 20% of methanol, their practical use will not be possible in the nearest situation, but probably only in a future perspective.

These results show that the use of methanol at 20% vol would enable the production of premium unleaded fuel (if the consequent reduction of MON would be accepted) without increasing the severity of the conversion processes for octane boosting and therefore without penalizing the situation in terms of energy consumption even when the lead content was 0.6 g/l.

The use of synfuels would therefore enable to remove one of the constraints of the traditional system i.e. the correlation "RON-PROCESS ENERGY REQUIRED".

The energy value of synfuels as high octane components has been derived for a synfuel of given composition:

Methanol	80%vol
Ethanol	4%vol
n-Propanol	6%vol
iso-Butanol	10%vol

This blend, successfully engine tested against pure methanol, results in a higher stability of the alcohol fuel-gasoline mixture in presence of trace water (fig. 3)/6/.

The calculation has been performed assuming a particular energy scenario projected to 1995 (tab. VI) by assuming to satisfy a given demand only with oil or with the contribution of the alcoholic synfuel.

TABLE VI. FORECAST OF DOMESTIC OIL PRODUCTS CONSUMPTION IN ITALY

	10^6 MTpY		% OF TOTAL	
	1985	1995	1985	1995
LPG	1.4	1.6	1.2	1.5
Premium gasoline	11.2	13.3	9.9	12.7
Regular gasoline	1.3	2.4	1.2	2.3
TOTAL GASOLINE	12.5	15.7	11.1	15.0
Petrochem. Feedstock	8.5	12.4	7.5	11.8
Kerosene, jet fuel	4.5	5.2	4.0	5.0
Diesel gasoil	10.0	12.2	8.8	11.6
Heating gasoil	18.3	18.3	16.2	17.4
TOTAL GASOIL	28.3	30.5	25.0	29.0
Fuel oil	57.8	39.6	51.2	37.7
Total	113.0	105.0	100.0	100.0

The most significant results, reported in tab. VII /6/, show that for the study case of total removal of lead, one ton of alcoholic synfuel can replace two tons of crude oil or, in other words, one Joule from alcoholic synfuel is equivalent to 4 Joules from oil.

TABLE VII. CRUDE OIL SAVING BY USING ALCOHOLIC SYNFUEL/GASOLINE BLENDS - YEAR 1995 (10^6 MTpY) /6/

Lead content in gasoline	g/l	0.6	0.4	0.15	0
Crude oil consumption					
- without alcoholic synfuel		112.3	112.98	113.65	118.35
- with alcoholic synfuel		-	109.82	109.92	111.92
Crude oil saving		-	3.16	3.73	6.43
Alcoholic synfuel input		-	3.10	3.10	3.11
Crude oil saved per alcoholic synfuel utilized	t/t	-	1.02	1.20	2.07
	J/J	-	2.0	2.4	4.1

Diesel Engine

The Diesel engine has a higher overall efficiency compared to the Otto engine because the specific engine fuel consumption is lower and the refinery energy consumption for the production of diesel fuel is also lower than that required for obtaining gasoline.

A comparison of gasoline versus diesel fuel in terms of energy consumption in fuel production is reported in table VIII /7/.

TABLE VIII. REFINERY ENERGY CONSUMPTION IN FUEL PRODUCTION (kJ/kg) /7/

Gasoline			
Pb g/l	0.6	0.4	0.15
RON			
- 90	4603	5148	6319
- 96	6575	7348	9094
- 98	7462	8424	10534
- 100	8508	9672	-
Diesel fuel			
S % wt			
- 0.5	1465		
- 0.05	2971		

The above figures would indicate that the energy rationalization in the transport sector implies a total dieselization of the car park.

Actually almost all commercial vehicles are equipped with diesel engines, while only few percent (tab. II) are the passenger diesel cars in the total car park.

Dieselization is however blooming in the present market of passenger cars.

From the point of view of the customer the comparison Diesel vs Otto for passenger cars is more complex because performance is an important factor to be considered besides the energy aspect.

The data in tab. VIII should however be given great attention.

It should be noted that these data refer to the present european refinery which provides the middle distillates only from topping which is a low energy process and the yield is approximately 30% on crude oil processed.

Should the diesel fuel demand increase in excess of a certain threshold, much more expensive conversion and upgrading processes would be needed.

Therefore from the energy point of view, the present system based on oil would not allow the dieselization to expand beyond a certain limit notwithstanding the favorable engine consumption.

The results of a study concerning the effect of dieselization on crude oil consumption in the present domestic energy scenario are of special significance (fig. 4) /7/.

If we assume a substitution ratio of premium gasoline to diesel fuel equal to 1 to 0.9, the replacement of Otto engine with Diesel engine cars would result in a reduction of oil consumption up to 12% dieselization ratio; above this value the crude oil consumption would start to increase.

The use of synfuels in the transport sector could render the rate of dieselization independent of the present refinery structure and take advantage of the higher efficiency of the diesel engine.

The synfuels could either directly substitute or extend the oil derived diesel fuels or provide an indirect benefit by substituting the middle distillate in the residential heating, thus making available a corresponding share of oil derivatives to be used as diesel fuel.

Due to their properties, particularly in terms of cetane number, synfuels to be used directly in diesel engines are:

- middle distillates from F-T synthesis
- middle distillates from shale oil up-grading

As a conclusion for the first objective, improved energy utilization in automotive sector can be achieved by the introduction of synfuels in both Otto and Diesel engine cars:

Otto engine : high octane components

Diesel engine : high cetane number fuels (or indirectly shifting the oil derivatives from domestic heating to diesel).

SECOND OBJECTIVE: Maximized use of the domestic energy sources

This issue is particularly important for Countries with lack of energetic raw materials.

So far the greatest efforts have been made both at R&D and industrial level in the "fixed energy" sector such as thermal power generation, residential heating and industrial energy consumption.

The domestic energy sources to which much attention is given are as follows:

- nuclear energy, which is regarded as a domestic source even if the fuel comes from abroad
- geothermal energy
- wind energy
- solar energy
- use of solid or slurry fuels for electric power generation
- use of biomass (biogas, etc.)

In the transport sector the most significant examples of commercial scale exploitation of domestic energy sources, alternative to oil and natural gas, are the Sasol in South Africa, the gasohol in USA and the alcohol project in Brazil.

R&D studies are in progress in many other Countries.

The activities in some european Countries aimed at producing and using synfuels for transport from domestic sources are briefly reviewed.

Italy

A feasibility study for the production of ethanol fuel grade from biomass with a preliminary evaluation of the availability of sugar, starch and cellulose containing raw materials was carried out in 1976 /8/.

The energy balance of the entire ethanol production cycle (cultivation, harvesting, fermentation, purification, byproducts, etc.) has been evaluated for the sugar beet.

The production of ethanol from rise leaves and corn cob is presently under investigation.

The feasibility of using the Sulcis coal for methanol production has also been studied /10/.

Important research has evaluated the use of methanol and alcoholic mixtures in automotive applications /5, 11, 12/.

A novel MAS (Methanol and higher alcohols) production process has reached the demonstration stage /13, 14/.

France

Beginning 1981 the Government introduced the "Carburol" program which is finalized in particular to the use of methanol (mainly from import raw materials) and a mixture of butanol and acetone (acetonbutilique) from biomass /15/.

This very wide program is conducted by the major oil companies and car manufacturers that have already supplied the Government with the technical indications to be followed when introducing low concentrations of the above high octane components in gasoline.

West Germany

The studies conducted in this country are probably the most advanced not only in Europe and are directed towards the use of methanol both at 15% in gasoline and as neat alcohol in engines tailored for this type of fuel. Although to a lesser degree the use of methanol as diesel fuel extender has also been investigated /16, 17/.

The demonstration program run in 1979-82 with over 1,300 vehicles was partly sponsored by the German Government which then promoted an agreement among the oil and car industries for the "voluntary" use of low (3-5%) alcohol concentration in gasoline. This is now current practice.

Another agreement has been reached between the Government and the car builders that, as of beginning 1982, produce cars suitable for this type of gasoline-methanol blends by selecting the appropriate materials.

Norway

The Norvegian program, coordinated by the Norvegian Methanol Group (NMG), has examined the use of methanol in gasoline at 4% and 15%v

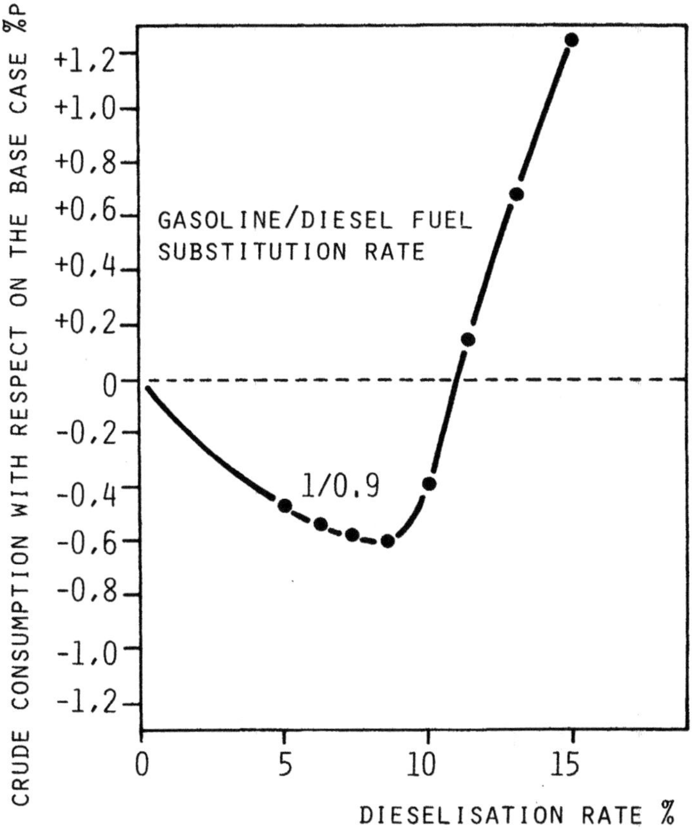

Figure 4. Impact of dieselisation rate on crude consumption.

concentration. The field trials with 70 cars have been conducted in the vicinity of Oslo (15%) and Bergen (4%). Methanol should be derived from natural gas produced offshore /18/.

Sweden

A wide program is underway since 1975 and is coordinated by SAAB, which is an organization created by the Government and the Volvo Company. This program, mainly concerned with the use of methanol at 15%, has as primary objectives to demonstrate the feasibility of supplying the gasoline-methanol blends in the existing distribution system and to suggest to the engine builders all the necessary technical measures to be taken for optimizing the cars running on this type of fuel.

As in Germany, also in Sweden the use of alcohol in diesel engines has been considered.

Special care has also been devoted to the examination of the toxicological implications posed by the use of the new fuels and their possible environmental impact at different concentrations (exhaust pollution in particular) /19, 20/.

U.K. and Ireland

These Countries have cooperated extensively with New Zealand which has considered the use of methanol as such and as a gasoline component.

Methanol in New Zealand is derived from the natural gas of the Maui offshore field and its favored use is through the "Methanol to Gasoline" MTG process of the Mobil Oil Company /21, 22/.

Ireland is also interested in ethanol production from crops.

THIRD OBJECTIVE: Diversification of the import energy sources

This is a very important objective of energy policy in Countries, like Italy, which are highly dependent on imports for meeting their energy requirements (tab. IX) /23/.

A sound energy policy should try to diversify as much as possible

(i) the geografic areas
(ii) the geopolitical and currency areas
(iii) the types of raw materials

A wider selection of technico-economical solutions for the raw materials (iii) will obviously lead to more freedom in the selection of (i) and (ii) which have a mainly political bearing.

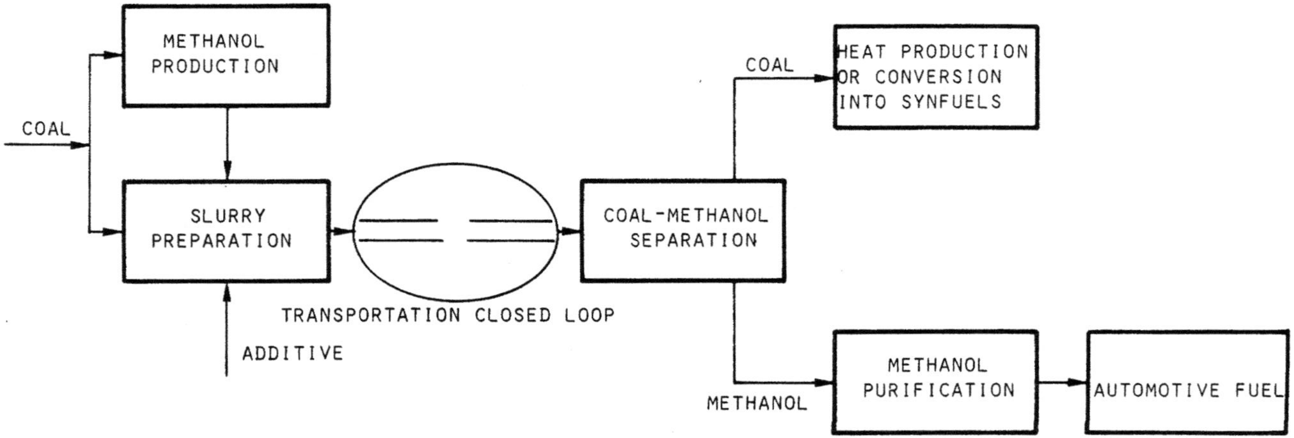

Figure 5. Coal-methanol slurry.

TABLE IX. PRIMARY ENERGY SOURCES (Italy) /23/

	1980			
	10^6 toe	%	Domestic produc. %	Import. %
Solid fuels	12.4	8.5	8.9	91.1
Natural gas	23.0	15.5	44.8	55.2
Crude oil	98.8	67.2	1.8	98.2
Hydro-geo electricity	10.9	7.6	100.0	-
Nuclear electricity	0.5	0.3	100.0	-
Import. electricity	1.3	0.9	-	100.0
Total	146.9	100.0	16.8	83.2

The entire spectrum of synfuels can effectively contribute to the diversification of the energy sources; this applies when the imported raw materials are converted in synfuels by the consumer Country as well as when the synfuels are imported directly due to economical advantages.

Intermediate alternatives are obviously possible when a partial conversion is accomplished in the producing Country, and the intermediate product is supplied to the consumer Country where the production cycle is completed.

Whatever solution is favored, an energy policy defined in close cooperation between the producing and the consuming Countries will be beneficial to both.

The methanol-coal slurry is a typical example of the contribution of synfuels to the diversification of the energy sources.

The methanol-coal slurry is produced as close as possible to the mine following the scheme of fig. 5.

The slurry is then moved through a closed system "pipe-vessel-pipe" to the consumers where the two components are separated and sent to the desired utilization.

Another example is given by the tar sands or the oil shales.

The recovery of tar and its upgrading to a stable finished product for transportation to the consumer Country is accomplished at the mine.

A list of synfuels that can contribute to the diversification of the energy sources in automotive sector (directly or indirectly) is reported in table X.

TABLE X. SYNFUELS

OIL PRODUCT	SYNFUEL	ALTERNATIVE SOURCE
1. GASOIL	1.1 Methanol	Coal
	1.2 Middle distillate from coal liquefaction	Coal
	1.3 Middle distillate from bitumen upgrading	Tar sand
2. DIESEL FUEL	2.1 Middle distillate from F.T. process	Coal
	2.2 Middle distillate from shale oil upgrading	Oil shale
3. GASOLINE	3.1 Methanol	Coal
	3.2 Ethanol	Biomass
	3.3 n Butanol	Biomass
	3.4 MAS (Methanol and higher alcohols)	Coal
	3.5 F.T gasoline	Coal
	3.6 Gasoline from MTG	Coal
	3.7 MTBE	Coal
	3.8 Gasoline from direct coal liquefaction	Coal

FOURTH OBJECTIVE: to enable the free circulation within the EEC in particular and within the European territory in general

To drastically modify the energy system of a Country but only relative to the fixed plants (electric power generation, residential heating, industrial heat) simply poses a number of technico-economical problems.

Conversely, the transport system in the european Countries has an important constraint posed by the need of free circulation of cars and trucks within the entire territory.

The Brazilian solution (100% ethanol) could therefore be adopted for well defined fleets operating only inside each Country or in selected narrow areas. This places a great obstacle to any change of the present system because any such change should be the same and generalized at least in all Countries of western Europe.

This situation is testified by the EEC proposed directive for saving oil through the use of synfuels presented to the Council by the Committee on August 5, 1982 /24/. This proposal allows the use of organic oxygenated compounds in gasoline not exceeding 10% vol

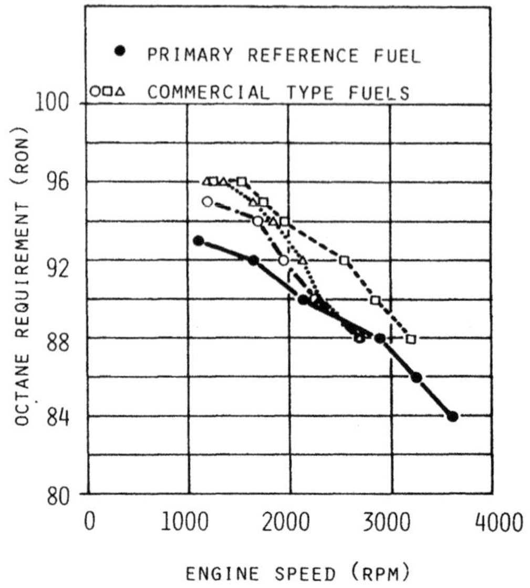

Figure 6. Typical trend of engine octane requirement vs speed.

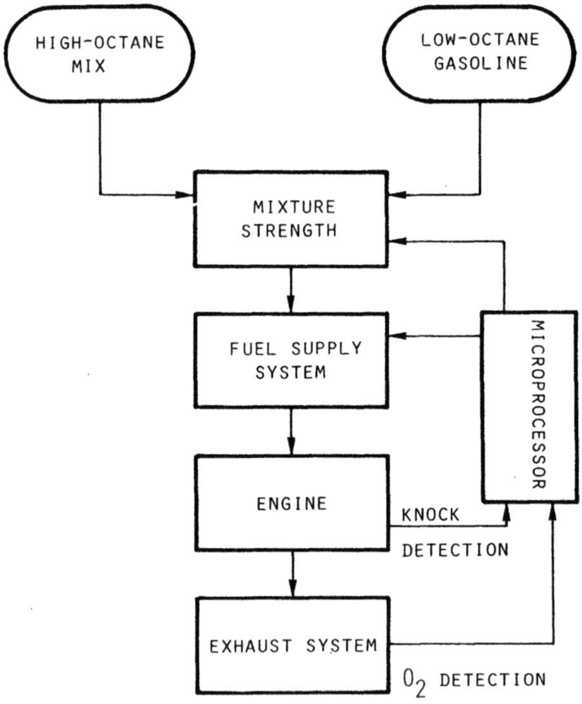

Figure 7. A new approach for the future.

overall, with a further limitation for methanol at 3% max. These constrainsts are dictated by the need of maximizing the use of oxygenated compounds without imposing any major modification of the present system: refining, distribution, engine, cars etc.

The acceptance of this principle does however limit the contribution of synfuels to transport and could considerably penalize their future development.

Even if the 3% of the total european gasoline production equals about $2.5 \cdot 10^6$ t/y, its production should be confined to a limited number of large plants with a consequent cost penalty in transportation and distribution or, conversely, be divided in small plants with loss of the large scale benefits.

A NEW APPROACH FOR THE LONG TERM

A new approach to the use of synfuels in transports is suggested as a discussion item because it could satisfy the four objectives proposed and enable each Country to plan an appropriate energy policy in this sector.

These possible benefits do however require facing high technological costs for several reasons that will be examined.

Briefly, this proposal involves a future transport system based on just one type of Otto engine for all of Europe having a high compression ratio capable of burning "any type" of fuel with optimized efficiency.

This solution is based on an interesting issue that could result in considerable energy savings even if at the expence of high technological costs; it involves a conceptual change from "maximum" octane requirement which imposes a certain fuel quality to "average" octane requirement.

It is well known that the octane requirement of an engine changes drastically with the running conditions.

An example is illustrated in fig. 6 /25/.

The car of the average customer probably necessitates high octane fuel for no more than 25% of its service, medium range octane fuel for another 25% and low octane fuel for the remaining 50%.

These data are merely indicative and would require a confirmation survey about the driving habit of European drivers.

However, if this assumption is verified, the private transport by gasoline engined vehicles could be supplied with:

25% of 99 RON
25% of 92 RON
50% of 86 RON

In this hypothesis the resulting "octane pool" would be 91.
The supply of "custom tailored" gasoline to the engine requires the installation of two tanks containing the high and the low octane fuels.

Depending on the instantaneous engine requirement the appropriate fuel is obtained by blending the two components in the proper ratio.

This operation would avoid waste of quality and therefore result in energy saving.

The car should have a high technological content because it would require:

- a high compression ratio engine
- a knock sensor
- an exhaust gas oxygen sensor
- a fuel system for each "gasoline"
- a microcomputer for actuating the entire system

A possible scheme of the system is reported in fig. 7.

The use of high technology available in each Country would enable great energy savings with consequent benefit for the balance of foreign payments and the use of alternative fuels with strategic advantages.

Such a car should in fact be the optimised solution in terms of Joules required per unit distance.
The high octane fuels to be used would be the different synthetic fuels available in each Country derived from both domestic raw materials (e.g. ethanol) or import raw materials (e.g. methanol).

This car could be supplied in service by anyone pair of high and low octane fuels because the system is designed to optimize the use of the available energy.

If the high octane component is methanol, this solution offers the double advantage of using an alternative component and obtaining a further saving in the refinery that would only produce the 80-85 RON low octane hydrocarbon component.

Flexibility is a great advantage of this system; methanol could in fact be used in the desired quantity, and the production of the hydrocarbon component in the refinery be planned accordingly.

In example, if an availability of 2 million tons of methanol is expected in Italy for 1991 then the hydrocarbon component should have a 90 octane rating and the engines would run on a 20/80 methanol/gasoline blend.

For 4 million tons in 1992 the gasoline octane rating would drop to 85 and the engine fuel would be a 40/60 methanol/gasoline blend.

Should methanol for any reason be unavailable for some time, the engine could be operated with premium and regular fuels blended in the correct ratio.

CONCLUSIONS

Europe should aim at finding a common strategy in the automotive transport that would enable to meet the future demand for mobility within the entire territory without penalizing the energy policy of any Country. The use of synfuels can play an important role in accomplishing this strategy.

The study of three component system is proposed for the Otto cycle engine, as follows:

- high technology car designed for the optimized use of a large spectrum of fuels
- use of several types of synfuels as high octane components
- rationalization of refining to meet the future synfuels availability

Such a system could enable each Country to meet the mobility needs within its territory and in the EEC territory with an optimized energy management.

As for the Diesel cycle engine, the direct or indirect introduction of particular types of synfuels could favor its growth.

This strategy in automotive transport is very complex and innovative, and as such it is worth of further development. To this effect a long term R&D program should be established under the sponsorship and cooperation of EEC for verifying the feasibility of the different steps as outlined below:

- 1st approximation technical feasibility study
- 1st approximation economical feasibility study
- engine testing program
- evaluation of the synfuels production from domestic and import raw materials
- 2nd approximation technical feasibility study
- 2nd approximation economical feasibility study
- engine testing program
- etc.

This program should evolve with concurrent cross check among the different operators and each subsequent step should be based and decided on the results of the previous one.

REFERENCES

/1/ 1982-Energy statistics - Eurostat 25.3.1983

/2/ ANFIA "Automobile in cifre 1982"

/3/ FEEMAS Report
"Sviluppo dell'auto diesel in Italia nello scenario energetico degli anni 80"
presented in Rome, february 24, 1983

/4/ F. Sezzi
Rational Use of Energy and Raw Materials
European Motor-Vehicle Symposium
Brussels, 9-12 December 1975

/5/ CNR - Progetto Finalizzato Energetica
"Uso del metanolo per autotrazione"
Libro bianco sulle ricerche condotte nell'ambito del sottoprogetto "Uso del metanolo per autotrazione"
Roma, dicembre 1982

/6/ G. Pecci, P. Garibaldi
"Methyl fuel and its effect on crude oil consumption"
Proceedings of 3rd International Symposium on Alcohol Fuels Technology - Asilomar, California, May 28-31, 1979, 11 pp.

/7/ G. Roncolini, A. Frondaroli, G. Pecci, P. Garibaldi
"Impact of partial diesel conversion of the automobile population on crude oil consumption in Italy"
Agelfi, European Automotive Symposium
Brussels, October 1977

/8/ "Etanolo per via fermentativa - possibilità e convenienza della produzione in Italia per l'uso nel settore dell'autotrazione"
CNR-PFE-LB-1
Milano, luglio 1979

/9/ Ambiente, Risorse, Salute
N° 18, Agosto 1983

/10/ V. Laganà, V. Arena
"La trasformazione del carbone del Sulcis per la produzione di metanolo"
Giornata di Studio "Il metanolo come vettore energetico - Ruolo e prospettive"
organizzata da FAST in collaborazione con ENI, Milano 24/6/82

/11/ F. Ciselli, S. Corso
"Esperienze nell'uso di miscele benzina-metanolo nell'autotrazione"
Giornata di studio "Il metanolo come vettore energetico - Ruolo e prospettive" - organizzata da FAST in collaborazione con ENI, Milano 24/6/82

/12/ M. Bidoli, R. Trerè, G.F. Zanoni
"Esperienze del metanolo in autotrazione"
Giornata di studio "Il metanolo come vettore energetico - Ruolo e prospettive" organizzata da FAST in collaborazione con ENI - Milano 24/6/82

/13/ A. Paggini, V. Fattore, F. Ancillotti, M. Sposini
"Processo per la produzione di metanolo ed alcoli superiori e loro impiego in autotrazione"
Giornata di studio "Il metanolo come vettore energetico - Ruolo e prospettive" organizzata da FAST in collaborazione con ENI - Milano 24/6/82

/14/ M. Sposini, R. Pea, F. Ancillotti, P. Ghezzi
"Engine performance and water tollerability of gasoline containing the alcoholic mixture obtained by the Snamprogetti process"
5th Int. Alcohol Fuel Technology Symposium, Auckland, N.Z., May 13-18, 1982

/15/ P. Le Prince, M.H. Ory
"Programme for gasoline substitution in France"
5th International Alcohol Fuel Technology Symposium, Auckland, N.Z., May 13-18, 1982

/16/ J. Bandel, E. Plassmann
"Research program "Alcoholic fuels for road traffic" of the

German Federal Ministry for R&T in the Federal Republic of Germany during the period 1979-1982"
3rd Int. Alcohol Fuels Technology Symposium
Asilomar, California, May 28-31, 1979

/17/ H. Quadflieg
"M-100 research and fleet tests in the German Federal alcohol fuels project"

/18/ J. Hancke
"Report from Norsk metanolgrupp 810917"
Project Group "Fuels and Distribution" of the COST 304 European Action

/19/ G. Svahn
"Methanol-gasoline mixtures in four stroke otto engines"
3rd Internation Alcohol Fuels Technology Symposium,
Asilomar, California, May 28-31, 1979

/20/ L.A. Bern
"The Swedish oil and fuel policy in the 1980's"
3rd International Alcohol Fuels Technology Symposium,
Asilomar, California, May 28-31, 1979

/21/ C.G. Nystrom
Chairman report to 2nd meeting of subgroup COST 304 "Alcohol-fuels and distribution 811028"

/22/ G.R. Cassels, W.G. Dyer, R.T. Roles
"BP New Zealand experience with methanol/gasoline blends"
3rd International Alcohol Fuels Technology Symposium,
Asilomar, California, May 28-31, 1979

/23/ "Piano Energetico Nazionale"
Ministero Industria Commercio e Artigianato
Roma, 31 luglio 1981

/24/ Gazzetta ufficiale delle Comunità Europee
C229 25° anno - 2 settembre 1982

/25/ J. Weissmann
"Carburants et combustibles pour moteurs à combustion interne".
Publications de l'Institut Français du Petrole.

ECONOMICS OF THE SYNFUEL PROCESSES AND THEIR PROSPECTS

Dr.-Ing. D. Wiegand
Bergbau-Forschung GmbH
Postfach 13 01 40
4300 Essen 13
Federal Republic of Germany

ABSTRACT. Coal can and will make an essential contribution to world-wide energy supply in the future. It can be used for direct heat generation but also for production of electricity, chemical feedstock, gasoline, liquid and gaseous fuels. Coal in the form of conversion products, therefore, is a universal-type source. Meanwhile coal conversion technologies have reached a far advanced state of development. Economic calculations on the production of SNG and hydrocarbon liquids from both Western European and imported hard coal have shown that economic viability depends on many parameters. Not only coal prices and investment costs, but also the financial conditions and last not least the market prices of crude oil derived fuels and natural gas dispose of economic attractiveness and prospects of synfuel processes.

The presentation will cover economic aspects of synthesis gas and SNG production of various gasification processes. The conversion of synthesis gas to ammonia, methanol and other hydrocarbons will be discussed.

The contribution of mineral oil to global energy consumption will decrease during the coming two or four decades mainly substituted by coal and nuclear energy. It is nevertheless expected that within a slowly expanding total energy consumption also oil will have a small increase. Future consumption growth will, however, be much less than assumed only a few years ago. At the World Energy Conference held in New Delhi in September last year were presented the results of a survey carried out by the Conservation Commission of the WEC. According to them the world energy consumption will be between 14 and 17 billion tce in the year 2000 and around 20 to 26 billion tce by 2020. This means first a substantial decrease of the energy demand forecast compared with previous estimates, since at the World Energy Conference in Munich four years ago the same body estimated energy consumption to be around 17 or 20 billion tce in 2000 and 26 to 33 billion tce in 2020. On the other hand these figures continue to represent a rise in consumption by 10 billion tce if referred to the consumption of 1980. Diverging growth rates are expected for the different energy fuels. The market shares of coal and

G. E. Beghi (ed.), Synthetic Fuels, 349–363.

nuclear energy are assumed to go up, against levelling out shares of natural gas and hydro-power. Taking into account new energy sources, e.g. solar energy, oil is reckoned to go down to a market share of 20 % by 2020. In absolute figures such percentual decrease for oil involves still, depending on the scenario, a rise from 3.9 billion tce in 1982 up to max. 5.1 billion tce by 2020.

At the 11th World Petroleum Congress held in London in September 1983, John H. Lichtblau, president of the Petroleum Industry Foundation Inc., stated that he felt oil consumption would go up in the long run, with the major growth share accounting for developing countries. According to Mr. Lichtblau there would be no problems up to 1990 to cover the oil demand of the western world.

But if a balanced energy supply structure beyond 1990 is to be ensured it is now already that new coal mines, nuclear power plants, and oil viz. gas fields have to be developed which start production during the nineties. To reduce the dependence of oil in Western Europe (the present share of crude oil in the EC amounts to 48 %, against as little as 24 % for coal) the preferred solution in view of her resources is coal. Coal has the biggest potential of contributing significantly and reliably to making energy supplies safe.

Expanding world coal trade in the recent years is an indication of a return towards coal. Slumping demand for coal during the last few years has to be attributed to the general economic depression in the industrialized nations. But in the medium and long run it can be expected that with economic revival the price advantage of coal over oil plus the considerations of safe coal supplies will force the consumption of coal.

Today, the image of coal still is largely one of an energy fuel of the past. And coal is predominantly used along with technology of the past. A task of importance therefore is to translate research and development results into viable technologies permitting either to burn coal cleanly or convert it in gaseous and liquid energy fuels acceptable to the environment.

Environmentally acceptable technologies burning coal for electricity and heat production have been developed. These technologies gradually conquer the market.

As to gasification and liquefation of coal, relevant processes have been developed and commercially used already several decades ago. Such processes to in many respects no longer comply with the requirements of modern technology. Many countries have, therefore, been working substantially on the improvement of these technologies and have also been developing novel ones.

To make different processes comparable one essential presupposition is to assume uniform basic data for calculation. The assessment of the results of economic calculations is only reasonable if one knows the parameters having served as a calculation basis. Fig. 1 shows the basic economic data for calculating the coal conversion processes under review. The costs are determined by a static calculation for the first operational year and for commissioning late in 1983. This paper will not enter into dynamic calculations with alternative parameters although such cost calculations are absolutely necessary for comparing processes representing mainly longterm projects (which is the case with gasification and liquefication of coal) in order to obtain stringent results.

DEBT/EQUITY RATIO	70/30
DEPRECIATION	20 A
DEBT FINANCING	9 %
EQUITY FINANCING	10 %
MAINTENANCE	3 %
TAXES, INSURANCE	1.5 %
OPERATING TIME	7,500 H/Y
LABOR COSTS	20,500 $/MAN-YEAR
OVERHEAD, ADMINISTRATION	16,400 $/MAN-YEAR
WESTERN EUROPEAN COAL	93 $/T.C.E.
IMPORTED OVERSEA COAL	60 $/T.C.E.

Figure 1. Basic Data for Cost Calculation

Coal gasification is the most versatile among available techniques as fas as process routing and utilization of products is concerned. There are numerous procedural approaches of which a few only have been advanced to commercial maturity so far. Today, production of synthesis gas as feedstock for ammonia going into the fertilizer industry or as raw material to gasoline synthesis (Sasol) is clearly in the focus of interest.

Synthesis gas is an energy fuel not occurring in nature like methane; it has to be produced by conversion processes. Many experts feel, therefore, that if coal gasification is to be applied on a large scale, syngas production would economically be more promising than SNG production since the latter has to compete with a natural energy fuel like methane or natural gas.

Presently synthesis gas is produced in large commercial plants by reforming of natural gas or naphtha, by partial oxidation of heavy fuel oil, an by gasification of coal. Synthesis gas, in turn, is only an intermediate product. The main applications for it in the Federal Republic of Germany are represented on Fig. 2. These are ammonia synthesis, methanol synthesis as well as different oxo-syntheses. In 1980 syngas consumption in the Federal Republic of Germany was around 8 billion m^3, excluding the internal consumption of hydrogen in the mineral oil and petro-chemical industry and excluding also coke oven gas for the heat market. The consumption is estimated to rise to 10.5 billion m^3 by 1990. Rising consumptions are assumed for all of the three sectors, with ammonia accounting for 1.5 %/year, methanol and oxo-syntheses for 4 %/year.

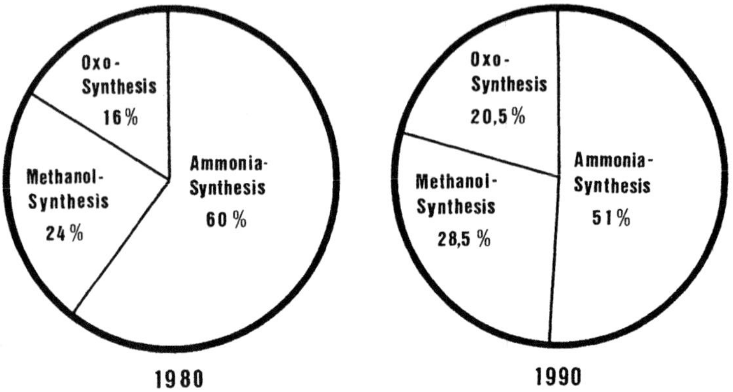

Figure 2. Synthesis Gas (CO; H_2) Consumption (FR Germany)

A production cost comparison for syngas on the four alternative fuels of respectively coal, heavy fuel oil, naphtha, and natural gas, gives some information on the economic attractiveness of the different raw materials. From Fig. 3 may be derived the syngas cost associated to determined raw material costs. The different curve rises are caused

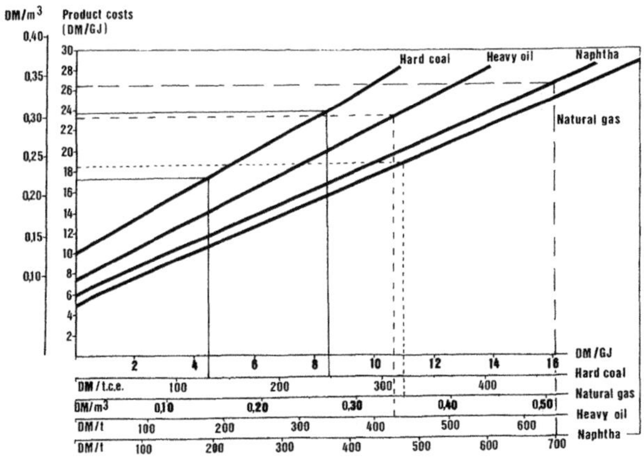

Figure 3. Production Costs of Synthesis Gas as a Function of Energy Prices

by the efficiencies of the different processes, whereas the differing intersections with the ordinate represent the unit amount of conversion cost and, as such, also the amount of capital investment. Using coal involves the highest conversion cost, and the steeply rising curve reveals a low conversion efficiency. This process can, therefore, only compete economically with other raw materials if the raw material coal is cheap enough. On Fig. 3 were considered the price levels valid in the Federal Republic of Germany at the end of 1983. The price of 132 $/t for heavy fuel oil is understood to be an internal refinery price, whereas the price of heavy fuel oil containing max. 1.5 % of sulphur amounts to 171 $/t in the Rotterdam market. Taking the latter pricing as basis, syngas cost increase from DM 0.26/m³ to DM 0.31/m³. Using naphtha does not appear to be an attractive alternative for syngas production, given the present price level. The average price of coal imported from overseas to Western Europe can be assumed with $ 60/tce which would mean barely DM 0.06/m³ of cost-saving for syngas production, reducing cost price to DM 0.24/m³. In the first quarter of 1984 we have some changes in energy prices. Heavy fuel oil goes up to 140 $ (internal refinery price) and to 200 $ in the Rotterdam market. This means an increase of 1 - 3 Pf/m³ syngas. On the other hand the price of coal imported from oversea decreases to about 50 $/tce leading to a price reduction of 2 Pf/m³.

Presently, syngas from coal is commercially produced according to the Lurgi and the Koppers-Totzek-processes. The Texaco process is in the phase of commercial introduction. Fig. 4 contains a comparison between the syngas production costs for the three alternatives on the basis of the economic parameters shown in Fig. 1. As to the Lurgi process, the methane recovered from gasification has first been cracked in a reformer integrated in the process, in order to maximize the syngas yield. If methane is sold as a by-product, the cost of the syngas strongly depends on the by-credit for methane.

	DIM	Process Lurgi	 Koppers-Totzek	 Texaco
Coal throughput	10^6 t/y	0.567	0.558	0.532
Gas production	10^6 m³/y 10^6 GJ/y	690.31 8.62	690.31 8.62	690.31 8.62
Investment Costs	10^6 DM	520	495	505
Capital costs Coal costs Operating costs	DM/GJ DM/GJ DM/GJ	9.86 16.21 3.57	9.38 15.95 3.30	9.58 15.19 4.24
Production costs	DM/GJ Pf/m³	29.64 37.04	28.63 35.79	29.01 36.26

Figure 4. Design Data of Synthesis Gas Plants

When comparing the production cost of the three alternatives it turns out that the differences in costs are only around 10 %. And when considering that the capital investment has been estimated by a cost calculation with a probable accuracy of 15 %, one cannot speak of significant cost advantages of one particular process, given the high investment share of any of them.

As already mentioned syngas is only an intermediate product applied for a multitude of subsequent conversion processes. The whole range of possible applications shown on Fig. 2 (60 % for ammonia synthesis, 25 % for methanol synthesis, 15 % for oxo-syntheses) does not apply to the Federal Republic of Germany but is the typical case with almost any industrialized nation. Main application for syngas is ammonia synthesis. 80 % of the ammonia is used either in numerous compounds or directly as fertilizer.

Commercial ammonia plants are nowadays built as single-stream units where one "stream" has a maximum capacity of approx. 1,300 t/d. The cost of ammonia production in a plant of 1,300 t/d during the first operational year has been calculated for different raw materials. The result of such a calculation is shown on Fig. 5. Depending on the raw material, production cost range between 200 and barely 300 \$/t of NH_3. The present market price for ammonia in Western Europe amounts to about 200 \$/t. The cost of ammonia on a West European price basis was calculated for a price of natural gas of 5 \$/t MMBTU, for a heavy fuel oil price of 170 \$/t and for the cost of imported coal of 60 \$/tce. Following such calculations, imported coal is an economic alternative to heavy oil. It appears, on the other hand, that none of the alternatives is cost-covering with the present market situation.

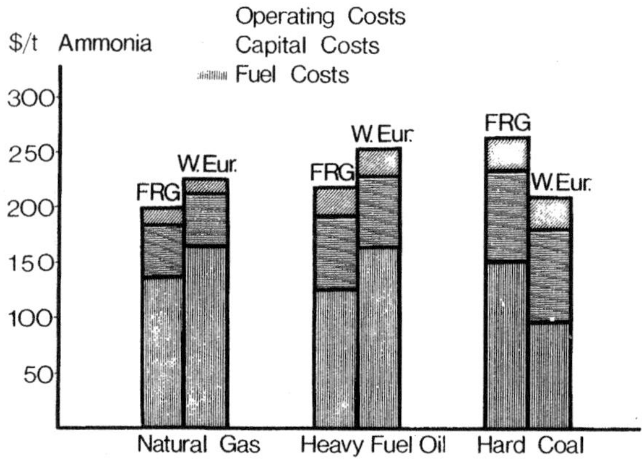

Figure 5. Production Costs of Ammonia

While a global consumption increase in nitrogen fertilizers of approx. 4 %/y is expected in the medium range, the situation in 1985 will likely see a production capacity of 113 million t versus a consumption of only some 67 million t. This means a worldwide utilization of barely 60 % of the capacity. On this background the construction of new ammonia plants particularly in Western Europe should be looked upon with a certain reserve. The OPEC nations of the Near East are clearly in a more favourable position with their cheap raw material sources, mainly of the associated natural gas.

Methanol production is, after that of ammonia, the second important market for syngas. Prospects for methanol appear to be substantially more favourable although it should be borne in mind that 1982 was consumed worldwide no more than some 11.3 million t of methanol against barely 60 million t of ammonia-containing fertilizers.

Present and expected consumption structures for methanol, by the example of the USA, is given on Fig. 6. The consumption sectors "motor fuel additive" and "acetic acid", considered separately from the 1979 general structure, represent the actual growth potentials of methanol. As can be seen from the right-hand figure, however, other new sectors will join them. They are the antiknock methyl-tertiary butly ether and, secondly, application of methanol for peak load electricity generation.

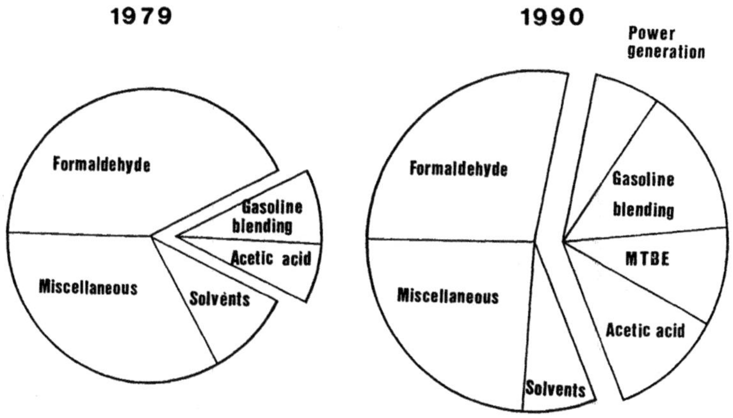

Figure 6. US Methanol Demand

Japan furthermore considers the possibility of a methanol-coal-mixture (MCM) as a modified coal-oil-mixture (COM) on the power plant sector. The reason is that methanol, being a clean sulphur- and ash-free fuel, would be an excellent feed to power plants in countries with severe environmental regulations. Another application consists in cracking for peak load town gas production, -- a technique which is at present practiced on a commercial scale in Berlin.

The unknown factor No. 1 with future applications for methanol is doubtlessly the motor fuel sector. If gasoline/methanol mixtures, e.g. M 15, or pure methanol as motor fuels were introduced on a large scale then methanol production would expand dramatically. There are, however, many indications of both technical and economic nature which, unlike some years ago, do not speak so much in favour of such option any longer. Another technology under development is the Mobil-M process allowing to produce, via methanol, either gasoline or diesel or, by other process alternatives, also ethanol or ethylene. If these processes were applied commercially methanol demand ought to soar dramatically. There is hardly any other product in the market for chemical feedstock -- whereof methanol is one -- whose products have been the subject of that much controverse discussions. There is a great number of factors which, especially in the European market, could be of critical importance to the eventual methanol production in Western Europe. The presently emerging export capacities in the USSR, North Africa, and the Arabic countries make prospects for West European methanol producers look quite gloomy. Raw material costs for syngas production are substantially lower in the aforesaid countries since, unlike in Western European, they do not depend on world market prices and can always be cheaper than methanol produced in Western Europe. Saudi Methanol Company has already started to export to the European market, and the big projects in Tomsk and Gubakha (USSR) as well as in Marsa el Brega (Libya) are about to be completed so that methanol prices are expected to get under strong pressure. The National Methanol Company of Saudi Arabia has scheduled the commissioning of a plant of 650,000 t annual capacity of methanol for 1985.

Today methanol is produced in integrated one-stream plants (production both of syngas and of methanol takes place in one plant) of economic size capacities between 1,000 and 3,000 t/d. One-stream plants of 5,000 t/d have already been designed, but not implemented so far.

Fig. 7 gives the costs of methanol production with different feedstock in a one-stream plant of 3,000 t/d. Have been considered first the energy prices as valid in the Federal Republic of Germany and, secondly, present Western European price averages for natural gas (5 $/MMBTU) heavy fuel oil (170 $/t) and imported coal (60 $/t).

The production costs range betwenn 190 $ and 270 $/t although the present market price is at a low level between 160 and 170 $/t. Under these conditions construction of new methanol plants would be uneconomic. The calculation results as per Fig. 7 are based on a 3,000 t/d plant which, if fully utilized, would yield barely 1 million t of methanol each year. Consumption in the Federal Republic of Germany presently amounts to 1.3 million t, 0.7 million t of which are produced in this country by three plants: BASF (220,000 t/y), UK Wesseling (350,000 t/y), and Veba (220,000 t/y). These plants are thus much smaller than those assumed by Fig. 7. The influence of plant size on methanol production cost is shown on Fig. 8 which offers two options for syngas production (Lurgi and Koppers-Totzek).

In the case of massive introduction of methanol in the motor fuel sector further cost reductions, mainly for syngas production, would be achievable. If German coal were used, methanol production cost could go down to 250 $, against 210 $/t with the use of import coal. Such level is, however, still considerably above current market prices of 170 $/t.

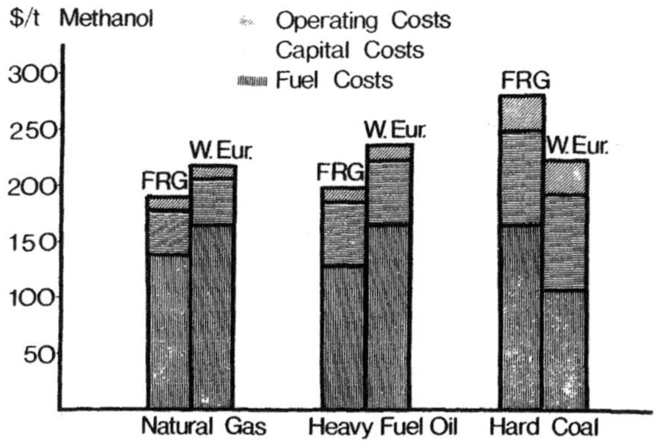

Figure 7. Production Costs of Methanol

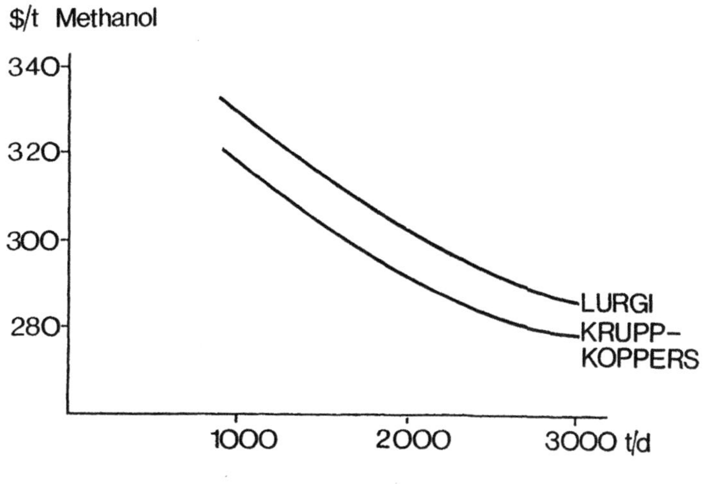

Figure 8. Production Costs of Methanol - a Function of Plant Size

Another and internationally discussed possibility of using syngas is the production of a high calorific gas by methanation. Such gas is described as synthetic natural gas. (SNG). SNG production may be of gaining importance in the heat market, particularly if cheap coal is available. Actually slumping or stagnating prices for natural gas as well as massive conservation have led to a suspension -- or even cancellation -- of scheduled projects for the erection of commercial plants.

Fig. 9 gives the results of an economic comparison for SNG produced from various coal gasification processes. Plant sizes are characterized by product yields of 200,000 m³ of SNG/h each. It is worthwhile noting that the process with the lowest coal feed (Lurgi) requires the highest capital investment, whereas the process requiring the highest coal feed (Koppers-Totzek) needs the lowest investment.

		Process		
		Lurgi	Koppers-Totzek	Texaco
Coal throughput	10^6 t.c.e./y	3.2	3.8	3.6
Gas production	10^9 m³/y	1.5	1.5	1.5
	10^6 GJ/y	56.0	56.0	56.0
Investment costs	10^6 DM	3,550	2,580	3,200
Capital costs	DM/GJ	10.57	7.68	9.52
Coal costs	DM/GJ	14.09	16.75	15.86
Operating costs	DM/GJ	2.07	1.62	1.83
By-credit	DM/GJ	0.68	--	--
Production costs	DM/GJ	26.05	26.05	27.21

Figure 9. Design Data of SNG-Plants

Considering capital investment and present prices for German coal, the results are practically idèntical product costs in the Federal Republic of Germany for all of the three processes in question, with deviations of the Texaco process remaining within calculation tolerances Changes in the cost of coal, e.g. by feeding imported coal, results in changes of gas production prices. Modifying the determinative parameters for calculation the different investment components also has a divergent effect on the process costs. The gas cost indicated with 10 $/GJ is, however, twice as high as the current price of natural gas between 4 and 5 $/GJ. Converting imported coal would, while reducing SNG cost to some 8 $/GJ, not bring us much closer to the break-even point though.

Another application for syngas is represented by the Fischer-Tropsch synthesis for the production of liquid hydrocarbons. Depending on the process route, could be yielded thereby preferential alternative fractions, e.g. gasoline, diesel, fuel oil. Present-time commissioning of a plant of 2 million annual tons of production capacity would require almost $ 2.6 billion of investment in the Federal Republic of Gemany. Taken into account the credit for by-products, production cost per ton of primary product would ammount to $ 720/t. By primary products are understood the gasoline fraction as well as the fractions C_1 trough C_4.

High production cost is critically determined by high coal prices.

Basing on a specific consumption of over 4 tce per ton of gasoline this comes already to something like 380 $ with German feed coal, i.e. to over 50 % of the production cost. Cheap imported coal, while reducing the product price by about 140 $/t, would not (costs being at 580 $/t of product) allow to attain the economic break-eben point. Following the calculations, with imported coal as feed, the price of crude oil would have to go up to at least 61 $/bbl and, for German coal, to 83 $/bbl before gasoline from Fischer-Tropsch synthesis could compete at all.

Another alternative of gasoline production is the Mobil-M process where methanol, in the presence of zeolithic catalysts, undergoes a practically quantitative conversion to gasoline. Separation of by-products and their upgrading, necessary with the Fischer-Tropsch synthesis, has not to be done, this is one main advantage of the process. Investment requirements are lower than for the Fischer-Tropsch synthesis, they are depending strongly on the type of gasification process preceding the methanol production. A first commercial plant of reportedly 0.5 million t of gasoline per year is under construction in New Zealand. The plant is based on natural gas as feed stock for methanol production.

A third process route to gasoline production is hydrogenation where as specific coal consumption of 3 tce per ton of product is considerably lower than with the Fischer-Tropsch synthesis. Capital requirements for the construction of such a plant are also lower than with Fischer-Tropsch synthesis. First calculations for a plant of a capacity of 2 million t of gasoline per year gave an investment cost of 2.2 billion $, with commissioning within 1983. The product costs are around 550 $/t, this means, barely 170 $/t lower than of the Fischer-Tropsch product. The economic attractiveness is reached by an increase of crude oil to 68 $/bbl in comparison with German coal respective 51 $/bbl with imported coal.

A cost comparison of the three alternatives of gasoline production (Fischer-Tropsch, Mobil-M, and hydrogenation) is given on Fig. 10. The capacity of all the plants is 2 million t/y of gasoline. As appears from the figure, gasoline from hydrogenation is cheapest (around 550 $/t), whereas the Fischer-Tropsch gasoline, costing around 720 $/t represents the most expensive alternative. Gasoline from the Mobil-M process should cost around 640 $/t based on identical assumptions. The current market price for crude oil-based gasoline is 312 $/t taking the refinery sales price as a basis. This means that gasoline production from German coal is not economic as yet.

When discussing the possibilities of an eventually increasing methanol consumption was also mentioned its direct use in the motor-car sector. Without entering into the details of infrastructural costs implied with a transformation to pure methanol-operated cars nor into the issues of motor design, a simple comparison of present-day methanol and gasoline prices with the costs of gasoline and methanol from coal derived earlier, is to highlight the economic prospects of methanol. The present gasoline price amounts to 312 $/t, against 170 $/t for methanol. Referred to the energetic basis the costs are as follows: petrol-based gasoline: 7.3 $/GJ; methanol: 8.7 $/GJ. At this moment methanol is thus more expensive than gasoline. Even when assuming a by 15 % higher motor efficiency with methanol -- which many experts put in question by now -- methanol would, at 7.4 $/GJ, still be slightly more expensive than gasoline. Fig. 11 shows the comparison of present gasoline and methanol costs (expressed as gasoline equivalent) against the to be expected cost for gasoline from hydrogenation and for coal-based methanol. While costs of distribution and trading have been considered, taxes and government duties were not included.

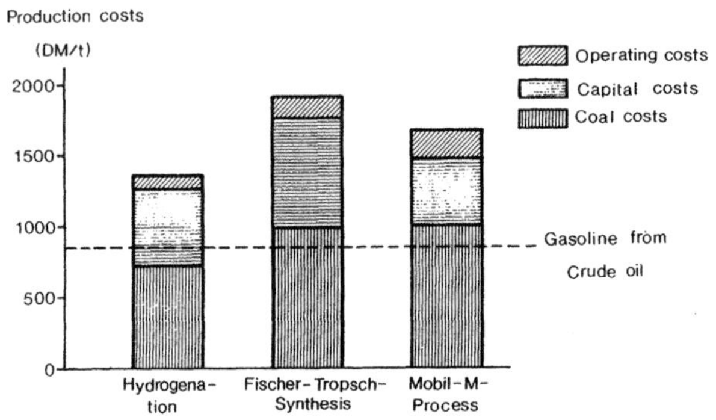

Figure 10. Production Costs of Gasoline

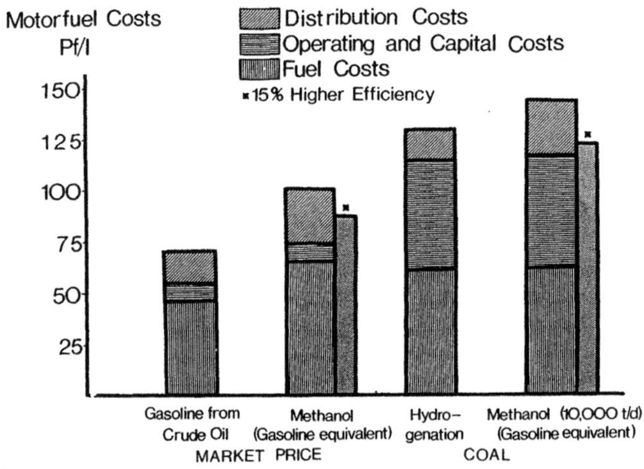

Figure 11. Motorfuel Costs of Gasoline and Methanol

The result is that direct use of methanol is more expensive than the application of gasoline from hydrogenation. Thence it appears logical that the Mobil-M process results in higher cost than hydrogenation.

Fig. 12 finally is a comparison of the results of all the processes under review versus current market prices. The production of gaseous or liquid energy fuels from coal is, at the present moment, not economic in any of the Western European countries. To be economic, the cost prices of coal would have to be still much below current import prices of 50 $/tce.

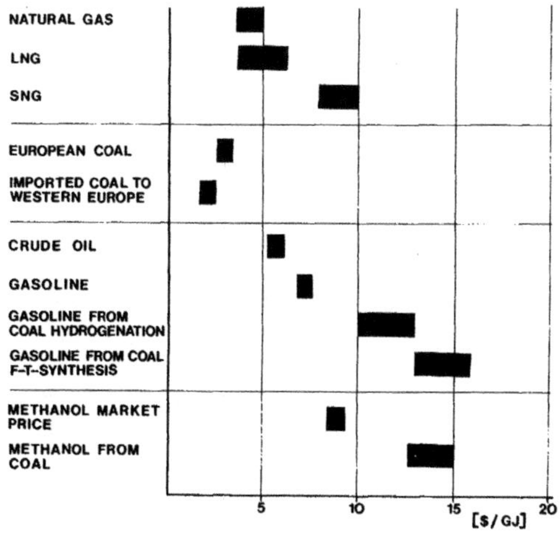

Figure 12. Energy Costs (1983)

At the same time we have to state that the industrial production of the above-mentioned products also on the basis of liquid and gaseous hydrocarbons in new commissioned plants is not economic attractive. Reasons are market prices not caused by production costs. The prices are determined by overcapacity and reduced supply.

About 40 percent of coal consumed in conventional coal gasification processes like Lurgi, Winkler, Texaco or Koppers-Totzek are used for process energies (electricity, steam of process heat). But this part of coal can be substituted by energy from a high temperature nuclear reactor. In this way it is possible to reduce the specific coal consumption per unit of gas produced. Since the beginning of the seventies this concept is under development in the Federal Republic of Germany in two parallel process routes, the hydro-gasification and the steam gasification. Since some years two semi-technical plants are under operation both with an coal throughput of 200 kg/h. For the hydro-gasification of lignite a first pilot plant with an throughput of 24 t/h is since 1979 under construction.

The comparison between conventional coal gasification processes and the process with the application of nuclear process heat shows a cost advantage of about 30 % for this new technology. More substantial cost advantages can be reached by the use of catalysts for increasing reactivity.

More important than the price comparison are the arguments of better coal utilization, independence of energy imports, saving of values and price regulating by domestic energy resources. By the application of steam gasification with nuclear process heat the total natural gas imports for the FRG in 1982 could be substituted by the gasification of 35 million tons of hard coal, by conventional gasification about 59 million tons would be needed.

Similar calculation could be done for liquid fuels and synthesis gas products. The commercial introduction of this sophisticated technology presuppose long-term development. For that reason conventional gasification processes habe to be realized in the meantime.

A comprehensive assessment of the economics of a process includes dynamic cost calculations, i.e. looking at the development of production costs over a period of several years after commissioning of the coal conversion plant.

The present state of development of coal conversion technologies is diverging. Whereas coal gasification according to the Lurgi and Koppers-Totzek processes habe worldwide found commercial application since many years and the Fischer-Tropsch synthesis for gasoline production is also applied commercially in the Republic of South Africa, other processes are still in the development stage. The first commercial plants based on the Texaco process are being built in the USA, in Japan, and the Federal Republic of Germany.

Progress in coal liquefaction in the USA suffered serious setbacks. The existing big pilot plants were closed down or sold. In the United Kingdom plans covering the following step of large-scale application were reduced drastically, above all due to the decision by BP not to share in the financing of the construction of a first demonstration plant. For that reason the troughput of the scheduled pilot plant of 25 t/d is to be cut back to 2.5 t/d. the situation in the Federal Republic of Germany is as follows; on the background of the present price situation and of the high capital requirements for a commercial coal liquefaction plant which could be operated exclusively with substantial government support, is examined how technological improvement of coal liquefaction can be continued at an as low as possible financial commitment. A hydrogenation plant of 200 t/d throughput is being operated in the Federal Republic of Germany. The next step could already be the design and construction of a large-scale industrial plant. Due to the fact, however, that with the present cost and price situation such plant cannot be operated on a commercial basis, it is envisaged to implement only one stream of such big plant where between 800,000 and 1.2 million t of coal per year are to be converted to 0.4 to 0.6 million t of liquid products. These will subsequently go to a conventional oil refinery for upgrading. Without the upgrading steps of refinement, hydro-cracking, and reforming, the investment requirements of such a liquefaction plant can be reduced considerably to about

0.6 billion $. This would allow, notwithstanding reduced financial expenditure, to demonstrate large-scale application and gain the know-how necessary for later construction and operation of such plants. In the present market situation this appears to be a viable solution to which the industry and the government should jointly contribute.

ENVIRONMENTAL PROBLEMS FROM SYNFUELS PRODUCTION AND UTILIZATION

P.P. Garibaldi
ASSORENI
Via Fabiani, 1
20097 San Donato Milanese (MI)
ITALY

ABSTRACT. The environmental science has grown in the oil era as a safeguard against the undiscerning use of this energy source in the different sectors of the energetic system (transports, heating, electricity etc..) and in the chemical industry. Numerous studies have been performed on the direct and indirect pollution by oil and derivatives. Methods of sampling and analysis and the relative instruments have been designed, realized and calibrated on the oil system. The synfuels technology developed in the last 10 years has shown new environmental problems sometimes different from the old ones, that could for many reasons require changes of the already experimented solutions applicable to the oil system. Other aspects of the synfuels technology could benefit from experience gained in the chemical industry; in example, the production, transportation, storage and general handling of methanol could involve the present structure of the chemical industry, even if scaled up accordingly. Conversely, some peculiar aspects of the synfuels will require special studies and the definition of appropriate measures. A general review of the various synfuels, from production to final use, has been fulfilled in this report giving prominence to the environmental aspects. Coal and some other sources such as oil shale, tar sand and biomass have been examined.

INTRODUCTION

It is not easy to precisely describe the "synfuels" category.

A good starting point is the definition given by the organizers of the 2nd Worldwide Synfuels Symposium /1/ that have considered as synfuels the liquid and gas derivatives of coal, oil shale and tar sand that can substitute the oil derived fuels in present use. Some

G. E. Beghi (ed.), Synthetic Fuels, 365–415.

types of biomass derived synfuels have also been considered besides those of fossil origin.

The 1982 production of the "european refinery" regarded as the overall output of the EEC refineries is reported in table I as an indication of the potential synfuel market /2/.

TABLE I. EEC REFINERY IN 1982 /2/

	million tons	%
Crude oil charge	440	100
Products		
LPG	10	2.3
Naphtha (to chemicals)	21	4.8
Gasoline	88	20.0
Kero and jet fuel	22	5.0
Gasoil and diesel fuel	137	31.1
Heavy fuel oil	107	24.3
Other products	23	5.2
Total products	409	92.7
Losses and refinery fuel	31	7.3
T O T A L	440	100.0

In the short or long term the synfuels will cause a reduction in the quantity and a change in the relative distribution of this refinery output and will consequently have an impact on the total quantity of the crude oil processed and on the refinery schemes.

This diversification will have economical, political, and social effects and also an environmental effect in the broad sense.

By assuming the hypothetical refinery of table I as the base line, the possible impact of each synfuel as petroleum substitute in the production and utilization stage will be reviewed, at least qualitatively.

The synfuels considered are those likely to substitute partially in the medium or long range the products of largest use in the EEC that amount to a total of 350 Mt/y:

Heavy fuel oil	107	Mt/y
Gasoil + diesel fuel	137	Mt/y
Kero + jet fuel	22	Mt/y
Gasoline	88	Mt/y

The types of synfuels covered in this paper are reported in table II.

Following the layout of table II, the possible synfuel alternatives to the five main oil products are examined and the environmental aspects relative to the production stage and utilization are briefly covered.

TABLE II. SYNFUELS

OIL PRODUCT	SYNFUEL	ALTERNATIVE SOURCE
1. Heavy fuel oil	1.1 Heavy oil from coal liquefaction	Coal
	1.2 Coal oil mixture	Coal
	1.3 Coal water mixture	Coal
	1.4 Low, medium, high BTU gas	Coal
2. Gasoil	2.1 Methanol	Coal
	2.2 Middle distillate from coal liquefaction	Coal
	2.3 Middle distillate from bitumen upgrading	Tar sand
3. Diesel fuel	3.1 Middle distillate from F-T process	Coal
	3.2 Middle distillate from shale oil upgrading	Oil shale
4. Jet fuel	4.1 Light-middle distillate from shale oil upgrading	Oil shale
5. Gasoline	5.1 Methanol	Coal
	5.2 Ethanol	Biomass
	5.3 n-Butanol	Biomass
	5.4 MAS (methanol and higher alcohols)	Coal
	5.5 F-T gasoline	Coal
	5.6 Gasoline from MTG	Coal
	5.7 MTBE	Coal
	5.8 Gasoline from direct coal liquefaction	Coal

1. HEAVY FUEL OIL

The synfuels that can be used as a substitute for this product are:

heavy oil from coal liquefaction processes

coal oil mixture

coal water mixture

low, medium and high BTU gas from alternative energy sources

1. Heavy oil from direct coal liquefaction

e direct coal hydroliquefaction technology is still at the pilot ant stage (disregarding the plants constructed in Germany during e second world war).

A listing of the main so called 1st generation processes is ven in table III /3/.

Some of these processes have led to light and medium distillates d to boiler fuel.

A general process scheme is reported in fig. 1 /4/.

As an example, the properties of a boiler fuel obtained through e SRC II process /5/ are reported in table IV; it is a heavy pro-ct with a high content of aromatics and nitrogen.

Toxicity tests run on this typical product have revealed a ncerogenic and mutagenic activity definitely superior compared to e corresponding petroleum cuts (tab. V) /6/.

The chemical "types" responsible for this mutagenic effect are bably molecules similar to those reported in fig. 2.

It is obviously possible to upgrade the product by removing the eroatoms with a reduction of the mutagenic activity, however the h cost involved would almost certainly prevent the possibility of stituting the fuel oil of petroleum origin.

This is the reason why studies are in progress for modifying the uefaction process schemes aiming at a complete elimination of this vy fraction by means of internal recycling and through its partial lization in the hydrogen production process (as illustrated under .1).

Besides technico-economical considerations, the toxicity of the-products has already dictated a resolution in the production stage.

. Coal-oil mixture

s is obviously a part-synfuel, half of the mixture being oil.

The coal-oil mixture (COM) can be regarded as a pseudo-liquid use its transport, handling and utilization can be assumed to be similar to those of the conventional heavy fuel oil.

TABLE III. MAJOR PILOT PLANTS FOR DIRECT COAL LIQUEFACTION /3/

PROCESS	SPONSORS	LOCATION PROJECT TIMING START UP	PROCESS DATA - CAPACITY - PRESSURE - TEMPERATURE
1G Farbenindustrie modified	RURHKOHLE AG, VEBA OEL AG	BOTTROP-NRW (RFT) 1975-1984 running since 1981	200 t/d coal 300 kg/cm^2 475°C
1G Farbenindustrie modified	SAARBERGWERKE AG	Wolkingen-Saar (RFT) 1975-1982 running since 1981	6 t/d coal 300 kg/cm^2 450-460°C
SRC II	US DEPT. OF ENERGY, PITTSBURGH & MIDWAY COAL MINING CO.	Tacoma (Washington, USA) 1975-1982 running since 1977	50 t/d coal 140 kg/cm^2 450-460°C
EDS	US DEPT. OF ENERGY, EXXON EPRI, JAPAN COAL LIQUE-FACTION DEV. CO., PHILIPS PETROLEUM, ATLANTIC RICH-FIELD CO., RUHRKOHLE AG. AGIP SPA	Baytown (Texas, USA) 1974-1982 running since 1980	200 t/d coal 100-150 kg/cm^2 450°C
H-Coal	US DEPT. OF ENERGY, ASHLAND SYNTHETIC FUELS INC., STANDARD OIL CO., CONOCO COAL, EPRI, COMMON-WEALTH OF KENTUCKY, RUHRKOHLE AG	Catlettsburg Kentucky, USA running since 1980	160-180 t/d coal 190 kg/cm^2 450°C

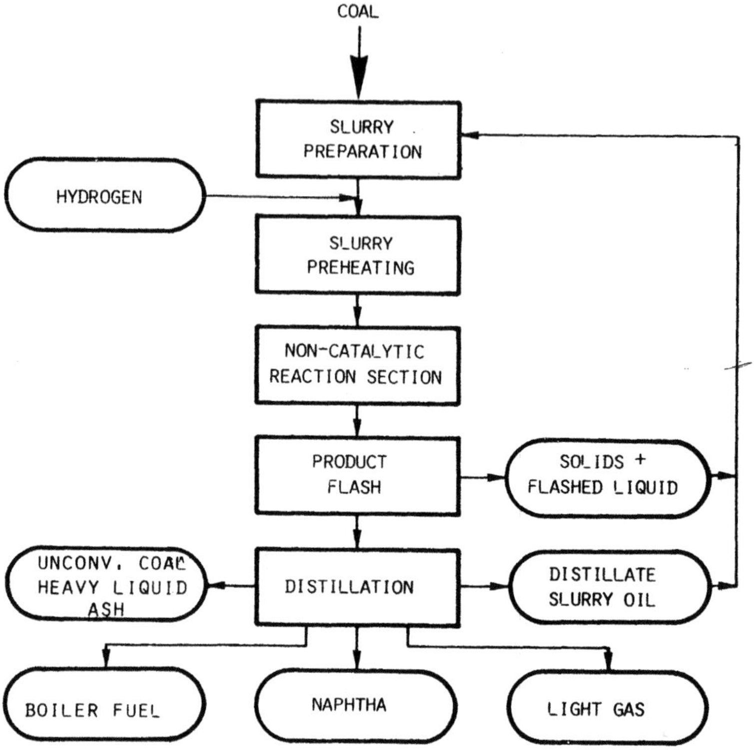

Figure 1. Hydroliquefaction of coal: 1st generation process scheme /4/.

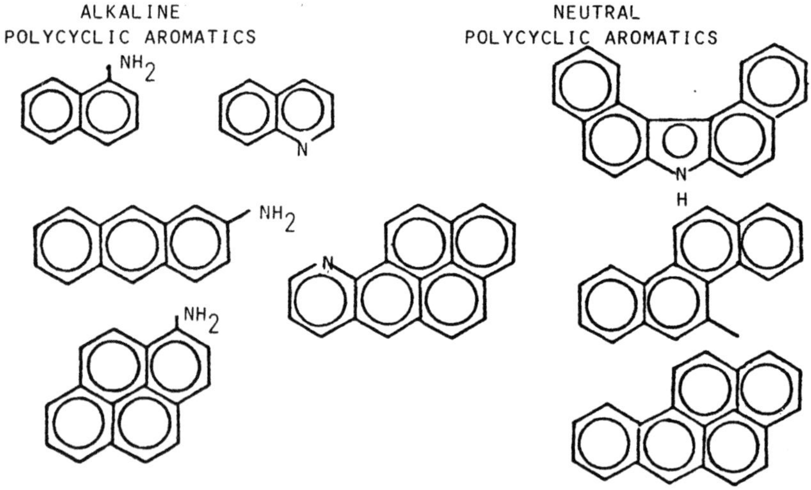

Figure 2. Chemical "types" responsible for bacterial mutagenicity of synfuels.

TABLE IV. TYPICAL PROPERTIES OF SRC-II DERIVED FUEL OIL /5/

		SRC-II FUEL OIL
Specific gravity	kg/l	1.02
Viscosity	cSt	
100°F		6.64
210°F		1.57
Pour Point	°C	-46
Nitrogen	wt %	0.98
Sulfur	wt %	0.24
Carbon residue	wt %	0.44
Total acid no.	mg KOH/g	1.2
Total base no.	mg KOH/g	7.0
Asphaltenes	wt %	0.04
Nickel	ppm	0.1
Vanadium	ppm	0.1

1.2.1. Production. The COM is produced by mixing coal and oil in the approximate ratio of 1 to 1.

Chemical additives, the use of very fine pulverized coal or other systems ensure mixture stability.

This technology has reached the demonstration stage, with several pilot and demonstration plants operating in various parts of the world.

A production scheme is reported in fig. 3 /7/.

No special environmental problems have been anticipated at this stage of production as one would expect upon considering that coal transport, handling, storage and grinding involve technologies common to the present coal utilization system.

1.2.2. Use. The environmental aspects posed by the use of COM involve the efficiency of fly-ash separation in the flue gas and its disposal (fig. 4).

A problem yet to be dealt with involves the analysis of the organic portion of the fly ash that escapes the separation unit.

As for gas pollution, there are no major differences between COM and petroleum fuel oil (tab. VI); only the NOx concentration is

TABLE V. TUMOR-INITIATING AND MUTAGENIC ACTIVITY OF FUELS DERIVED FROM PETROLEUM AND COAL /6/

SAMPLE	N° OF MICE WITH TUMORS LASTING 30 DAYS*	AVERAGE N° OF TUMORS PER TUMORBEARING MOUSE	MUTAGENIC ACTIVITY (WITH ACTIVATION) TA98	TA100
Positive control				
benzo(a) pyrene	26	6.0	++	++
South Louisiana Petr.				
naphtha	2	1.0	--	--
light gas oil	3	1.6	--	--
mid gas oil	3	1.0	+	+
residue	5	1.6	+	+
crude	5	1.4	--	+
H-coal fuel oil mode				
atmospheric still overhead	2	1.0	--	--
atmospheric still bottoms	14	2.4	++	+
vacuum still overhead	24	4.5	++	++
vacuum still bottoms	13	2.4	++	++
Exxon donor solvent				
naphtha	0	0	--	--
solvent	12	2.5	++	+
heavy gas oil	18	3.6	++	++
vacuum gas oil	20	3.3	++	++
residue	17	2.4	++	++
SRC-II				
naphtha	3	1.0	--	--
fuel oil	9	2.3	++	+
heavy gas oil	18	3.1	++	++
stripper tower bottoms	15	2.5	++	++
vacuum tower bottoms	10	1.9	++	++
residue	21	3.0	++	++

* based on 30 mice per group

-- $<$ 0.01 revertant/μg

\+ 0.1 to 0.01 revertant/μg

++ $>$ 0.1 revertant/μg.

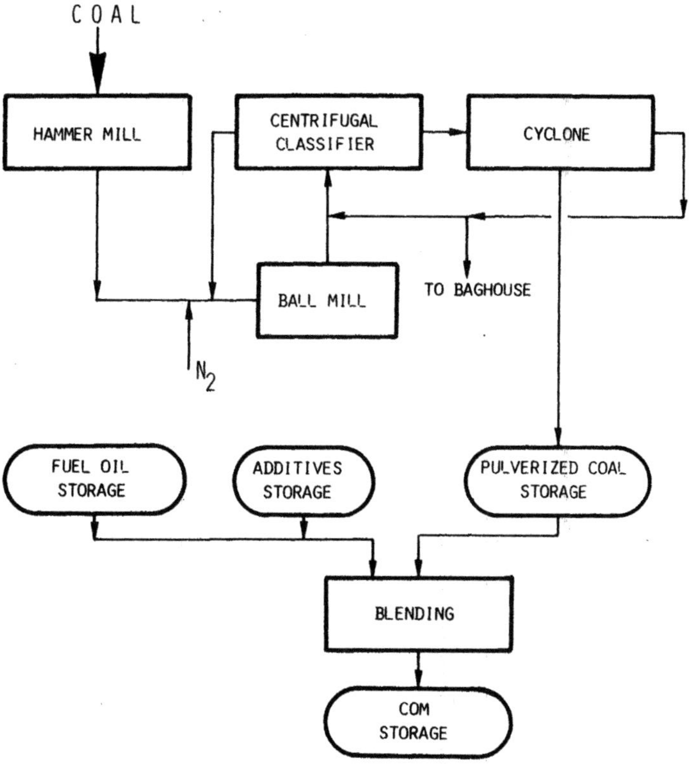

Figure 3. Production scheme of coal-oil mixture /7/.

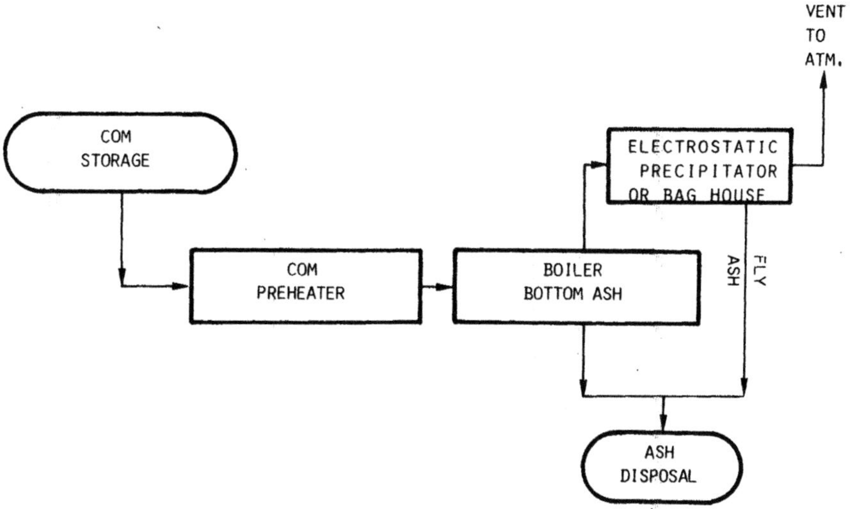

Figure 4. Coal oil mixture utilization.

TABLE VI. COMBUSTION OF COAL-OIL MIXTURES IN A 700 HP WATERTUBE BOILER /8/

FLUE-GAS ANALISIS	50% COM			40% COM			30% COM		n° 6 FUEL OIL		
O_2 (%)	2.99	2.35	2.97	3.33	2.75	2.40	2.76	2.39	3.00	2.43	3.02
CO_2 (%)	14.49	15.14	14.48	14.02	14.59	14.82	14.36	14.68	13.84	14.22	13.77
CO (ppm)	7.7	615	25.3	35	52	91	37.4	71	17.8	119	69
NO_x (ppm)	414	366	344	363	366	350	321	305	223	207	209
THC (ppm)	3.0	14.1	1.1	.88	3.3	11.0	0.6	2.7	0.94	2.34	4.05
Opacity (%)	36.5	54	32.5	27.7	29.5	33.5	29.5	31.5	20.2	20	9

larger, which is probably due to the higher nitrogen content of coal compared to oil /8/.

The SO_2 emissions should be smaller due to the lower S content of coal compared to oil (on an energy content basis) and because the SO_2 partly reacts with the alkaline metals of the ash.

The lower vanadium content of coal, relative to oil, could also decrease the oxidation of SO_2 to SO_3 with a relief of the environmental impact.

The addition of chemical additives for stabilization purposes could contribute to atmospheric pollution with emissions different from those found in conventional coal oil combustion.

However their concentration will be approximately 0.2% and should not contribute specifically to emissions.

Most additives are composed of organic materials which are combustible and should not present unusual problems.

1.3. Coal water mixture

Also this product is a pseudo-liquid whose typical composition is as follows /9/:

70%	coal
29%	water
1%	additives (flow improvers and stabilizers)

1.3.1. Production. The technology has reached the demonstration plant scale.

A process scheme is reported in fig. 5 /9/.

The very nature of CWM requires to provide for coal beneficiation by means of gravimetric methods or based on the different surface properties of macerals and the mineral phase.

The production of CWM with ash below 2-3% and sulfur below 1% seems therefore to be technically and economically feasible in the next future.

The CWM should therefore be regarded as a clean fuel to be suitably used also in highly populated areas.

1.3.2. Use. As the CWM is a pseudo-liquid, its utilization (transport over long distance, storage, handling) can be accomplished in closed systems with a negligible environmental impact.

The combustion tests performed so far have not anticipated special environmental problems, besides of course the separation of the fly-ash /10, 11/.

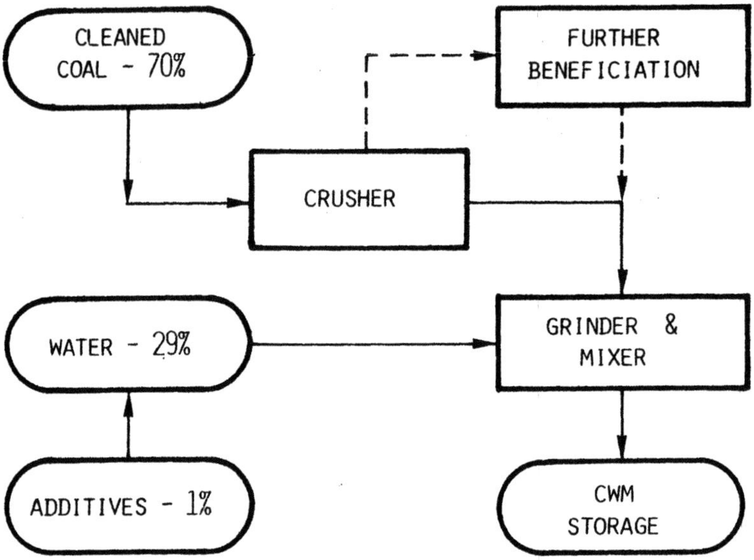

Figure 5. Coal water mixture: general production scheme /9/.

The tests underway and those planned in the near future should however give a confirmation to this effect.

Additives are usually present at high concentration (1%) and their effect on gas and particulate emissions should therefore be checked.

1.4. Low, medium, high BTU gas from alternative energy sources

Coal is the most important raw material even if other materials can be exploited for the production of gas.

Coal gasification is a well known process widely used in industrial scale.

Technological improvements are continuously developed for achieving more efficient processes, each tailored for a given type of coal and for obtaining a gas composition appropriate to the final use.

1.4.1. Production. A generalized scheme of a gasification plant is reported in fig. 6.

At this stage several environmental aspects common to other technologies (e.g. emission of particulate during coal milling) and others typical of the gasification process must be considered as shown in fig. 6.

In this connection the different problems can change, expecially quantity-wise, from plant to plant and their control from the environmental protection point of view is accomplished on the basis of long established technologies common to other productions.

By referring to specific processes and plants, a better characterization and definition of the environmental problems can be achieved.

The potential pollutants associated with the plants operating a fixed bed gasifier are:

- gas leaks from the coal feeding line which contain all components of the raw gas like tar, incondensable hydrocarbons, particulate, sulfur and nitrogen compounds;
- wet ash and polluted water from quenching operations, containing coal and other organic and inorganic compounds;
- flash gas coming from the quenching water and entraining raw gas and solid particulate;
- dust mainly containing fine coal contaminated by organic compounds;
- contaminated water from the condensation process, containing small amounts of tar, condensable hydrocarbons, sulfur compounds, ammonia and other raw gas components.-

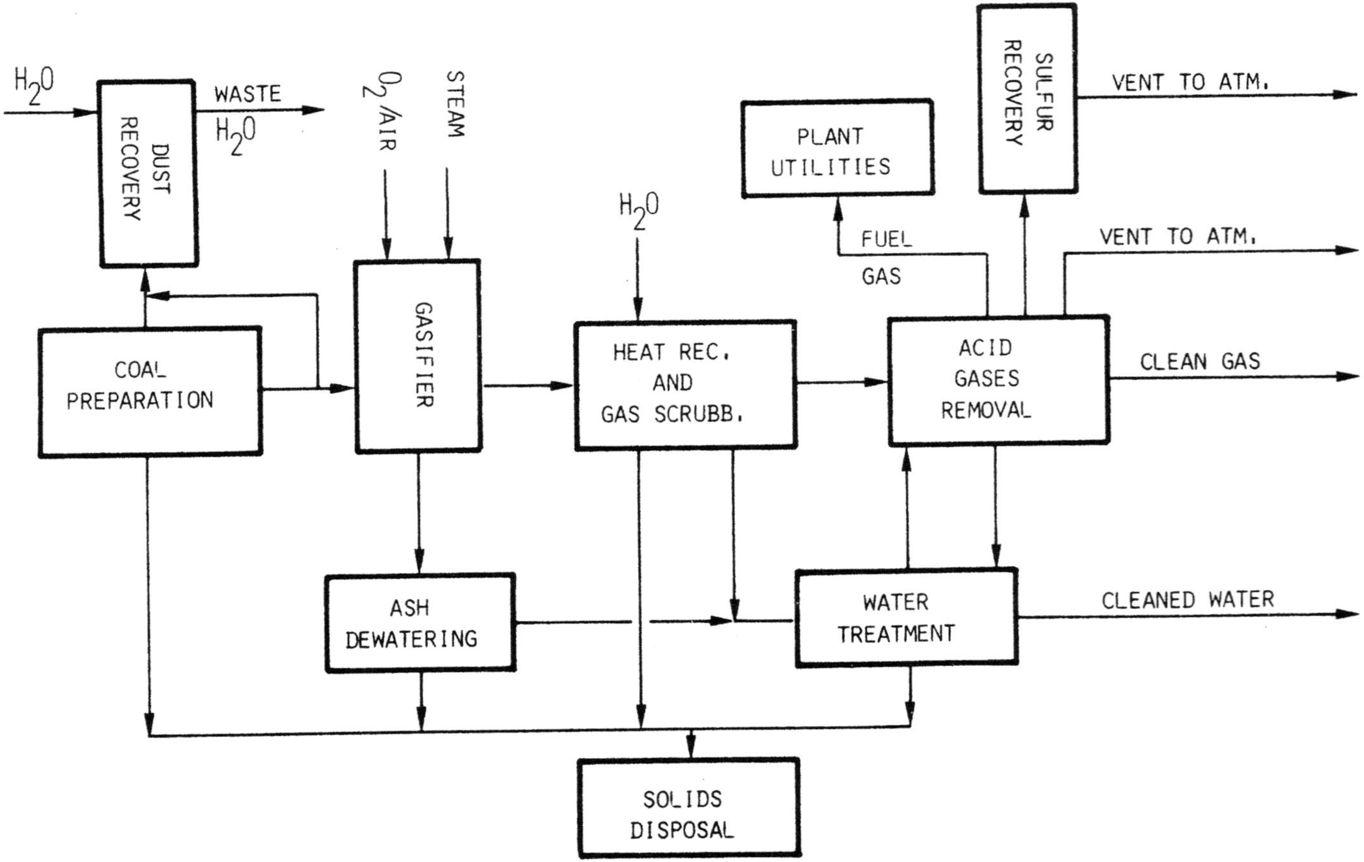

Figure 6. Block flow diagram of a coal gasification process.

The novel technologies developed recently will be further verified after the start-up and operation of the Tennessee Eastman coal gasification plant (Texaco technology) /12/ designed for the production of acetic anhydride from synthesis gas, via methanol.

1.4.2. Use. The utilization of fuel gas in place of heavy fuel oil is of course technically feasible, well accepted by the users, and does not pose any environmental problem.

The possibility of accurately controlling the combustion parameters enables to minimize the CO and NOx emissions.

Sulfur can be totally absent.

In practice the only pollutant is CO_2, this being the unavoidable price of the fossil energy.

2. GASOIL

This terminology defines the middle distillates (boiling range 200 to 400°C) mainly used as fuel in residential heating and in the production of electricity via turbogas.

Possible synfuels envisaged at the present time are:

- methanol
- middle distillates from direct coal liquefaction
- middle distillate from tar sand

2.1. Methanol

This is a well known chemical with wide applications (tab. VII)) /13/.

TABLE VII. METHANOL WORLD CONSUMPTION IN 1979 /13/

Formaldehyde	49.2%
Solvent	9.5%
Methyl halides and chloroform	6.8%
Methylamine	5.2%
Methylacrylate	4.9%
Dimethylterephthalate	4.9%
Acetic acid	4.2%
Others (for example MTBE)	15.3%

Total methanol consumption in 1979 = 11 million tons

The world production reached 11 Mt in 1979 and it is expected to increase considerably in the coming years as the demand for energetic applications develops.

2.1.1. Production. The schematic flow diagram of a 11 kt/d hypothetic methanol production plant is reported in fig. 7.

The environmental problems relative to the gasification units are illustrated under 1.4.1.

No special problems are anticipated for the synthesis units.

2.1.2. Use. Combustion tests performed in Italy within the PFE-1 CNR project /14,15/ have demonstrated the excellent suitability of methanol as fuel for residential heating.

The resulting emissions are very low, with aldehydes being absent despite the general expectation.

The handling problem remains to be solved, because methanol is associated with a high toxicity level.

2.2. Middle distillate from direct coal liquefaction

2.2.1. Production. In addition to the so called 1st generation processes outlined under 1.1, more advanced processes are being developed in smaller capacity pilot plants. The process which is gaining more favor is a multi-step type because it offers a greater plant flexibility and enables to optimize the production of middle distillates (fig. 8) /16/.
The estimated investment and operating costs indicate that the coal derived liquids will only be competitive with petroleum products towards the end of the century.

The present experience indicates that the coal liquefaction plant can pose considerable environmental problems and appropriate solution must be pointed out for each stage of the coal liquefaction process.

As to the pollution related to coal processing i.e. waste water, sour gas, ammonia, particulate, ash, hydrocarbons, the problems posed can probably be solved using solutions and technologies well known and already widely tested in other processes.

The experience gained at the pilot plant stage has indicated also a number of potential dangers for the manpower of the future industrial plants (tab. VIII) /17/.

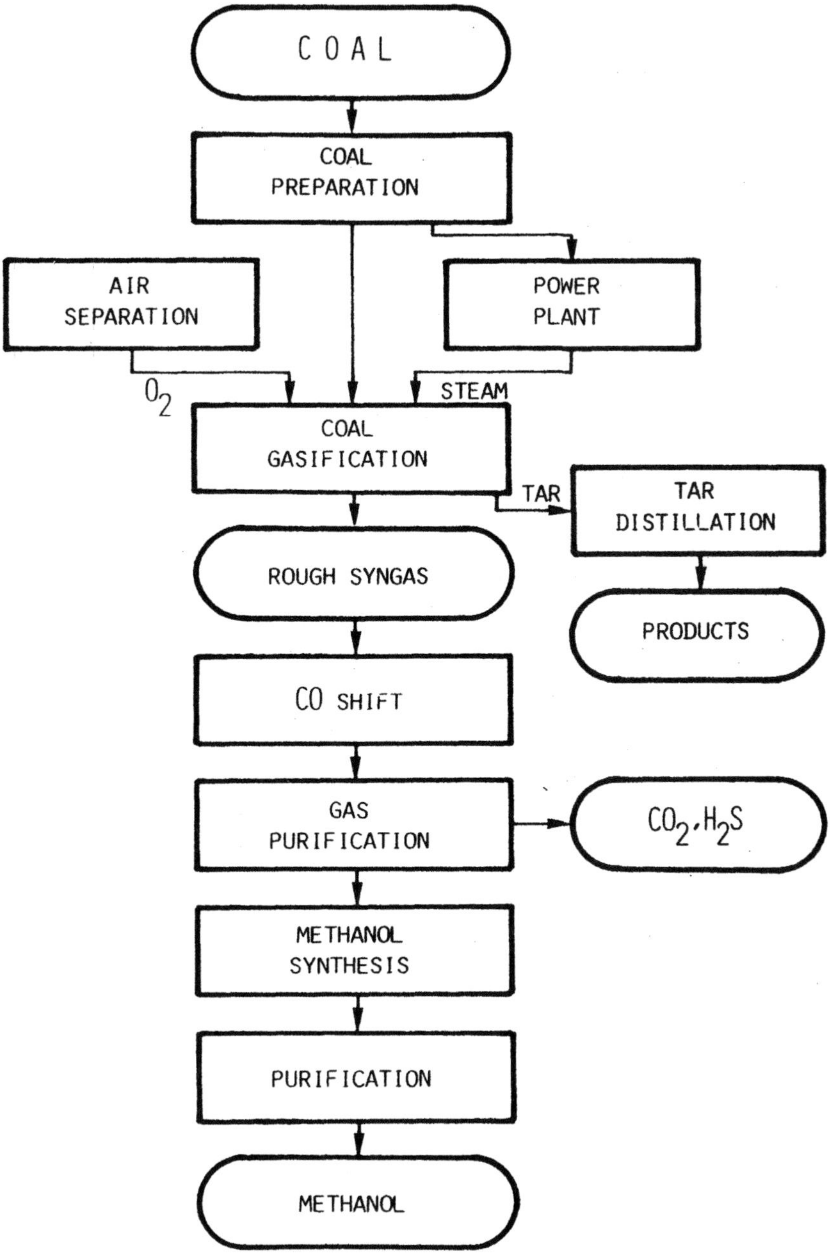

Figure 7. Schematic flow diagram of methanol production from coal.

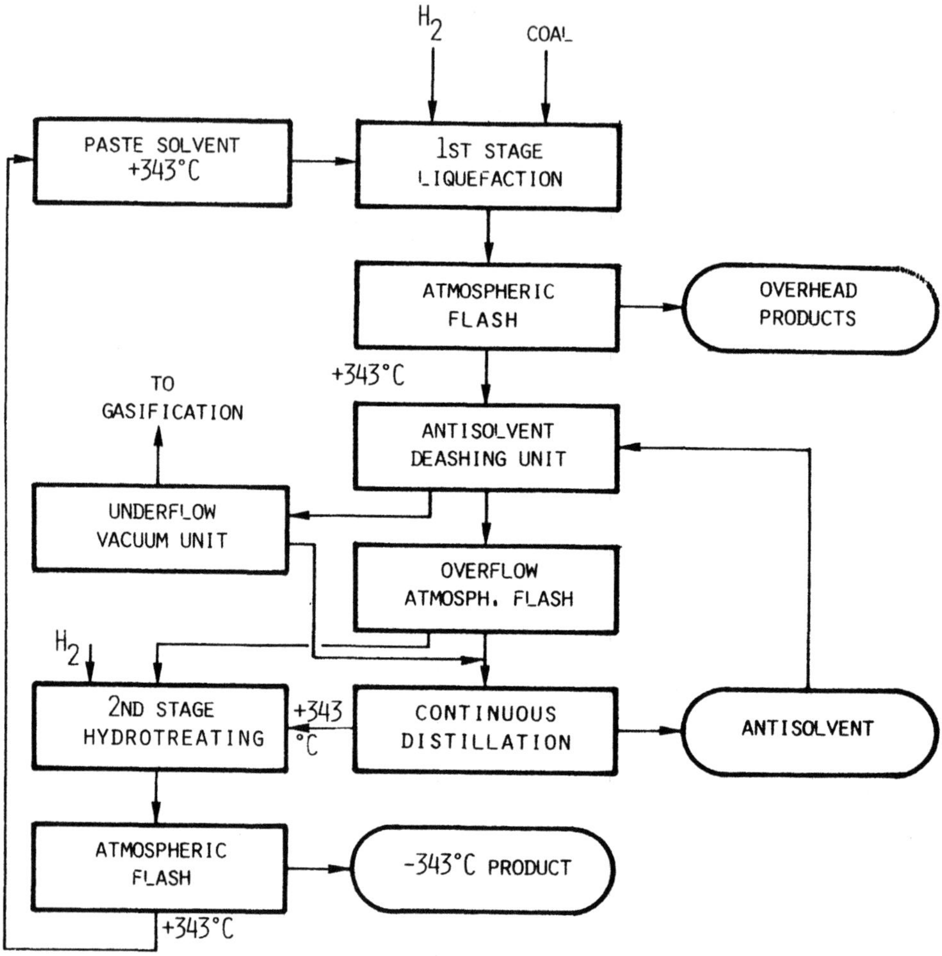

Figure 8. Hydroliquefaction of coal: adavanced two-stage process scheme /16/.

TABLE VIII. POTENTIAL OCCUPATIONAL HAZARDS IN COAL LOQUEFACTION PLANTS /17/

UNIT OPERATION/UNIT PROCESS	POTENTIAL HAZARDS
Coal handling and preparation system	Coal dust, noise, fire, explosion, asphyxia, (nitrogen and carbon monoxide gases), burns
Liquefaction system	Phenols, ammonia, tars, thiocyanates, pah, carbon monoxide, hydrogen sulfide, hydrocarbons, fires, explosions, burns, high pressures, noise, ash, slag, mineral residue, spent catalyst
Separation system	Oils, phenols, hydrogen cyanide, ammonia, hydrogen sulfide, burns, fires
Upgrading and gas purification	Light hydrocarbons, phenols, ammonia, hydrogen sulfide, carbon dioxide, carbon monoxide, burns, fire, explosion, high pressures
Hydrogen production	Tar, naphtha, hydrogen cyanide, fire, catalyst dust, burns, hot gases (carbon monoxide, hydrogen)
Waste treatment facilities	Hydrogen cyanide, phenols, ammonia, particulates, hydrocarbon vapors, sludges, spent catalyst, sulfur, thiocyanates

2.2.2. Use. The gasoil produced by direct coal liquefaction processes has a content of aromatics and nitrogen and oxygen containing compounds much higher than the corresponding petroleum products.

The sulfur compounds can instead be present at a low level.

A typical product analysis is given in tab. IX /18/.

TABLE IX. MIDDLE DISTILATE FROM COAL HYDROLIQUEFACTION /18/

Specific gravity	kg/l	0.946
Aniline Point	°C	-32
Bromine Number	mg/g	2.0
Color ASTM		3
Flash Point	°C	57
FIA Analysis vol %		
Aromatics		77.3
Olefins		5.5
Saturates		17.2
Carbon	wt %	88.86
Hydrogen	wt %	10.97
Sulfur	ppm	15
Nitrogen	ppm	185
Oxygen	ppm	1300
CON-C 10% Bottom	wt %	0.05
Corrosion	Cu strip	2C
Distillation ASTM °C		
10%		172
50%		210
90%		289
Viscosity at 100°F	cSt	1.9
Smoke Point	-	7.8

The raw product from the liquefaction plant contains PNA and aromatic amines that make these products cancerogenic and mutagenic (tab. V) /6/.

As a result of the subsequent hydrogen treating for the product upgrading required for meeting the specification of the market, there is a considerable reduction of the toxic products to the extent that the finished product is comparable to the present oil derivatives (tab. X and XI) /19/.

The combustion of these products for residential heating should not increase the pollutant output with a probable exception for the NOx /19/.

Due to the higher aromatic content of the fuel, also PNA emissions should be carefully verified.

TABLE X. CATALYTIC HYDROTREATING OF MIDDLE DISTILLATE FROM SRC-II PROCESS CONDITIONS AND RESULTS /19/

Sample	LIQUID PRODUCT H%	N%	S%	O%	H_2 cons. SCF/bbl	T °C	LHSV
(FEED)	8.42	1.057	.25	3.29	-	-	-
1	10.81	.352	.01	.59	1,930	325	0.50
2	10.06	.631	.02	1.65	1,270	325	1.0
3	11.54	.033	.01	.17	2,670	350	0.50
4	9.80	.700	.02	1.58	970	310	0.50
5	12.24	.001	.01	.09	2,880	375	0.50
6	12.39	.001	.01	.03	3,250	400	0.50
7	12.82	.001	.01	.06	5,010	400	0.35
8	9.68	.742	.01	2.00	980	325	1.0

TABLE XI. CATALYTIC HYDROTREATING OF MIDDLE DISTILLATE FROM SRC-II TOXICOLOGICAL DATA /19/

Sample	Concentration, µg/plate	Number of revertants without/with metabolic activation TA 98	TA 100
FEED	250	30/1822	144/322
FEED	100	34/1122	163/348
FEED	50	27/547	150/306
Background	0	21/60	144/142
8	100	15/131	134/257
4	100	19/111	176/232
2	100	23/178	151/226
1	100	16/64	158/218
3	100	10/40	161/194
5	100	22/64	162/194
6	100	19/70	128/186
7	100	17/65	153/218
Background	0	25/66	169/220

2.3. Middle distillate from tar sand

2.3.1. Production. The production of syncrude from tar sand reached the commercial stage since 1967 in Canada where two plants for an overall capacity of 200,000 BPD are in operation. Short term developments in the USA are possible /20/.

The surface extraction process scheme presently employed in the Canadian plants is shown in fig. 9 (hot water process) /21/.

The extracted tailings are sent to the tailing pond that consists of an artificial lake with the banks constructed with the processed sand.

After a certain period of time, much longer that anticipated, a relatively clear water for recycling becomes available from the top layer in the lake. The environmental impact of this technology is quite evident and the Canadian authorities do not regard it as acceptable any more.

These environmental problems should hopefully be overcome by means of recovery technologies under study and development in Canada and USA.

One problem that can be resolved, but yet to be resolved, is the disposal and re-vegetation of the processed sand whose volume is larger than that of the original sand.

2.3.2. Use. The middle distillate originated from the Canadian tar sands has an aromatic content higher than the corresponding petroleum cut.

As a result of the severe hydrogenation process the heteroatom content is however much lower (tab. XII) /22/.

Its use in residential heating should not therefore give additional problems with respect to the oil products.

3. DIESEL FUEL

The diesel fuel has a specified cetane number quality requirement (CN ≥ 40) which defines the hydrocarbon composition of the product.

By definition the cetane number of a given fuel is represented by the composition of the binary mixture of n-hexadecane (CN = 100) and -methyl naphtalene (CN = 0) that results in the same ignition delay in the CFR engine.

A synfuel should therefore meet this requirement in order to be an acceptable fuel for the present and the short and long projected car population.

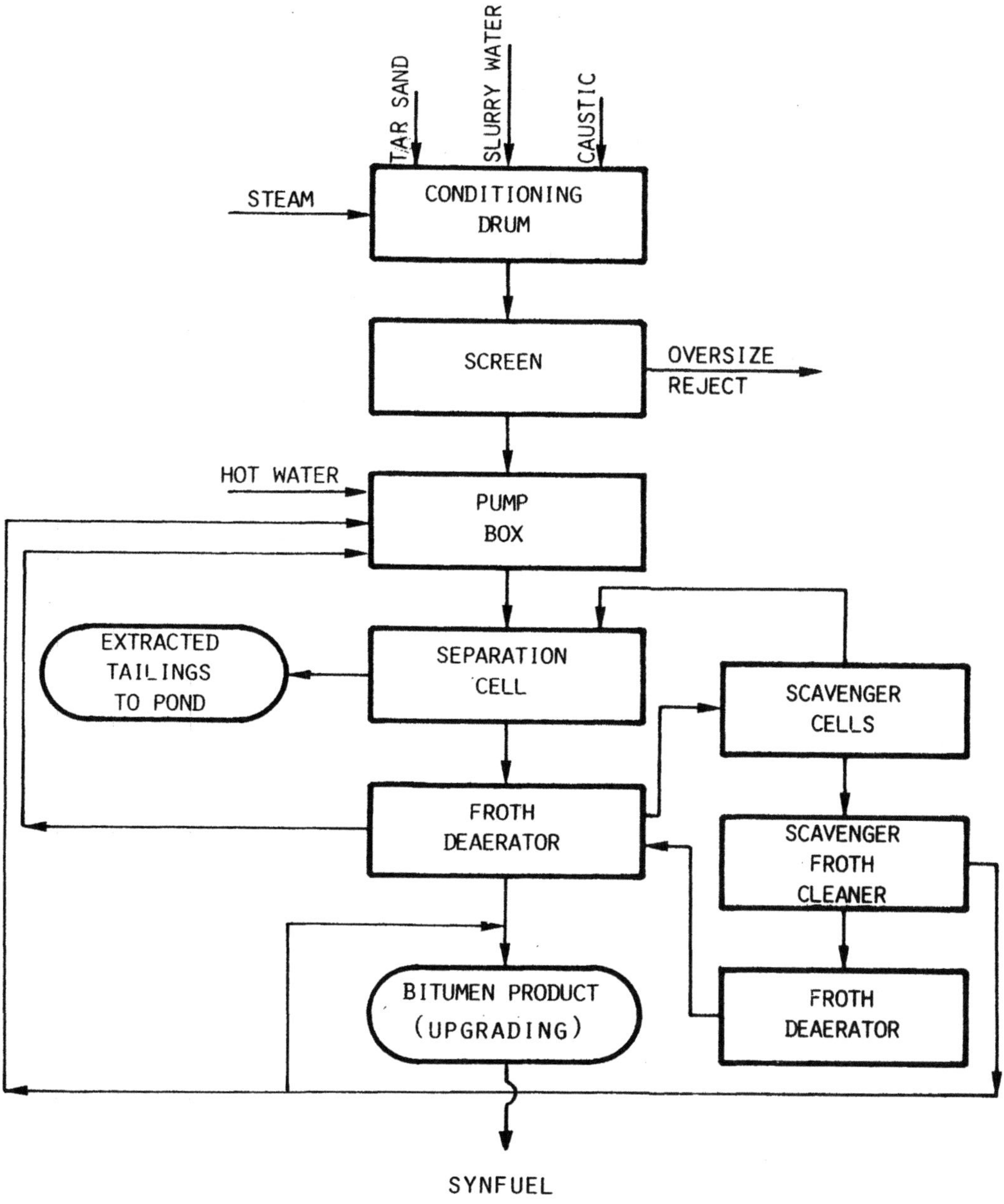

Figure 9. Bitumen extraction from tar sand: hot water process - block diagram /21/.

TABLE XII. MIDDLE DISTILLATE FROM TAR SAND

Specific gravity	kg/l	0.8974
Viscosity at 40°C	cSt	3.3
Cloud point	°C	-27
Pour Point	°C	-50
Sulphur	wt %	0.03
Nitrogen	ppm	25
Aromatics	vol %	44
Cetane no.	-	34
Cetane index	+	42
Distillation ASTM	°C	
IBP		157
10%		192
50%		274
90%		329
FBP		334

The synfuels that could at present be a substitute for the oil derived diesel fuel are:

- the F-T middle distillate
- the middle distillates from oil shale.

3.1. F-T middle distillate

3.1.1. Production. The synthesis of hydrocarbons from CO and H_2 is well known since long and is accomplished in several plants (tab. XIII) /3/.

Three plants are presently operating in South Africa (Sasol 1, Sasol 2, Sasol 3).

The Sasol production schemes are reported in fig. 10 /3/.

The environmental problems are mainly confined to the gasification step (see item 1.4.1).

The waste water and gas treatment processes derived from a study conducted in the USA /23/ are reported in fig. 11 and 12.

3.1.2. Use. The properties of the diesel fuel obtained via this process are reported in tab. XIV /24/.

The total absence of sulfur and nitrogen makes this product preferable to the current diesel fuel from environmental consideration.

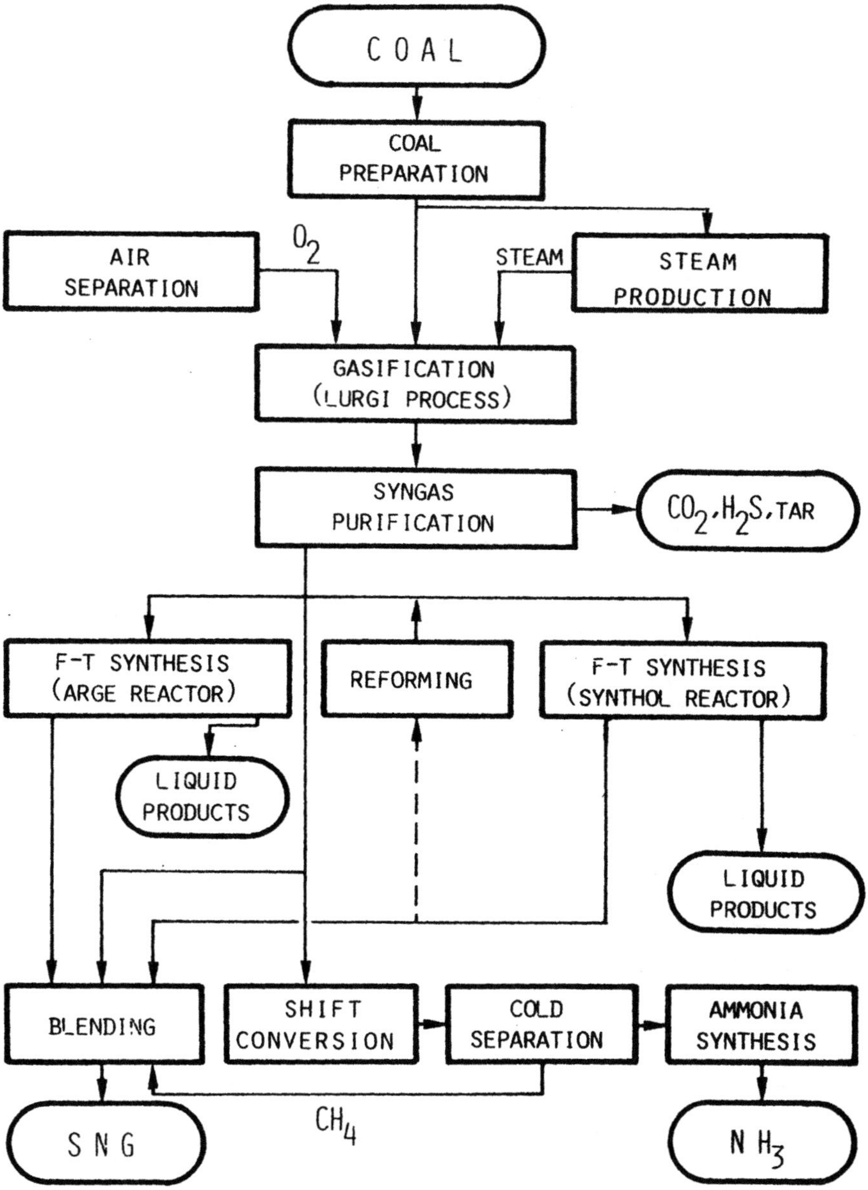

Figure 10. Simplified "Sasol" process: block diagram /3/.

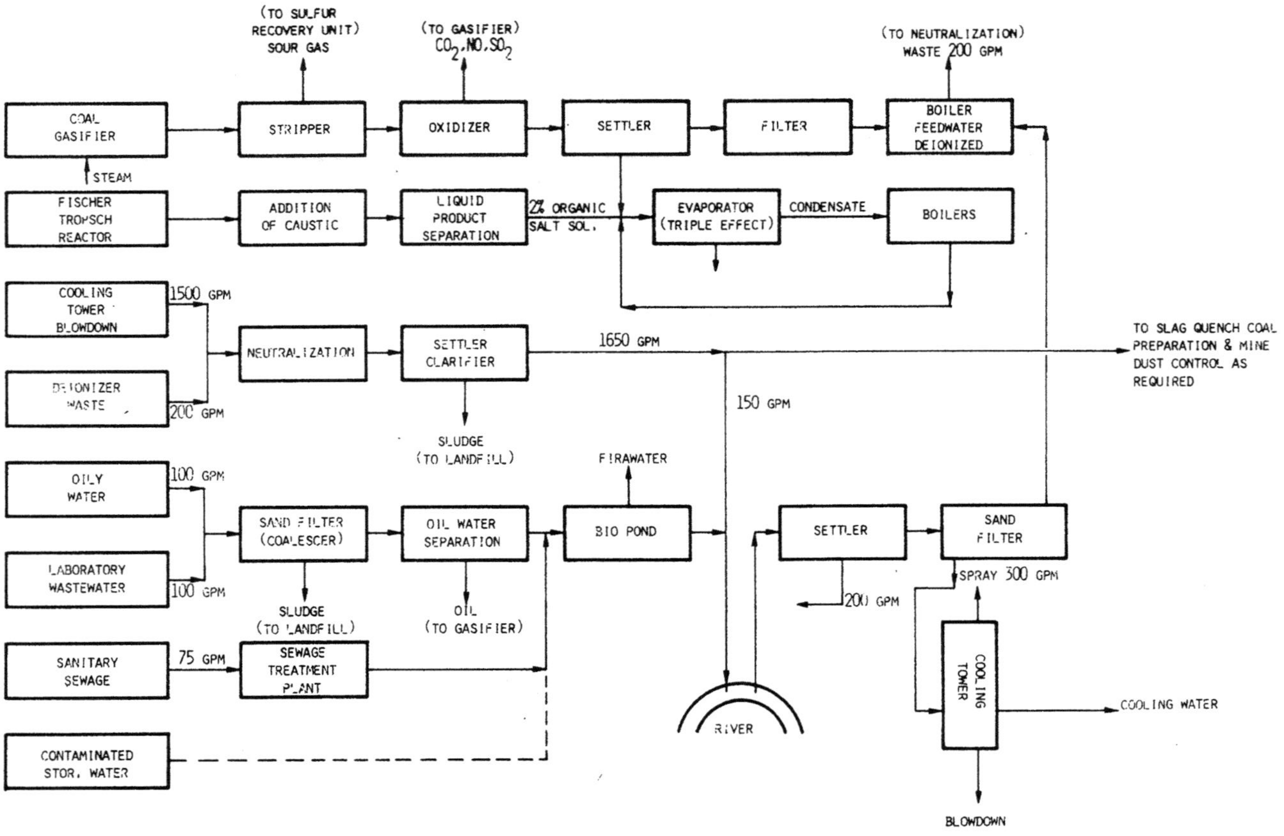

Figure 11. Water treatment scheme for Fischer-Tropsch process /23/.

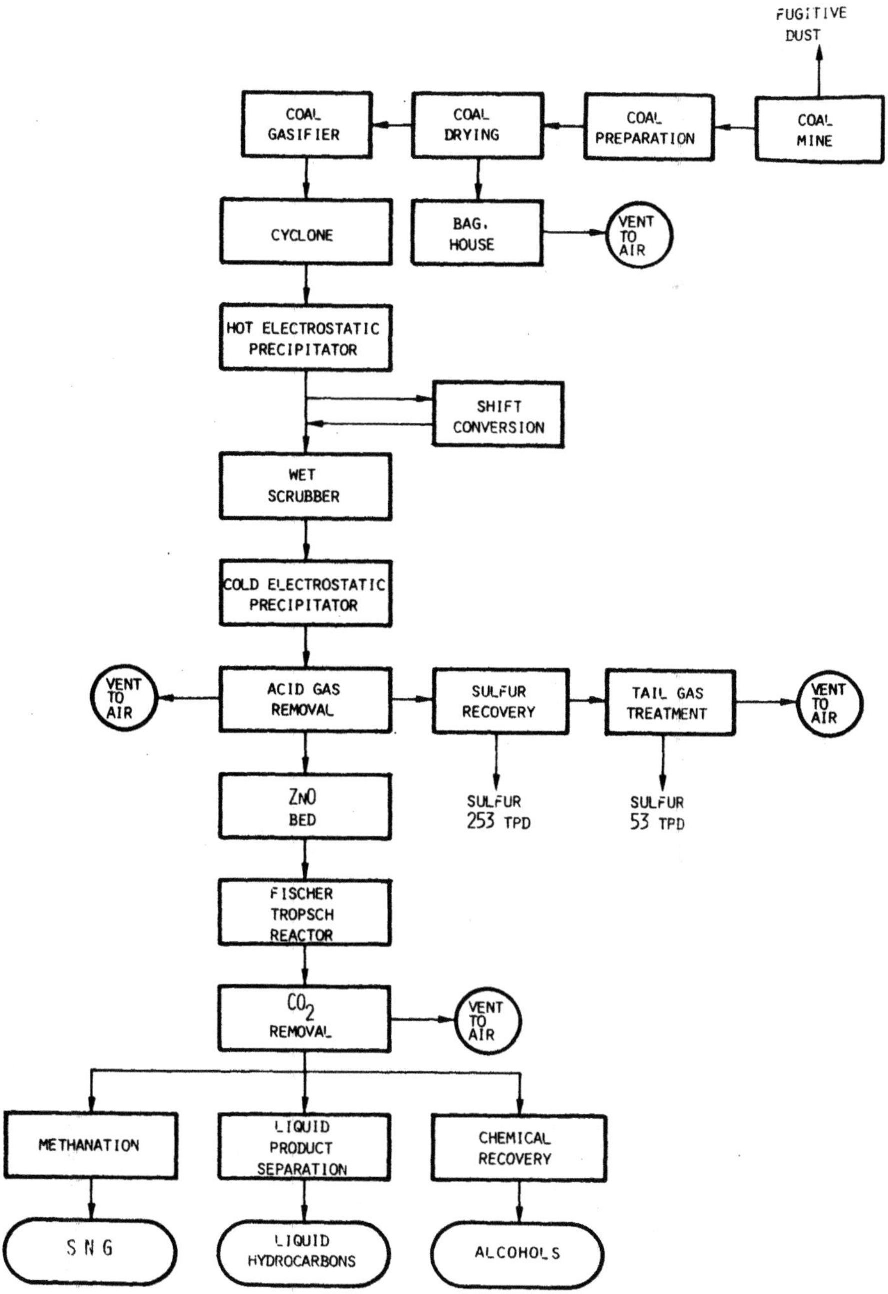

Figure 12. F-T plant: air pollution abatement /23/.

TABLE XIII. FISCHER-TROPSCH SYNTHESIS PLANTS INSTALLED UP TODAY /13/

Year	Country	N° of plants	Plant capacity (10^3 t/y of liquid products)*
1935-39	Germany	9	740
1937	France	1	30
1939-43	Japan	4	260
1939	Manchuria	1	50
1950	USA	1	360
1965	South Africa (Sasol I)	1	260
1980	South Africa (Sasol II)	1	1.700**
1982	South Africa (Sasol III)	1	2.100**

(*) 10^3 t/y are equivalent to about 20 B/D
(**) Gasoline, gasoil, jet fuel

TABLE XIV. DIESEL FUEL QUALITY FROM F-T PROCESS /24/

Specific gravity,	kg/l	0.770
Cetane number	-	50
Pour point	°C	-7/-12
Sulfur	wt %	NIL

3.2. Middle distillate from oil shale

3.2.1. Production. The possibility of oil recovery from oil shale is well known since over a century. In the recent past small capacity industrial productions were accomplished in several countries (Sweden, China, USSR).

At the present time the major interest in this field can be found in the USA where a 10.000 BSD plant is due to start-up in 1983 /25/.

A production scheme with a detailed description highlighting the streams with environmental implication is given in fig. 13 /26/.

The ecological problems related to the shale industry are considerable and will therefore influence to a large degree any commercial development.

The environmental impact is important expecially during mining

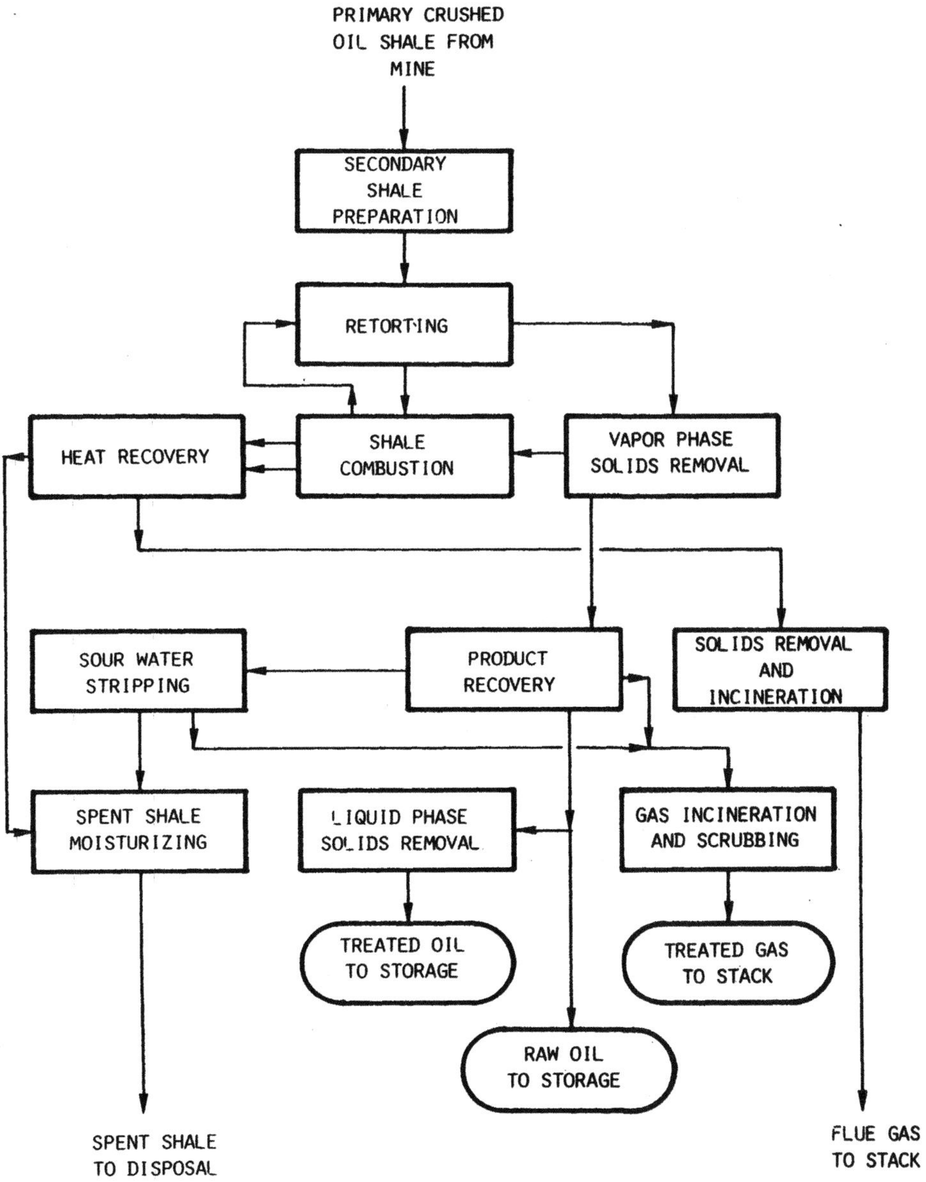

Figure 13. Block flow diagram of an oil shale plant /26/.

and retorting both on the surface and, even if to a lesser degree, in situ.

The problems posed are three fold:

- air pollution (particulate, sulfur and nitrogen compounds)
- water pollution (fluorides, chlorides, sulfates, ammonia, nitrates, cyanides, phenols, arsenic, sulfides, trace elements)
- disposal of the processed shale with the related problems of retorting it in situ and probably revegetating the area; this issue is complicated by the fact that the spent shale has a volume 20% larger than the original material.

3.2.2. Use. The properties of the diesel fuel obtainable are excellent (tab. 15) /27/ and its use should alleviate the problems connected with the use of the present oil derived products on account of the low content of heteroatoms and aromatics (high cetane index).

TABLE XV. DIESEL FUEL FROM SHALE OIL /27/

Specific gravity,	kg/l	0.828
Distillation ASTM	°C	
50%		266
90%		331
End Point		345
% Recovered	vol %	98
Sulfur	ppm	5
Cloud Point	°C	-18
Viscosity at 100°F	cSt	2.5
Flash Point	°C	58
Cetane Index	-	55

4. JET FUEL

This product is very important and its demand is bound to increase with the development of the civil aviation. It is also a strategic product of vital importance in case of war.

The quality specifications are very tight and limit the composition in terms of both hydrocarbon types and boiling temperature ranges.

Up to now the only synfuel appropriate as a replacement of jet fuel is a medium-light distillate from oil shale.

4.1. Jet fuel from oil shale

4.1.1. Production. The considerations reported under 3.2.1 apply.

4.1.2. Use. The product has excellent properties: low in aromatics, high smoke point, high thermal stability (tab. XVI) /27/.

Environmental problems different from those of petroleum derived JP4 are not anticipated.

TABLE XVI. JP-4 FROM SHALE OIL /27/

Specific gravity	kg/l	0.776
Distillation	°C	
20%		143
50%		194
90%		229
EBP		248
Olefins	vol%	0.0
Aromatics	vol%	9.5
Sulfur	ppm	5.0
Mercaptans	ppm	3.0
Nitrogen	ppm	2.0
Vapor pressure	psi	2.2
Smoke point	-	30.7
Freezing point	°C	-60
Existent gum	mg/100 ml	1
Thermal stability JFTOT		
Pressure drop,	mm Hg	0
preheater deposit	-	1

5. GASOLINE

The fuel for Otto cycle engines has been given so far the greatest attention in the search for possible alternative products.

Synfuels can be used either as components in a blend with oil derived hydrocarbon products or as finished products.

This fuel could be used in the present car population, in a more or less modified car population or in a totally new car population designed to meet the future mobility needs.

Synfuels that can potentially substitute or extend the present gasolines are:

- methanol
- ethanol
- acetone, butanol, ethanol mixtures (ABE)
- alcohol mixtures
- FT gasoline
- M gasoline
- MTBE
- gasoline from direct coal liquefaction

5.1. Methanol

This is the synfuel most widely studied in the last ten years that offers the best chances.

5.1.1. Production. The production of methanol has been examined under 2.1.1.

5.1.2. Use. The areas of application of methanol so far examined are as follows:

- as a gasoline component for the present car population up to 3% concentration. This solution has already been put in practice in some markets.
  On account of the very small concentration of methanol, the exhaust emissions effect can be regarded as negligible.
  A positive effect may come from the possibility of reducing the lead alkyl content of gasolines without increasing the energetic cost of production of the hydrocarbon base.

- as a gasoline component at 10-20% concentration for the present car population duly modified in carburation and in the intake system.
  In this case the effect of methanol on exhaust emissions is larger.
  If compared to the standard gasoline the M10-M20 blend is associated with the following changes:

  - decrease of the CO content

- small increase of the HC content
- no change in the NOx emissions
- large reduction of the PNA emissions
- large increase of the aldehyde emissions

As an example the results of a test run on 7 cars fueled with a gasoline/methanol 90/10 mixture are reported in fig. 14 /28/.
The increase in aldehyde emissions affects the concentration of oxidizing compounds in the atmosphere with a bearing on the photochemical smog.
This is a very important aspect presently under investigation in many laboratories and research institutes.
The methanol emissions resulting from evaporation losses due to the high vapor pressure of the blend are another possible source of pollution.
The antiknock rating of methanol in gasoline at 15-20% concentration makes it acceptable from an energetic point of view for reducing or even completely eliminating lead in gasoline (tab. XVII) /13/.
This would help the introduction of catalytic mufflers with a clear benefit for atmospheric pollution.

TABLE XVII. CLEAR RON OF BASE GASOLINES FROM T+CR+CC+A REFINERY SCHEME (1) /13/

Lead content	g/l	0.6	0.4		0.15		0	
Methanol	vol %	zero	10	20	10	20	10	20
Base gasoline clear RON		91.2	88.8	81.3	91.5	84.5	96	92.1

(1) Final blend RON is 98-99 in all cases

- as neat compound (100%) for fuelling specially designed engines. The considerable advantages in terms of emissions (total absence of sulfur and lead, drastic reduction of the PNA) are however associated with an increase of formic aldehyde. The adoption of the catalytic muffler would make this solution highly desirable in terms of environmental protection.
  One aspect to be carefully considered is the possible loss of methanol in liquid or gaseous state during the distribution cycle.

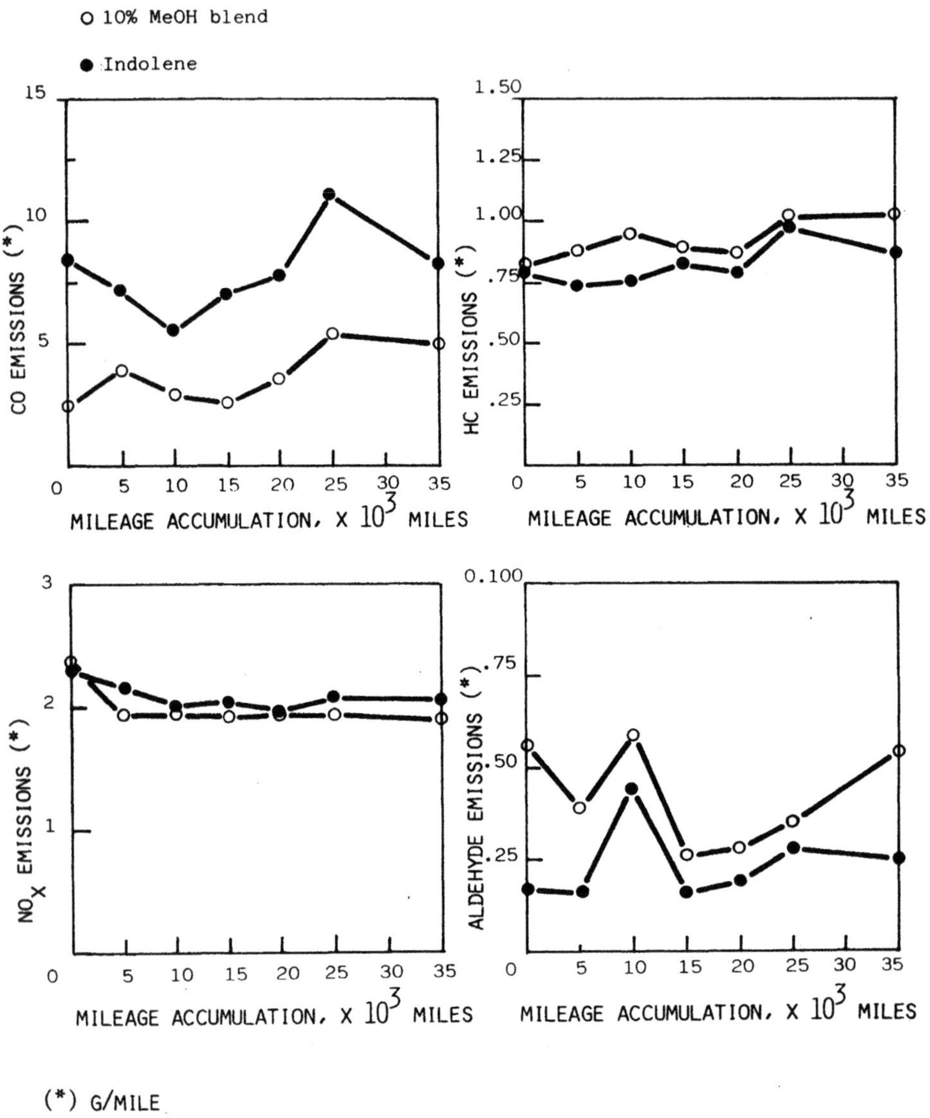

Figure 14. Exhaust emission comparison between methanol containing blend and a conventional gasoline /28/.

5.2. Ethanol

Ethanol from biomass has already been adopted as fuel in transports on a commercial scale in

- Brazil E20 and E100
- U.S.A. E10 (Gasohol)

5.2.1. Production. The production is presently based on the alcoholic fermentation process of sugar and starch containing materials (sugar cane in Brazil and corn crops in the U.S.A.)

The production of ethanol from cellulose via enzimatic hydrolisis followed by alcoholic fermentation is still at the research stage.

The production process of ethanol from biomass is schematically reported in fig. 15 /29/.

The major environmental problem is the disposal of the distillery residue unloaded from the bottom of the rectifier.

Fermentation processes (anaerobic and aerobic) can solve the problem with a recovery of energy (biogas) and by products.

5.2.2. Use. The same considerations relative to the use of methanol apply; however ethanol has a toxicity much lower than methanol as a millennial experience has proved.

5.3 Acetone, Butanol, Ethanol (ABE) mixtures

These three compounds are co-produced by fermenting sugar in presence of some species of bacteria (clostridium). This process reached in the past the commercial stage (production of acetone and n-butanol) but was later abandoned when low price oil became available.

5.3.1. Production. A large scale production of this mixture for energy use would be favored by the type of cellulosic raw materials involved, which however require a preliminary stage of acid or enzimatic hydrolisis for the production of sugar.

A production scheme from cornstalks is reported in fig. 16 /30/.

5.3.2. Use. In the french "Carburol" project the anticipated use of this product as such or as pure n-butanol is as cosolvent for methanol because it stabilizes very effectively the gasoline-methanol mixture in presence of water (fig. 17) /31/.

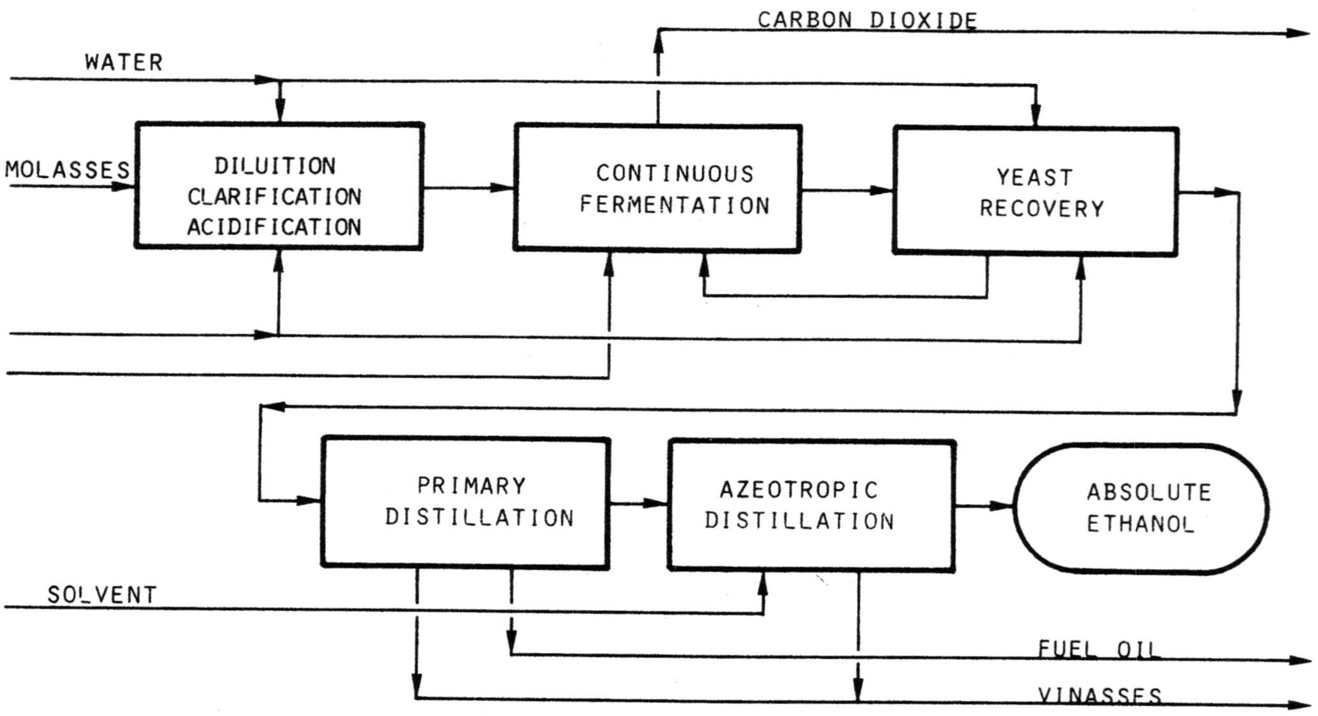

Figure 15. Process scheme for the production of automotive ethanol /29/.

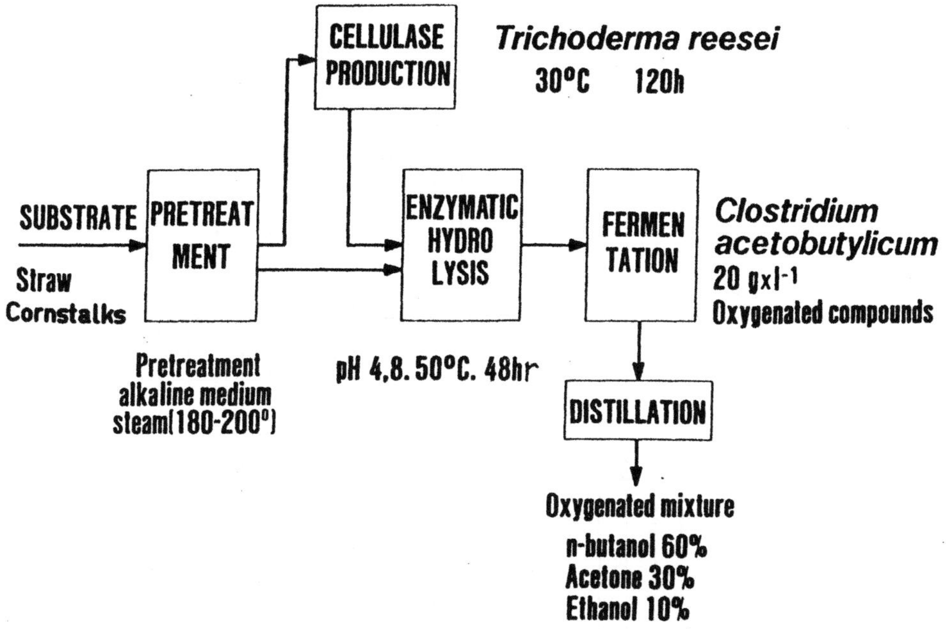

Figure 16. Acetone-butanol fermentation of cellulosic substrate /30/.

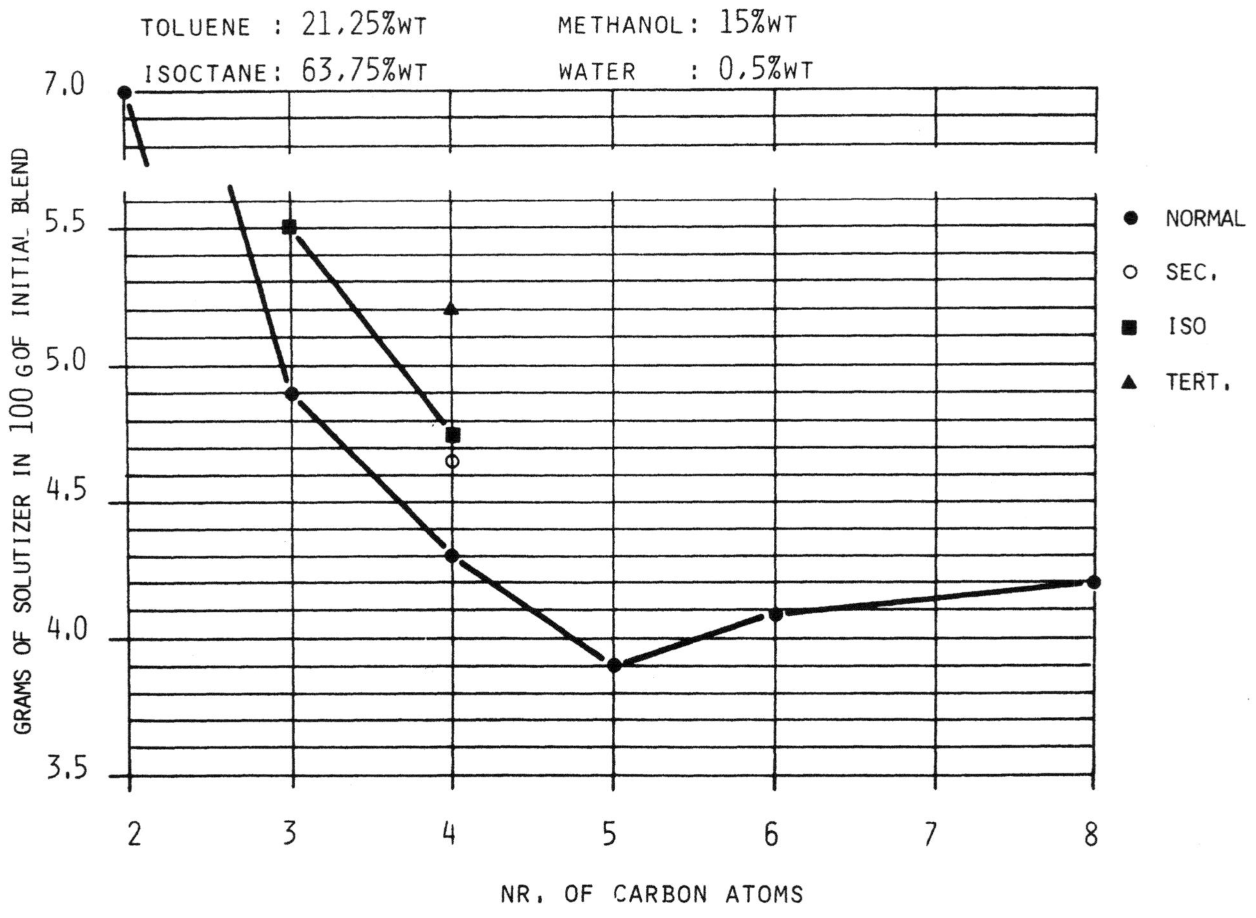

Figure 17. Solutizer required by blend to be omogeneous at 20°C /31/.

5.4. Alcohol mixtures

In the last years interest has increased for the direct synthesis of methanol and higher alcohols from coal derived syngas.

The higher alcohols (C_2-C_4) in fact increase considerably the stability of the methanol-gasoline mixtures and reduce the risk of phase separation at low temperatures or in presence of water in the fuel distribution system (fig. 17).
The addition of higher alcohols to methanol also increases the calorific value and improves the volatility characteristics, even if the antiknock rating is slightly lower.

5.4.1. Production. The direct synthesis of methanol and higher alcohols can be achieved by modifying the composition of the catalysts that are normally employed for the conversion of syngas to methanol /32/.

5.4.2. Use. As an example of the type of alcohol mixtures obtainable, the properties of methanol are compared with those of the MAS produced according to the Snamprogetti process, which is presently at the pilot plant stage (tab. XVIII) /33/.

TABLE XVIII. CHARACTERISTICS OF "MAS" VS METHANOL /33/

		"MAS"	Methanol
Density at 15°C		0.806	0.796
RVP	kg/cm^2	0.240	0.356
E70/E100	vol %	60/92	-
Lower heat value	kJ/kg	24572	19955
Flamm. at 50°C			
upper/lower	vol %	32.2/3.5	38.5/4.4
Carbon	wt %	46.1	37.5
Hydrogen	wt %	12.8	12.5
Oxygen	wt %	41.1	50.0
BLENDING PROPERTIES			
RON		117-132	123-139
MON		90-100	90-102
E70	vol %	120-185	160-220
E100	vol %	78-97	85-104
RVP	kg/cm^2	1.5-2.7	1.8-3.5

Note: Oxygenate in blends (vol %): 5÷15
Lead content (g/l): 0-0.4
Aromatics & olefins: typical W. Europe concentration

When this product is used in engines as a gasoline component, the regulated emissions decrease whilst the unregulated emissions, and the formic aldehyde in particular, tend to increase (tab. XIX) /33/.

The last feature is however commonly associated with the use of alcohols as fuel, methanol in particular.

TABLE XIX. CARBONYL COMPOUNDS EMISSION /33/
Example of experimental results on 1 of the 13 test cars (hot phase of ECE 15 cycle)

Carbonyl Compounds (g/test)	Conventional Gasoline	SP 10(*)	M 10(°)
Aldehydes:			
C 1	0.11	0.23	0.29
C 2	0.06	0.09	0.10
N C 3	0.04	0.04	0.05
ISO C 4	0.02	0.04	0.04
N C 4	0.02	0.03	0.03
Ketones:			
Acetone	0.07	0.10	0.11
Mek	0.09	0.14	0.17

(*) Blend containing 10% "MAS"
(°) Blend containing 10% methanol

5.5 F-T gasoline

5.5.1. Production. The products obtained by the Fisher-Tropsch process are mainly linear paraffins and alfa-olefins; the aromatic hydrocarbons are practically absent. As a consequence, the production of premium grade gasoline requires subsequent steps for octane upgrading (catalytic polymerization or alkylation of the C_3/C_4 olefins, C_5/C_6 isomerization, hydrogenation and catalytic reforming of naphtha).

5.5.2. Use. Some properties of the finished FT gasoline are reported in tab. XX /34/.

The high gum content reveals instability problems that should however disappear by adopting more severe production schemes and process operating conditions.

The use of FT gasolines should not pose additional environmental problems to those related to the present oil derived gasolines; the S and N heteroatoms content should be zero.

TABLE XX. PROPERTIES OF A GASOLINE FROM FISCHER-TROPSCH LIQUEFACTION PLANT /34/

Density at 15°C	kg/l	0.738
RVP	kg/cm^2	0.589
RON + 0.3 g/lPb		94.6
MON + 0.3 g/lPb		83.8
Aromatics	vol %	23
Olefines	vol %	0
Actual gums	mg/100 cc	3.4

5.6. M-gasoline

The Mobil Research and Development Corp. announced in 1976 to have finalized a catalytic system for the selective conversion of methanol to gasoline. The pilot program, partly sponsored by the USA Department of Energy, has led to the development of the process called MTG (Methanol to Gasoline) or M-gasoline (Mobil-gasoline).

5.6.1. Production. The scheme of the synthesis unit of the MTG plant is reported in fig. 18. The only environmental problem could be created by the water co-produced in the methanol conversion reactor.

In the process schemes evaluated this water is however recycled in the reactor for the syngas production.

5.6.2. Use. The Mobil gasoline has properties very comparable to those of oil derived gasolines with the advantage of being sulfur and nitrogen free, as shown in table XXI /35/.

5.7. MTBE

Other products energetically interesting can be obtained by means of coal liquefaction process via syngas and the conversion process of the intermediate products. The synthesis of methyl-ter-butyl ether (MTBE) is presently being given much attention. This compound, obtained from methanol and iso-butene, has been quite successful in recent years as a high octane component of gasolines /36/.

(GASIFICATION STEP-LURGI PROCESS)

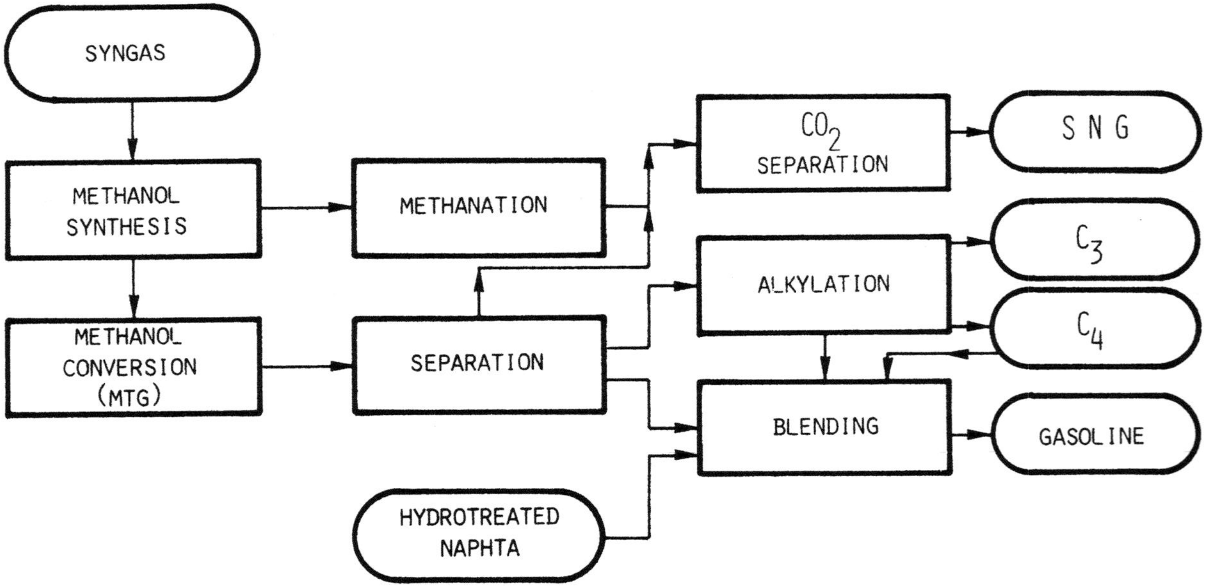

Figure 18. Methanol to gasoline process (gasification step-Lurgi process).

The present production capacity is nevertheless rather limited due to the availability of small quantities of iso-butene in the petroleum and petro-chemical industries.

TABLE XXI. PROPERTIES OF M-GASOLINE (FLUIDIZED BED PROCESS) /35/

Specific gravity	kg/l	0.731
RVP	kg/cm^2	0.78
IBP/EP	°C	30/216
RON (unleaded)		96.8
MON (unleaded)		87.4
Existent gum	mg/100 ml	1.3
Potential gum	mg/100 ml	2.5
<u>Composition</u>	vol %	
Aromatics		27
Olefins		6
Saturates		67
Sulfur		Nil
Nitrogen		Nil

5.7.1. <u>Production</u>. The production of MTBE from coal would offer a viable alternative; a research program in this direction is being completed by Suntech Corp with the support of the Department of Energy /37/. The proposed production process consists of five main steps (fig. 19):

a. coal gasification
b. purification and shift of the syngas
c. synthesis of methanol and iso-butanol
d. dehydration of iso-butanol to iso-butene
e. etherification of iso-butene with methanol to MTBE

5.7.2. <u>Use</u>. MTBE is an excellent gasoline component because it is "hydrocarbon" in nature and has a high blending octane rating (blending RON of 115-135).

MTBE is completely soluble in hydrocarbons and no phase separation occurs even in presence of water /38/. This product is used at present at 5 to 15% vol concentration in gasoline.

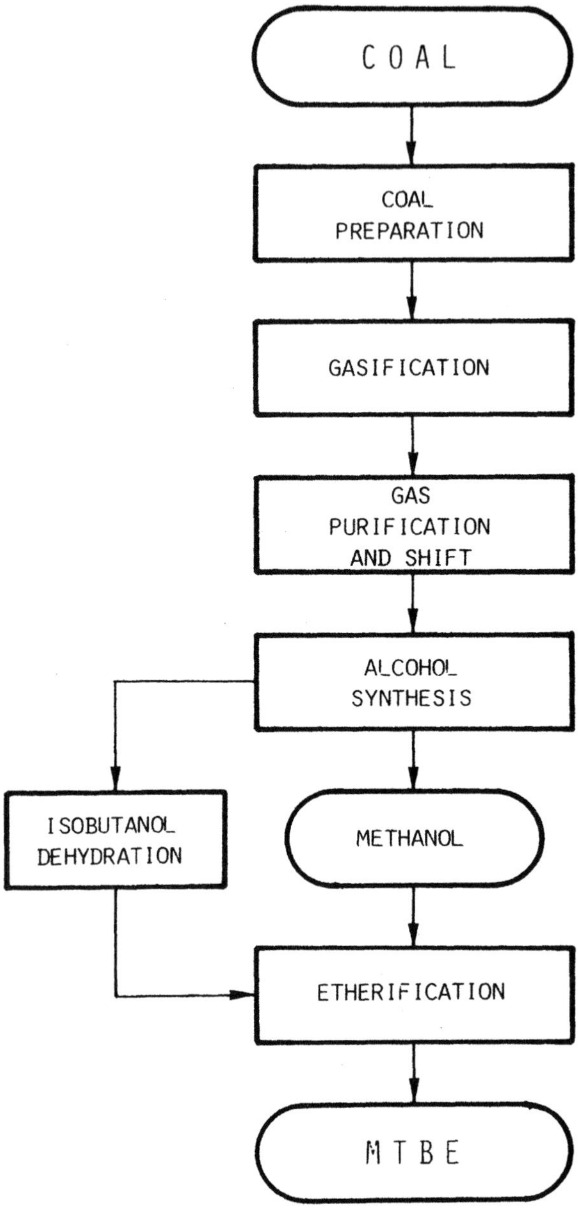

Figure 19. Proposed "MTBE from coal" production process.

Before entering the commercial stage MTBE has been successfully tested in a variety of toxicological tests /36/.

The large scale utilization of this product over the last ten years has not revealed any environmental problem.

5.8. Gasoline from direct coal liquefaction

5.8.1. Production. The production is described under 2.2.1.

Naphtha from coal has higher concentrations of cyclic compounds, olefins and heterocyclic compounds than the oil derived naphtha.

Coal naphthas can be turned into premium gasoline with processing similar to that currently used in oil refineries.

The basic reforming process is favored by the large presence of aromatic hydrocarbons precursors.

5.8.2. Use. The properties of a liquid coal derived gasoline should not differ from those of a conventional gasoline; however, the density and the concentration of aromatics should be higher.

According to literature, the exhaust emissions should be similar to those of indolene. Other authors believe that the higher concentration of condensed aromatic hydrocarbons in the catalytic reformate (tab. XXII) would lead to an increase in PNA emissions; at any rate, the reduction of the final point of the reforming feedstock would make the concentration comparable to that of oil derived catalytic reformates /18/.

CONCLUSIONS

Energy and the environment are always associated because, with very few exceptions, the use of energy has a direct or indirect impact on the environment.

The oil cycle (production, transportation, refining, use) has obviously a number of "constant" environmental effects (engine exhaust emissions, etc.) and accidental ones (oil spill, etc.).

Considerable research and development efforts and investments have been made and are in progress aimed at controlling this environmental impact.

The introduction of synfuels as partial or total substitutes for the petroleum fuels will affect somehow this impact.

From the general survey conducted, the general picture for the synfuels is not of great concern.

As for oil, the synfuels too pose a number of problems.

TABLE XXII. COMPARISON OF CONDENSED AROMATICS LEVELS IN REFORMATES /18/

	PETROLEUM		EDS NAPHTHA			
Feedstock	Paraffinic	Naphthenic	Wyodak 370°C EP		Illinois 340°F EP	Wyodak 325° EP
RON CLEAR	100.2	100.2	100.3	104.6	105.3	105.2
Concentration, WPPM						
Acenaphthene	540	880	990	1070	430	410
Fluorene	910	1330	1160	1510	1110	1040
Phenanthrene	670	1220	920	1360	890	710
Fluoranthene	470	740	720	1070	540	320
Benz(a) Anthracene	50	95	130	170	44	23
Dimethylbenz(a) Anthracene	1	11	9	27	5	4
2-Dimethylbenz(a) Anthracene	4	11	4	16	6	4
Coronene	-	1	1	4	1	1
TOTAL	2645	4290	3935	5230	3025	2510

The different environmental aspects can be taken care of by solutions already existing or developed as appropriate.

Indeed, the environmental aspects should be carefully considered in the economical evaluation of synfuels for commercial scale production.

REFERENCES

/1/ "Synfuels' 2nd Worldwide Symposium"
Brussels, Oct 11-13, 1982

/2/ "Quarterly Oil Statistics - Fourth Quarter 1982"
OECD 1983 No.1

/3/ Garibaldi P. - Assoreni
"Conversione del carbone in prodotti liquidi"
Giornata di Studio "Le nuove tecnologie del carbone nella politica energetica nazionale: Obiettivi per una strategia di ricerca e sviluppo" FAST
Milano, 13 luglio 1981

/4/ Matar S.
"Synfuels-Hydrocarbons of the future"
Penn Well Books 1982

/5/ "Electric utility use of coal-derived fuels: health, personnel protection, and regulatory considerations"
AP-2288, EPRI, Final Report, March 1982

/6/ Calkins K.H., Deye J.F., King C.F., Hartgrove R.W., Krahn D.F.
"Syntetic crude oils from coal: mutagenity and tumor initiation screening test. Coal conversion and the environment"; pag. 462-470
Technical Information Center U.S. Department of Energy DOE Symposium Series 54 - 1981

/7/ Calvi G., Pistone L., (Assoreni) Graziadio M., Ligasacchi S. (Enel-CRTN Pisa)
"Sperimentazione di un sistema di combustione bench-scale per miscele olio-carbone"
37° Congresso Nazionale ATI
Padova, 27/9/82-1/10/82

/8/ Pan Y.S., Bellas G.T., Mathur M.P. and Bienstock D
Pittsburgh Energy Technology Center
United States Department of Energy
"Combustion of coal-oil mixtures in a 700 HP watertube boiler"

/9/ Passman R.A. - Atlantic Res. Corp.
"Coal water slurry pilot facility"
Fifth Int. Symp. on coal slurry combustion and technology
Tampa, Fla., April 25-27, 1983

/10/ Garibaldi P. - Assoreni
"Overview of the italian program on coal slurry combustion"
Fourth Int. Symposium on coal slurry combustion
Orlando, Fla., May 10-12, 1982

/11/ De Michele G., et al. (Enel-CRTN Pisa); G. Salvi (Snamprogetti)
"Large scale CWS firing test at the S. Barbara Power Station"
Fifth Int. Symposium on coal slurry combustion and technology;
Tampa, Fla., April 25-27, 1983

/12/ King O.G., Ushiba K.K., White T.E.
"Make oxygenated chemicals from methanol"
Hydrocarbon Processing - November 1982

/13/ CNR - Progetto Finalizzato Energetica
"Uso del metanolo per autotrazione"
Libro bianco sulle ricerche condotte nell'ambito del sottoprogetto "Uso del metanolo per autotrazione"
Roma, dicembre 1982

/14/ Progetto Finalizzato "Energetica" - Sottoprogetto "Metanolo"
Tema: Impiego del metanolo nel riscaldamento
Rapporto riassuntivo sul lavoro effettuato da:
- Stazione Sperimentale per i Combustibili
- CNPM - Politecnico di Milano
- Assoreni
nel periodo giugno 1976/giugno 1980

/15/ Troyli M. - Assoreni
"Esperienze e prospettive del metanolo nel riscaldamento domestico e nelle turbine a gas"
Giornata di studio "Il metanolo come vettore energetico: ruolo e prospettive" organizzato da FAST in collaborazione con ENI - 24/6/823

/16/ Moroni, E.C.; Burke, F.P.; Winschel R.A. and Wilson B.W. "Integrated two-stage liquefaction solvent quality effects"
ACS 185th National Meeting, Seattle, Washington March 20-25, 1983

/17/ "Occupational hazard assessment - Coal Liquefaction" Vol. II Assessment
U.S. Department of Healt and Human Service - Public health service centers for desease control
National Institute for Occupational Safety and Health, March 1981

/18/ "EDS Coal Liquefaction Process Development Phase V "
EDS Product Quality FE-2893-68 Interim Report March 1981

/19/ Sutterfield D., Lanning WC "Hydrotreatment and biological test of SRC-II Coal liquid
US Dept Energy, Energy Tech. CTR
ACS Symposium series, Vol. 1981, N156, 1981

/20/ Chemical Engineering, 6/19/78, pg 67-69

/21/ Schumacher M.M.
"Heavy oil and tar sands recovery and upgrading"
Noyes Data Corp. 1982

/22/ Steere, D.E.; Mac Donald W.A. and Stone D.H.
"Tar sands products bring changes"
Hydrocarbon Processing, September 1981

/23/ O'Hara, J.B.; Loran, B., I.; Bela, A. and Jeutz, N.E.
"Environmental factors for Fischer-Tropsch coal conversion technology"
Coal Processing Technology, AICHE, Vol. 5, (1979)

/24/ Schreiner, M.
"Research guidance studies to assess gasoline from coal by methanol-to-gasoline and SASOL - type Fischer Tropsch technology"
DOE Report FE-2447-9, Washington DC, 1978

/25/ Private communication

/26/ Tamm, P.W.; Bertelsen, C.A.; Handel G.M.; Spars B.G. and Wallman P.H.
"The Chevron STB oil shale retort"
46th Midyear Refining Meeting, Session on Synthetic Fuels Processing - Part I, March 13, 1981, Chicago

/27/ Miller W.; Harvey C. and Hunter M.
"Premium syncrude from oil shale using Union Oil technology"
1982, NPRA Annual Meeting, March 21-23, 1982, San Antonio, Texas

/28/ Stamper K.R.
"50,000 Mile methanol/gasoline blend fleet study - a progress report"
3rd Alcohol Fuels Technology Symposium
Asilomar, California, May 28-31, 1979

/29/ A. Busacca
"Processo Finalizzato per la produzione di alcool etilico"
I.C.P., marzo 1980 pag. 39

/30/ Leprince P., Ory M.-H
"Programme for gasoline substitution in France
5th Int. Symposium Alcohol Fuel Technology
Auckland, N.Z., May 13-18, 1982

/31/ Terzoni G., Pea R., Ancillotti F.
"Improvement of the water tolerability of methanol-gasoline blends"
3rd Alcohol Fuels Technology Symposium
Asilomar, California - May 28-31, 1979

/32/ Ricci, R.; Paggini, A.; Fattore V.; Ancillotti F. and Sposini M.
"The production of methanol and higher alcohols from synthesis gas"
Seminar on Chemicals from Synthesis Gas, Geneva, 27 June- 1 July, 1983

/33/ Sposini M., Pea R., Ancillotti F., Ghezzi P.
""Engine performances and water tolerability of gasoline containing the alcoholic mixture obtained by the Snamprogetti process"
Proceedings Fifth International Alcohol Fuel Technology Symposium - Vol. II
Auckland, New Zealand, May 13-18, 1982

/34/ Bowden, J.N. and Brinkman D.W.
"Stability of alternate fuels"
Hydrocarbon Processing, 59, July 1980

/35/ Harney, B.M. and Mills, G.A.
"Coal to gasoline via syngas"
Hydrocarbon Processing, 59, February 1980

/36/ Garibaldi P., Pecci G.; Vicenzetto F., Razze S.
"Methanol and ethanol as raw materials for the synthesis of high octane components"
2nd International Symposium on Alcohol Fuel Technology
Wolfsburg, November 2123, 1977

/37/ AFUP
"Project planning document highway vehicle alternative fuels utilization program"
DOE report CS-0093, Washington DC, July 1979

/38/ Ancillotti F., Pescarollo E.
"Fuel blends create solubility problems"
Hydrocarbon Processing, November 1977

SYNFUEL PROGRAMMES IN THE FEDERAL REPUBLIC OF GERMANY AIMS AND RESULTS

H.J. Neef, N. Stump
Projektleitung Energieforschung
Kernforschungsanlage Jülich, Postfach 1913
D-5170 Jülich

INTRODUCTION

During this Ispra Course on "Synthetic Fuels" a number of development projects which have the aim of converting carbonaceous materials - such as coal, biomass, municipal wastes etc. - into syngas and synthetic fuels have been presented. The contributions from Germany were rather numerous. The reasons for this are manifold: The main reason is that hard coal and lignite are the only significant domestic energy sources of the Federal Republic of Germany (FRG). The hard coal deposits are unfavorable from a geological viewpoint so that mining costs are very high. The technically minable hard coal reserves are estimated at about 25,000 million tons. In addition, there are lignite reserves of 35,000 million tons which are economically minable. This, amounting to about 35,000 million tons of hard coal equivalent, compares with only about 400 million t oil and gas reserves.

One other reason for the numereous German contributions to coal technologies is that Germany has a history in industrial coal conversion processing, which was revived after the oil crisis by both Government and industry. Therefore in the non-nuclear energy programme of the German Federal Government coal was always given priority over all other primary energy programmes.

As in other countries, which depend on oil imports, one of the objectives of energy policy is to reduce the share of imported oil by better exploitation of domestic energies, use of renewable energies, and last but not least by energy conservation. For Germany, better exploitation of domestic sources of course means the efficient and environmentally acceptable use of coal. In addition the optimum use of imported energy is another goal.

The reduction in the dependence on crude oil between 1975 and the eighties can be seen in figure 1. In this, the imported proportion of the

G. E. Beghi (ed.), Synthetic Fuels, 417–434.

1975: 10^6T HCE	(%)	1980: 10^6T HCE	(%)		1983: 10^6T HCE	(%)
7	(2)	14	(4)	URANIUM	21	(6)
49	(14)	64	(16)	NATURAL GAS	54	(15)
181	(52)	186	(48)	OIL	154	(43)
67	(19)	77	(20)	HARD COAL	79	(22)
34	(10)	37	(10)	LIGNITE	38	(11)
10	(3)	12	(3)	HYDRO AND OTHERS	12	(3)
348	(100)	390	(100)		358	(100)

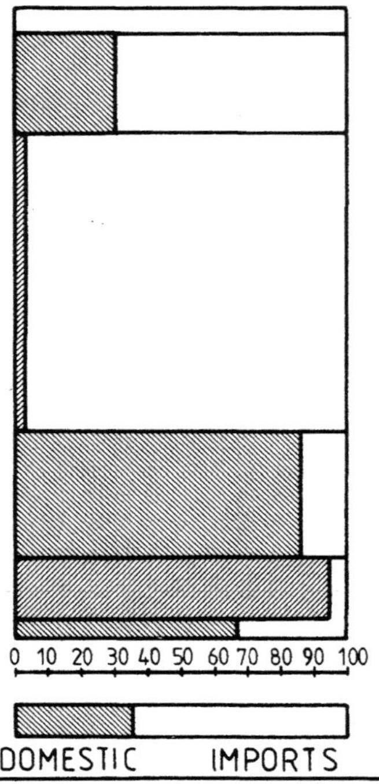

PRIMARY ENERGY DEMAND IN THE FEDERAL REPUBLIC OF GERMANY

various sources of primary energy is also shown graphically for 1980. At this point it must be clearly stated that the reduction in the dependence on imported crude oil is less a result of the synfuel programme, but is essentially a result of economic development and legislation in resent years as well as an efficient use of the source of primary energy, oil. At present, the synfuel programme is intended to maintain the high level of knowledge of industry and research establishments and to put this to use when necessary.

The Federal Governments' energy policy has strongly reduced the use of mineral oil for power production. The large proportion of coal used for electricity generation accounted for the fact that the Federal Republic's electricity supply was practically not affected by the oil crisis in 1970 and 1974.

85 % of the lignite production of about 130 million tons is used in base load power stations. Open cast mining of lignite is economical in the Federal Republic of Germany but for several reasons the production capacity cannot be increased significantly. The production of hard coal is heavily subsdized. The production capacity of the hard coal mining industry was fixed at 90 million tons of hard coal per year. In 1980 legislation was passed in order to increase the use of hard coal for power generation from 33 million tons per year to 45-50 million tons per year by 1995. In 1982 actually 36 million tons were used for power generation.

Within the framework of the Ispra Course the processes of coal gasification and direct coal liquefaction have already been treated in detail by German authors (Hedden, Romey). Indirect coal liquefaction was also reported on (Frohning, Reimert, Keim). For this reason, in the following report, a summary of the overall development is to be given.

COAL GASIFICATION

During the first half of this century three coal gasification processes were developed in Germany up to commercial utilization. These were fluidized bed gasification according to the Winkler process, fixed bed gasification according to the Lurgi process, and entrained bed gasification according to the Koppers-Totzek process. These processes have specific advantages and disadvantages with regard to coal quality and grain size, to flexibility of operation, to consumption of gasification agents, to the necessity for raw gas treatment and to product gas quality.

With the start of the Non-Nuclear Energy Research Programme in the mid-70s different coal gasification technologies have been further and newly developed in the FRG, partly based on the experience of the above mentioned "old" technologies. Nearly all of these process developments aim at increasing the reaction pressure to increase specific gas production and to decrease the dimensions of the equipment. This of course requires more sophisticated equipment for coal feed and ash removal.

Further objectives (Fig. 2) of the coal gasification development are:

- to increase gasification temperature for reduction of tars and phenoles in the product gas
- to obtain higher contents of CO and H_2 in the synthetic gas
- to increase the heating value of the product gas
- to increase the content of methane for generation of substitute natural gas
- to reduce the specific coal consumption by incorporating heat from nuclear reactors
- to be able to use a broad spectrum of coal qualities and coal grain sizes
- to integrate coal gasification into power production processes
- to reduce investment and operation costs

The development of a coal gasification process needs money and time. From laboratory scale tests to commercial application typical development costs of about 100 million DM over a time of 10 and sometimes 20 years are necessary. During this time period the economic perspectives and the international and national competition might change so that not all developments which have been started will be honoured by commercial success.

Within the German Energy R+D Programme of the Federal Ministry of Research and Technology research for several coal gasification processes has been and still is being supported. The State Government of Northrhine-Westphalia - which is the main coal area of Germany - also subsidizes some developments of coal gasification. One process development has been carried out by private industry without governmental support. Demonstration of two processes is being supported by the Commission of the European Communities under the Demonstration Program in the Energy Sector. During the R+D phase a number of pilot plants have been operated for the different gasification processes

Increase gasification pressure

Increase gasification temperature

High CO and H_2 content (Syngas)

High CH_4 content (SNG)

High heating value of product gas

Reduce specific coal consumption (nuclear heat)

Use of a variety of coal qualities

Use of different coal grain sizes (fines)

Integrate coal gasification into power production

Achieve high efficiency

Reduce investment and operating costs

Fig. 2: Development goals of coal gasification

Bergbau-Forschung Steam Gasification Process

Carbon Gas Technology Process

Klöckner Molten Iron Bath Process

KHD Humboldt Wedag MIP Process

Kohlegas Nordrhein Process

Korf KR Process

Krupp-Koppers Prenflo Process

Rheinbraun High Temperature Winkler Process

Rheinbraun Hydrogasification Process

Ruhrgas/Ruhrkohle/Steag Ruhr 100 Process

Ruhrkohle/Ruhrchemie Texaco Process

Saarberg-Otto Process

Shell SK Process

VEW Coal Conversion Process

Fig. 3: Coal Gasification R+D Projects (Pilot Plants)

Some of them are still under operation, some of them have demonstrated the technology in a way which allows continuation on a larger scale, but some have also shown that technical problems are not so easy to solve. It cannot be the purpose of this report to go into a detailed evaluation of all the processes. A list of activities related to test and pilot facilities in the coal gasification field is shown in Fig. 3.

ADVANCED POWER PLANTS WITH COAL GASIFICATION PROCESSES

Conventional coal power plant technology has reached a high standard in Germany. Primary and secondary measures are necessary to reduce the emission of pollutants to the required levels:

- primary measures: burner design, combustion area layout, application of fluidized bed combustion
- secondary measures: technologies to remove SO_2, NO_x, dust from flue gases.

R+D effort has been increased since 1977 in the FRG in support of these very important environmental tasks.

An important alternative for power generation from coal could be the combination of a power station with a pressurized coal gasification plant, as shown in Fig. 4. The principle of combining gas and steam turbines is known from natural gas-fired power production. SO_2, NO_x, dust and waste heat emissions can be reduced to very low levels without secondary measures. The integration of a coal gasification process into power generation requires R+D efforts for the following components:

- the coal gasification process
- raw gas treatment on an elevated temperature level
- gas turbines for high gas inlet temperatures and low NO_x output
- waste heat recovery

In addition a close integration of all components has to be achieved for a good overall efficiency which can go beyond 40 % and can even reach 50 % if the development potential of gas turbines is taken into account. Apart from the already mentioned coal gasification pilot plants, several studies are underway to investigate the optimum concepts for combined cycle power plants.

A process under development specially for power generation is the VEW Process. In this concept coal is gasified only partially. The gas will be used in the gas turbine and the char will be burnt in the boiler furnace. This combination allows a very high carbon conversion which is necessary to reach the overall efficiency.

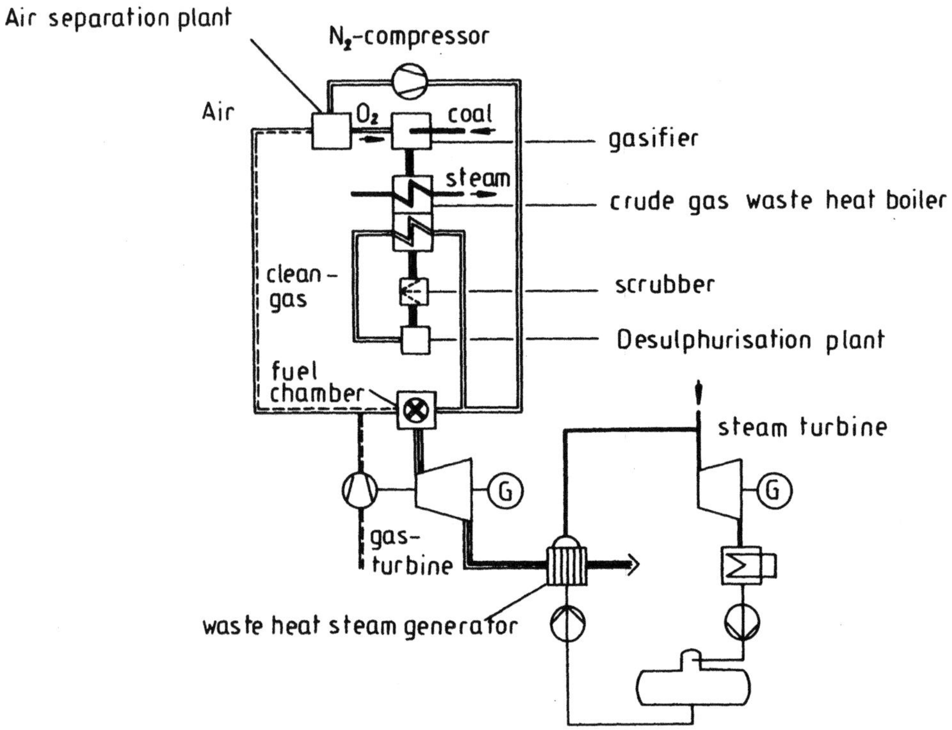

Coal gasification combined cycle system (source: KWU)

COAL LIQUEFACTION

German efforts in the field of coal liquefaction can follow up on the scientific and technical experience accumulated in earlier years. Aside from pyrolytic processes (carbonization, coking), in which only small amounts of liquid products are generated, there are basically two ways of generating liquid products from coal: on the one hand, direct hydrogenation of the feed coal and, on the other hand, gasification into carbon monoxide and hydrogen with subsequent synthesis to liquid products, the so-called indirect liquefaction. The last process - the Fischer-Tropsch synthesis - is being applied on a commercial scale in South Africa.

Some industries and university institutes in the Federal Republic of Germany carry out development work on the Fischer-Tropsch synthesis, most of it designed to improve the performance of catalysts with the aim of a narrower palette of products which can be used as chemical raw materials.

For the production of gasoline by synthesis the classical Fischer-Tropsch process has a competitor in the Mobil-process. Whereas the Fischer-Tropsch synthesis results in a large variety of hydrocarbons the Mobil-process converts methanol (which can easily be produced from synthesis gas) nearly completely to gasoline with a high octane level.
The so-called MTG-process (Methanol to Gasoline) is being operated in a pilot plant, subsidized jointly by the U.S. Department of Energy and the Federal Ministry of Research and Technology. In the pilot plant a fluidized bed is used for the zeolite catalyst.

In addition, efforts are being made to use zeolites in order to produce gasoline directly from synthesis gas. If experiments are successful, the intermediate step of producing methanol would no longer be required.

Direct hydrogenation of coal has a tradition in Germany. In 1943-1944 the capacity of all hydrogenation plants in Germany amounted to about 4 million tons of end products.

Today's hydrogenation processes have been developed from the IG-Farben processes applied in these plants. The major advantage of the improved versions of this process is the removal of those components which are difficult to hydrogenate, but which can well be used to produce, in a pressurized gasifier, the hydrogen required. Since these components are removed, the reaction pressure can be reduced from 700 bars to 300 bars.

Two laboratory facilities for the hydrogenation of hard coal have been operated by Bergbau-Forschung and by Saarbergwerke for several years.

On the basis of the results obtained in these facilities, Ruhrkohle and Veba have built a pilot plant with a throughput of 200 t/d. A smaller pilot plant with a throughput of 6 t/d is operated by the Saarbergwerke. The Rheinische Braunkohlenwerke has operated a laboratory facility for the hydrogenation of lignite; plans to build a pilot plant have been postponed.

A test facility (Salzgitter/Imhausen) for hydrogenation under very high pressure in a tube reactor is under construction. It will have a capacity of 65 kg/h hard coal and will start operation in 1984. Expected results are increased reaction rates, low gas production, the possibility of using a variety of coals, and better plant control, which should lead to economic improvements.

COMMERCIALIZATION OF COAL CONVERSION

The successful operation of the different pilot plants resulted in a programme of the Federal Government to continue the development of coal conversion technologies on an industrial scale. This programme was announced on January 30th, 1980 and German industry was asked to propose projects which should be promising from both energy-political and industrial points of view. In October 1981 the Federal Government published the industrial proposals. Seven large coal gasification projects and four coal liquefaction projects had been planned on the basis of preliminary feasibility studies.

For the gasification projects the Federal Government offered to carry up to 40 % (50 % if necessary) of the investment and up to 60 DM/t for the use of domestic coal. (Note: At the end of 1983 the costs were 240 DM/t for domestic and 150 DM/t for imported hard coal).

The proposed gasification projects are summarized in Fig. 5.

Rheinbraun has started construction of the first train of a High-Temperature Winkler Plant - without funding by the Federal Government, but with some assistance of the EEC. The synthesis gas produced will be used for methanol production thus substituting the synthesis gas generated so far from crude oil products.

In German newspapers one could find on Nov. 15, 1983 the announcement that Ruhrkohle and Ruhrchemie have authorized Uhde GmbH to be general contractor for a gasification plant for synthesis gas production based on the Texaco-coal gasification process. The investment costs are reported as 220 mill. DM and will be subsidized with 50 % by the Federal Ministry for Economic Affairs. Commissioning is expected for spring 1986, operation at full capacity for the fall 1986.

The proposal of Klöckner to produce a coal-based gas for the ironworks at Bremen has been withdrawn. The reason for this is the difficult position of the steel-industry where high investments are difficult to make.

	Operator	Process	Coal Capacity 10^6 t/a	Product Gas $10^6 m^3/a$	Usage 10^6 t/a
1	Ruhrkohle Ruhrchemie	Texaco	0,25	Syngas 400	Chem. Industry
2	Shell	Shell-Koppers	0,324	Syngas	Methanol 2,15
3	Rheinbraun	HT-Winkler	1,5[1)]	Syngas 880	Methanol
4	Klöckner	Iron Bath	0,85	Syngas 2000	Iron Works
5	Ruhrkohle Ruhrgas	Lurgi	1,5	SNG 230 + Syngas	Methanol 0,49
6	Gelsenberg Krupp	Koppers-Totzek	1	Syngas	Ammonium
7	PVC-Flick	Fixed Bed	0,25	Syngas 421	Heating Gas

1) Raw Lignite ≙ 0,6 Dry Lignite

Fig. 5: Proposal for commercialization of coal gasification (Status Oct. 1981)

Projects 2, 5, 6 and 7 as proposed in 1981 will not be realized.

Within the Coal Conversion Programme there are proposals to build and operate large coal liquefaction plants. In Fig. 6 the proposals received by the Federal Ministry for Economic Affairs in autumn 1981 are given.

In the meantime Rheinbraun has indicated that the commercialization of lignite liquefaction will be introduced later than expected in 1981. With regard to the hard coal liquefaction the investment costs of such demonstration plants are very high and during operation financial support might still be needed. Therefore it is foreseen to build only one hard coal liquefaction demonstration plant and the companies were invited to propose one joint project only. This process is still underway.

UNDERGROUND COAL CONVERSION

Long term research projects are devoted to the investigation of methods which would allow coal to be extracted underground without the use of human labour, and exploiting existing coal deposits to a great extent. This could be achieved by chemical fragmentation, gasification and liquefaction of coal underground.

Underground gasification offers the best chances of realization among these techniques. For this reason it has been chosen as the starting point for this long term development work.

In the past years Belgium has also taken up studies of underground gasification. In view of the similarity of interests, and also of geological conditions, an agreement on co-operation was signed between Belgium and the Federal Republic of Germany in 1976. The first joint field test started in Thulin (Belgium) in 1979. Money from EEC is also being provided for work within this project.

A study group "Kohlegewinnung Zweite Generation e.V. ("Second Generation Coal Production e.V.") was formed by the hard coal industry, research institutes and companies involved in the energy economy. This study group is to investigate basic data for the in-situ conversion of coal in an interdisciplinary manner. Projects initiated by the group are being supported to a considerable extent within the framework of the Energy Research Programme.

UPGRADING OF HEAVY OIL

In addition to coal conversion, the upgrading of residual oils from refinery technology and heavy crudes also belong to the German synfuel programme. If one considers in which field the substitution of oil by coal, nuclear energy and natural gas takes place then one finds that this is the area of heavy oil fractions. This means a large supply of top residues on the market. These can be converted into light products using processes of direct liquefaction. In principle these processes are similar to direct coal liquefaction and can also be applied to naturally occurring heavy crudes. The importance of heavy crudes for the future energy supply of the Federal Republic of Germany is estimated as being highly significant in spite of the present stable situation on the oil market.

This is made clear by the following data on crude oil reserves worldwide:

	10^9t
light crude	
- primary and secondary reserves	90
- enhanced oil recovery reserves	90
heavy crude/bitumen	150

(Source: Veba Oel AG)

The need of the FRG for a guaranteed supply of fossil sources of energy can be satisfied through close co-operation with producing countries whereby responsibilities for the technical development of heavy crude production processes and up-grading are borne jointly so that the co-operation is beneficial to both sides.

An example of such a co-operation is being carried out with Venezuela, where heavy crude deposits in the Orinoco belt amount to more than 25,000 mill. t according to present estimations. Co-operation within the R+D-sector covers the

- production of heavy crude
- upgrading of heavy crude to syncrude
- energy and hydrogen generation from residues occurring during the heavy crude conversion processes

In a comprehensive study it was proved that a concept of a plant that integrates these three tasks and which uses the German liquefaction technology is technically and economically meaningful for a location in Venezuela. Figure 7 shows the flow diagram for such an integrated plant. The realization of an "integrated" plant depends less on technological problems than on the development of the costs and the availability of liquid and gas sources of primary energy worldwide. For this reason the joint German-Venezuelan efforts concentrated on an upgrading plant for heavy and residual oils in Germany. These activities are of a commercial nature and are being accompanied by R+D projects which

	Operator	Process	Coal Capacity 10^6 t/a	Products
1	Ruhrkohle	Modified IG	4 or 8	Light/Medium Oil 1.5 or 3
2	Veba	Modified IG	3.9 or 5.2 1)	Liquid Products + LG 1.8 or 2.4
3	Saarberg-werke	Modified IG	2.3	Gasoline 0.68 other hydrocarbons 0.32
4	Rheinbraun	Modified IG	3.5 2)	Raw product for fuel and chemical industry 0.4

1) Alternative: Refinery residues or heavy oil
2) Raw Lignite

Fig. 6: Proposals for commercialization of coal liquefaction (Status Oct. 1981)

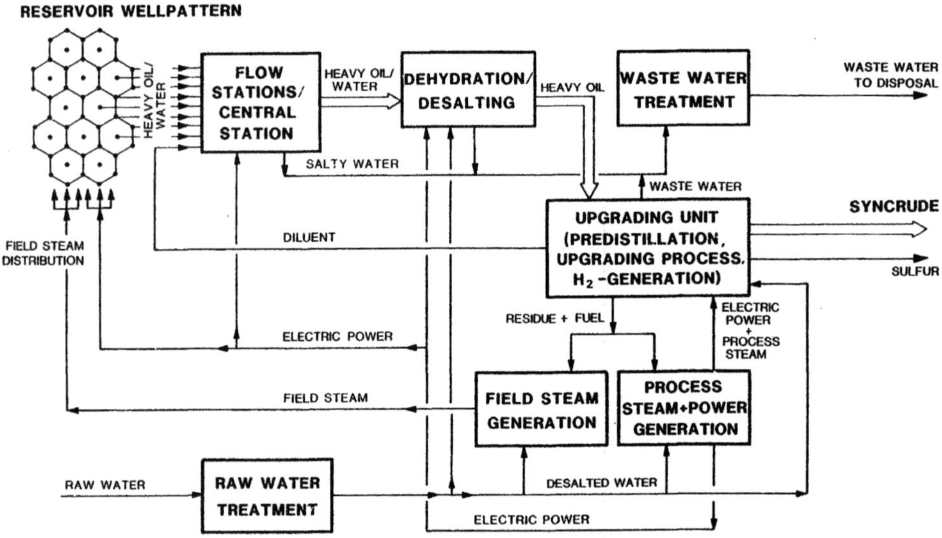

Fig. 7: Simplified scheme of an integrated system for heavy oil production and upgrading including energy supply.

are supported by the Federal Ministry of Research and Technology and to some extent by the EEC. In 1983 a pilot plant built by the partnership Veba Oel/Lurgi went into operation with a capacity to convert 1 t residue into syncrude per hour by the so-called VCC-(Veba-Combi-Cracking) process. The operation of this plant is primarily intended to confirm design data for a commercial plant in Germany.

In connection with the work on the upgrading of heavy crudes, it should also be mentioned that considerations are moving towards testing coal-liquefaction with the help of the heavy crude upgrading plant on the next larger scale. Whether such considerations can be realized and to what extent they are possible must still be carefully checked.

BIOMASS

The utilization of biomass for energy production does not really belong to the synfuel programme of the FRG as this concentrates more on coal conversion. However, since biomass is being dealt with in several reports within the framework of this ISPRA-Course, something should be said here about R+D efforts in the fields of biogas production as well as gasification and the combustion of biomass.

The further development of biogas technology concentrates on the practical demonstration of various plants. This is accompanied by scientific measuring programmes, the central point being the optimization of plant technology and the logging of the most important output and cost data. The considerable costs, inherent in this technology, still limit its potential utilization and, in addition, it is the technical weaknesses which put a question mark over the profitability. Above all the problem areas of hydrogen sulphide, corrosion, the remaining products (residues, condensates, slurries) as well as the efficiency of the plant should be mentioned.

In the case of the gasification of biomass work, which is being supported by the Federal Minister of Research and Technology, is concentrated on gasification of wood. In comparison with coal, oil and gas (and also nuclear energy), wood is not a very important source of energy for the Federal Republic of Germany. The utilization of wood for the production of electricity and heat is characterized by an expensive plant technique and by pollution problems. Central difficulties in the gasification of wood are the condensates which occur, gas purity, efficiency of the plant and finally the production costs for the gas. The latter are decisively influenced by the costs for the disposal of the condensate (50-300 DM/m^3). The costs for electricity generation vary, depending on the gasification process, between 25 and 50 Pf/kWh for 6000 hours of operation per year and a plant size of 500 kW_{el}. The high production costs are considerable and set the question of whether the production of electricity/heat using the steam process is not more cost-effective in the given output range than by using gasification.

Problems on the emission side resulting from combustion can be seen in the case of incomplete combustion in the CO content of the flue gases. Furthermore, the SO_2-concentration (120 mg/Nm^3) must be taken into account as they can represent a problem on the side of immissions because of the low stack.

Regarding the evaluation of the technology on the side of emissions it has been determined that the problem area for combustion is clearer; the energetic as well as the cost expenditure for removing all pollutant-loaded products is easier to estimate than in the case of gasification. Using the results from existing plants it can be concluded that electricity generation using combustion is economic in the range above 1 MW_{el}. Considering cost degression as well as the technical feasibility it has also been determined that according to present knowledge wood gasification is more favourable regarding costs in the range up to 200 kW_{el}, even though the profitability remains at least questionable.

Reliable results from research for the steam process are not yet available in the remaining output range 200 kW - 1000 kW, so that no statement is possible on whether the gasification principle is to be preferred to combustion.

Finally, an important aspect of the utilization of biomass should be mentioned. The importance of biomass does not lie only, or perhaps not primarily, in the aspects of energy-economy. It is more likely to be the disposal aspect, or applications in particular agricultural operations, which allow the utilization of this source of energy to appear meaningful in the Federal Republic of Germany.

In the case of developing countries the processes of biomass utilization are significant possibilities for the substitution of other primary sources of energy. Here, an optimum use of biomass for the production of energy can lead to reducing the load on the energy balance and to minimizing the consumption of fuel-producing plants. Several projects on the use of biomass in developing countries are therefore being financed by the Federal Ministry of Research and Technology.

GOVERNMENTAL FUNDING

It is always difficult to present meaningful data on available governmental funding. Either the information is complete, then too many details have to be reported because of the complexity of funding procedures and the involvement of many industries, universities and agencies on a European, federal, state, and local level; or the information is condensed, then the audience requiring information is somewhat frustrated.

	(10^6 DM)
Efficient Uses of Energy and Secondary Energy	1050
Coal Mining	530
Coal Utilization	1820
Oil and Gas	240
New Sources of Energy	1100
Total	4740

Fig. 8.1: Non-nuclear Energy Research Funding by the Fed. Ministry of Research and Technology

Summary 1972-1984

	(10^6 DM)
Efficient Use of Energy and Secondary Energy	90
Coal Mining	40
Coal Utilization	250
Oil and Gas	65
New Sources of Energy	115
Total	560

Fig. 8.2: Non-nuclear Energy Research Funding by the Federal Ministry of Research and Technology

Budget for 1984

Nevertheless, Fig. 8.1 gives a summary of the available funds for the Non-Nuclear Energy Research Programme of the Federal Ministry of Research and Technology. The summary indicates the high priority of coal research.

It is worthwhile mentioning the following breakdown of the 1.820 mill. DM for the use of coal:

Power plant technology and emission control	650 mill. DM
Coal liquefaction	630 mill. DM
Coal gasification	470 mill. DM
Coal use for heat generation	70 mill. DM

Fig. 8.2 shows the available budget for 1984.

CONCLUSIONS

In spite of the fact that synfuels from coal and biomass conversion are not competitive with oil products the energy R+D programme of the Federal Republic of Germany undertakes all efforts to decrease the dependence on oil. The technology is available and will be improved and demonstrated in years to come. The future oil prices in comparison with the price of other primary energy sources will determine the appropriate ways and means of achieving the commercialization of these technologies.

LIQUEFACTION AND GASIFICATION OF SOLID FUELS - DEMONSTRATION PROGRAMME OF THE E.E.C.

Dr.Ing. S. Furfari
Commission of the European Communities
Directorate-General for Energy - Coal Directorate
200 rue de la Loi
1049 Brussels - Belgium

1. INTRODUCTION

The aims of the energy strategy for the Community are to reduce oil dependense and to reduce oil consumption to a level of 40% of the raw primary energy consumption. The ways to reach these objectives are based on a more rational energy use and giving a broader variety of the supply of the energy sources.

Among these, solid fuels (coal, lignite and peat) will play a considerable part due to their strategic characteristics: wide world reserves, reserves in the Community, broad range of exporting countries, interesting prices.

The estimates of solid fuels consumption foresee that for the medium-term the main utilisation will continue to be for electricity production but for the long-term, although the forecasts are particularly unsettled, solid fuels conversion into gas and liquids should represent a large part of their consumption. The substitution products formed by the conversion of solid fuels will be used for energy purposes but also as a source of raw material for the chemical industry.

Indeed, solid fuels transformation into gaseous and liquid hydrocarbons represents a source of alternative energy making it possible to alleviate the dependence of the Community on the hydrocarbon importation. This will have a favourable effect both on industrial development and employment creation.

Conscious of the considerable role that the gasification and liquefaction of solid fuels will play in the future, the Commission has always kept in mind this use of upgraded coal. Since 1958, the Commission has supported research and development in this field on the basis of Article 55 of the ECSC Treaty. The ECSC aid granted to this domain until now is around 200 MioECU which covers 60% of the budget of the research supported. The support to such work has given the Community the possibility to maintain in this field active and qualified research cells, and this in the middle of the oil boom. In due time, the structure was in place to ensure a scientific development in the field of solid fuels conversion. Demonstration projects in this field were established thanks

G. E. Beghi (ed.), Synthetic Fuels, 435–454.

to - among others - the knowledge developed by the ECSC research. The necessity for support for demonstration projects in this field being perceived, regulations allowing the Commission to grant aid to demonstration projects have been created. A Community action in the field of liquefaction and gasification of solid fuels was justifiable and is still justifiable for the following main reasons:

- heavy investments only supportable with great difficulty by Member States;
- concentration of the effort in order to maintain European technology at this level.

2. FIRST DEMONSTRATION ACTION (1978-1982)

2.1 Generalities

Action at Community level launched by the Commission concerning demonstration projects relating to the liquefaction and gasification of solid fuels was covered by:

a) Council Regulation (EEC) No.1302/78 /1/, which concerns alternative energy sources, together with Council Regulation (EEC) No.728/79, which concerns the liquefaction and gasification of solid fuels;

b) Council Regulation (EEC) No.726/79 /2/, which lays down the maximum amount of aid for projects to exploit alternative energy sources.

According to these Regulations, demonstration links the R&D stage, sometimes on pilot plant, and the later investment stage. It differs from the R&D and pilot stages in the industrial scale of the projects, the requirement of having prospects of economic viability, and from the normal investment stage in that the inherent risks are still considered by the entrepreneurs to be too high.

Demonstration is an essential stage at both national and Community level. The European Parliament, for its part, has emphasized in several Resolutions /3/ the need for demonstration projects to develop alternative sources of energy and energy saving measures. When the Community regulations concerning demonstration projects were adopted in 1978, only two Member States had introduced support measures for demonstration projects. Initially, therefore, the Community programme also served to ensure a minimum level of demonstration activity in all Member States.

Since that time all Member States, often prompted by the Community programme, have set up national programmes to support demonstration projects. The Community programme - chiefly a stimulant until now - will in future, as in the case of the Community R&D programme, be required to coordinate and supplement national programmes as well.

The demonstration programme is not only the logical extension of Community R&D activities: because it provides the opportunity for exploiting projects from national R&D programmes at Community level, it also opens up a European scale market for them.

This Community action in the field of liquefaction and gasification of solid fuels was one of the fields of the demonstration programme of the Community. The other fields of application were:

a) for energy saving:
   - buildings
   - supply and use of process heat and of electricity in industry
   - energy industry
   - transport

b) for alternative sources:
   - exploitation of geothermal fields
   - exploitation of solar energy
   - liquefaction and gasification of solid fuels
   - exploitation of wave, tidal and wind energy

For the alternative energy sources, several considerations, including budgetary factors, have led the Commission to limit action to the first three sectors.

The Regulations which have enabled the Commission to carry out this action in the field of liquefaction and gasification of solid fuels, pointed out - among other things - that:

-"The Community may grant financial support for demonstration projects to exploit alternative energy sources in the Community which by their nature may serve as examples and which prior studies and research have shown to offer prospects of industrial and commercial viability. For the purposes of the Regulation, alternative energy sources means any potential source of energy, with the exception of nuclear energy and fossil fuels, exploited by conventional means.
- All projects must relate to the exploitation of alternative energy sources by means of new techniques or technologies which might encourage the establishment of other installations of the same type. With regard to the liquefaction and gasification of solid fuels in particular, the Regulations state that:
  . the design, construction, putting into operation, operation of or improvement to pilot or demonstration plant,
  . in the event of underground gasification, the preparation and execution of the borings and the creation of the linking between boreholes, as well as the recovery of the gases produced.
- Such support shall constitute only a minority share of the total financing of the project. Support may not in general exceed 40% of the total cost of the project.
  Half of the support shall be repayable in cases where, after the first year of operation, the plant or process is subsequently used for industrial or commercial purposes.
  The repayable part of the support shall be repaid over a maximum period of eight years, starting from the second year after the date on which exploitation of the installation or process brings in revenue. Arrangements for repayment shall depend on the nature of the project and shall be laid down in the contracts to be concluded with the recipients.
- Every project submitted for consideration by individuals or undertakings in the Community further to a call for submission of projects published in the official journal of the European Communities shall be examined by the Commission on the basis of the following information to be supplied by the applicants:

. a detailed description of the project and the timetable for its execution;
. a description of the preliminary work and research;
. the importance of the project for reducing energy dependence and increasing diversification of supply;
. the financial situation and technical capabilities of those responsible for the project;
. the nature and extent of the technical and economic risks involved in the project;
. the cost of the project, its potential economic viability and the proposed arrangements for financing it;
. whether at an earlier stage of research and development the project has been given financial support by the Community or by Member States;
. details of any other financial support for the project to be provided by, or expected from Member States or the Community;
. any other information which justifies the Community support requested;
. how it is proposed to publicize the results.

- The Commission shall decide whether to grant or refuse support for projects after consulting the Advisory Committee on the Management of the projects concerned, composed of representatives of the Member States."

During the first three years of the validity of the Regulations, the Commission issued two invitations to submit projects in the field of gasification and liquefaction of solid fuels. 35 proposals were submitted, involving a total investment cost of one thousand million ECU. In view of the limited amount of funds available, 13 projects were selected. The Commission was assisted in selecting projects by Advisory Committees whose members should be well-acquainted with national programmes because of their position in national administrations.

One of the selected projects was withdrawn by the applicant. The motives for a withdrawal can be:

- difficulties encountered by the applicant for financing the part of investment not covered by the Community grant;
- the proposers realized only later that the Community system, unlike national systems, imposes repayment requirements;
- the Community programme aims to ensure a good dissemination of knowledge acquired by the achievement of the project. It is indeed specified in the contract to be concluded between the Commission and the proposer that the latter will remain the owner of the patentable inventions and the know-how obtained during the course of the project but must, however, undertake to disseminate directly, with or without the collaboration of the Community, the results of the project. Several beneficiaries were reluctant to accept the clause.

Table 1 sums up, for all the sectors and for that of liquefaction and gasification of solid fuels, the main data of the invitations to submit projects. It appears from Table 1 that the projects relating to liquefaction and gasification of solid fuels are much more expensive than those of any other sector since 7% of the contracts signed represent 26% of the total investment of selected projects.

TABLE I - Main figures for the invitations to submit projects (period 1979-1982)

	Total	Liquefaction/ gasification	
Proposals submitted	1,432	35	2%
Total investment of proposals submitted MioECU	4,143	1,000	24%
Projects selected	331	13	4%
Projects withdrawn by the proposer	47	1	2%
Contracts signed	169	12	7%
Total investment of selected projects MioECU	878.6	226.3*	26%
Average investment per project MioECU	2.6	17.4	544%
Financial support granted MioECU	206.9	75	36%
Projects withdrawn MioECU	22.2	4.3	19%

* First contractual phase only

The commitments for all the sectors and for that of liquefaction and gasification of solid fuels have been, for the period 1979-1982: 183.2 MioECU and 71.4 MioECU, respectively. The breakdown by year is given in Table 2.

TABLE 2 - Commitments for all the demonstration sectors and for the "liquefaction and gasification of solid fuels" sector

Year	All sectors MioECU	Liquefaction/gasification MioECU	%
1979	27.3	11.6	42
1980	66.2	38.9	59
1081	34.2	1.5	4
1982	55.5	19.4	35
Total	183.2	71.4	39

It should be borne in mind that the projects in question are long-term ones entailing fairly considerable investment. Consequently, and in view of the cash limits imposed, it was thought preferable to support individual stages rather than entire projects.

Table 3 gives the ten projects that have been the subject of the twelve contracts named in Table 1. The Commission participation has been 40% of the cost of the supported phases but for one project it was only 10%. This Table 3 shows that even with limited available funds, the programme covers most of the main technological fields.

TABLE 3 - "Liquefaction and gasification of solid fuels"
Projects 1979-1982

Beneficiary	Project	EEC contribution MioECU
Fiat-Ansaldo (I)	Westinghouse gasification + combined cycle	6.7
Agip Carbone (I)	Texaco gasification of Sulcis coal	5.2
BGC (UK)	Lurgi slagging gasifier	12.3
Veba Oel (D)	Production of hydrogen for the hydrogenation of heavy oil and coal	6.1
Kloeckner (D)	Molten iron bath gasification	0.2
IDGS (B+D)	Underground gasification	11.3
GEGS (F)	Underground gasification	9.9
NCB (UK)	Direct liquefaction of hard coal	12.1
RBW (D)	Direct liquefaction of lignite	4.3
Haldor Topsøe (DK)	Indirect liquefaction	1.9

The ten projects supported in this first action will now be reviewed.

2.2 Projects Receiving Support

2.2.1. Fiat T.T.G./Ansaldo (I).

Fluidized bed coal gasification demonstration plant for the generation of electric power in a gas-steam combined cycle. The project covers the design, construction and operation of a fluidized bed coal gasification demonstration plant integrated with a gas-steam combined cycle for electric power generation. The gasification system allows the utilization of different feedstocks, including low rank coals. The integration with a combined cycle results in an increased power plant efficiency and environmental compliance. This project, through the operation of the demonstration plant, has the following main objectives:

- verify the simultaneous and integrated operation of the two systems (gasification and combined cycle) and the capability of the plant to follow the operating conditions required by the electrical network, such as: start-up sequences, partial loads operation, load-following capability, shut-down sequence, and emergency situations;
- verify the reliability of the processes and of the equipment used;
- verify the expected demonstration plant performance as regards: raw materials consumption, conversion efficiencies, capability of utilizing coal of different sources, and compliance with environmental standards;

- collect data and experience on which to base the future development of commercial plants, from both design and operational standpoints.

The programme established for the development of the demonstration project is structured in phases, of which Phase I (preliminary design) has been already carried out. In this Phase I, different options were analysed and the basic principles for the project selected (such as selection of coals to be used and their characterization, degree of integration between systems, more suitable size of main components, etc;).

On the above basis the preliminary design of the demonstration plant (heat and mass balance, flow diagrams and more representative drawings), the evaluation of the overall performance, the estimate of plant costs and of manpower requirements for operation and maintenance, and the general schedule of future activities for the subsequent phases of the demonstration project were carried out.

The gasification plant consists of two trains, each train including coal pressurization and feeding, one gasifier with a nominal capacity of 30 t/h of coal, ash removal, particulate recovery, heat recovery, gas quench, acid gas removal.

The power generating section of the plant (combined cycle) consists of two gas turbine units, each unit including one 100 MWe gas turbine and the relevant heat recovery boiler, and of one 90 MWe steam turbine. The gasification plant is designed to convert different types of coal in a low-BTU fuel gas to feed one of the two gas turbine units, the other unit being fed by natural gas or distillate oil. The net power of the demonstration plant is about 270 MWe while the efficiency ranges between 36.7 and 37.7% depending on coal.

The economics, carried out on Phase I, led to an evaluation of plant capital cost of about 300 billion Italian Lira (220 MioECU); time required for detailed design and construction of the coal gasification plant is approximately three years. Due to the high investment cost the decision to go ahead with the project has not yet been taken; hence the decision to suspend the project during 1984 in the hope that a decision to continue will be taken promptly.

2.2.2. Agip Carbone (I).

Coal gasification demonstration plant. The aims of this project were the design and construction of a Texaco gasification plant having a coal throughput of 1000 t/d to produce synthesis gas from the high ash and high sulphur coal of the Sulcis area. Gasification trials with Sulcis coal have been carried out at the Montebello (USA) and Oberhausen (D) plants and have shown that the Texaco gasifier can gasify such types of coal successfully. The investment cost of the construction is estimated to be 120 MioECU. However, due to the fact that this project is closely dependent on the development of the Sulcis mine and that the economics have not yet been assessed, this project has been suspended during 1983. The future prospects are that this project will be withdrawn.

to be commissioned in April 1984. Further phases projected by British Gas for the new gasifier include an endurance run and further optimisation culminating in operation to full commercial standard. The product gas can be purified by known methods to give clean medium calorific value gas or synthesis gas, combined cycle power generation (Cf.3.2.8) or upgrading to substitute natural gas by the British Gas HICOM process (Cf. 3.2.9).

2.2.4. Veba Oil (D).

Production of hydrogen for the hydrogenation of heavy oil and coal. The objective of the project is the construction and operation of a demonstration plant for the entrained bed gasification of solid fuels, such as hard coal, lignite and low temperature carbonization coke from residues of the hydrogenation of coal and heavy oils. The main emphasis of this project is on the industrial scale testing of a special feeding system for introducing solid fuels into pressure vessels. This system has been developed on a pilot plant scale. The main component of the feeding system is a twin screw feeder. In the feeder the powdered fuel is mixed at atmospheric pressure with about 15 wt% water or oil, compressed above the reactor pressure, and is then introduced at a precisely controlled rate via a specially designed burner into the gasification reactor.

The feeding system is characterized by the fact that it operates with a mixture which has a water content of only 15 wt%. This compares favourably to the alternative of feeding suspensions by slurry pumps, which requires a water content of 35 - 40 wt% minimum. The result is a considerable reduction in oxygen consumption and an increase in the specific gas yield.

Extensive pilot-scale tests at a throughput of 2.5 t/h have demonstrated that the feeder can be run without any problems and that no significant erosion has occurred.

Besides this, by means of these tests the basic data required for engineering the demonstration plant have been evaluated. The main parameters for fixing the process concept were determined. The final process concept takes into account the high reliability and availability of the demonstration plant. The gasification section will consist of two trains, having a capacity of 15 t/h each, at a gasification pressure of 60 bar. The sub-units such as high temperature shift conversion, acid gas removal, pressure swing adsorption, waste water treatment and grinding plant were designed as single train units. The hydrogen generated amounts to 60,000 Nm^3/h with a purity of 99.8 Vol.%.

The documents for the authorization for this process concept have been prepared. The further time schedule is as follows:

- completion of basic engineering: mid-1984;
- completion of detail engineering: mid-1985;
- start-up of the demonstration plant: mid-1986.

The total cost of this project is in the range of 90 Mio DM (40 MioECU).

2.2.3. British Gas Corporation (UK).

Demonstration of a commercial-scale slagging gasifier. This contract comprises two phases of a long-term coal gasification project in progress at the British Gas Corporation's Westfield Development Centre in Scotland. The British Gas process is a modification of the commercial process developed by Lurgi in order to allow the gasifier to work at a higher temperature and to recover the ashes as a slag. This has the advantage of reducing the steam consumption by 15%.

The British Gas/Lurgi slagging gasifier operates at 25 - 30 bar and the coal (6 - 50 mm) is introduced via a lock hopper into the top of the reactor. There it is gasified with steam and oxygen injected through tuyères near the base of the reactor. Countercurrent clow of coal and product gases result in modest outlet temperatures of about 450°C with no necessity for expensive high temperature heat recovery equipment. The low temperatures at the top of the bed result in liquid by-products such as tar and oil. Molten slag collects in a hearth and is removed through a slag tap.

The process achieves complete conversion of coal and has low steam and oxygen consumptions, high throughput and high thermal efficiency. Low steam consumption results in low production of aqueous effluent. The gasifier is insensitive to coal reactivity, can handle highly swelling and coking coals and coals with refractory ashes can be modified by the addition of flux.

The overall objective of the programme is to acquire the design specifications and performance data to enable commercial slagging gasifiers to be designed and operated to meet predicted engineering and performance guarantees. Pre-contract work on an existing gasifier, 1.8 m diameter, 350 t/d, had established the capability of the gasifier to operate on a range of coals from highly caking Pittsburgh No.8 t Wyoming sub-bituminous, in addition to eleven UK coals.

The first contract phase was carried out on the existing gasifier modified to incorporate design developments from the earlier tests. The specific objectives were to establish acceptable operational life of key components, extending the ability of the gasifier to accept fines (less than 6 mm) by injection through the tuyères and to eliminate the liquid by-products from the top of the bed by recycling them to the tuyères.

The first contract phase has been completed bringing the total operational hours since 1975 to 9000, including two long duration runs of 500 and 2000 hours with a total of 100,000 tons of various coals gasified. The process accepts coals with up to 25% fines to the top of the gasifier and up to about 25% via the tuyères. Various other options such as pelletised fines through the top of the gasifier (112 tons), slurries injected through the tuyères (30% water/70% coal) have been demonstrated. The total recycling of liquid by-products has been achieved.

The second contract phase covers the construction, commissioning and proving of a commercial scale gasifier (2.4 m diameter) with a capacity of 700 t/d. This gasifier, which will incorporate improvements resulting from the earlier phase, is under construction and is scheduled

2.2.5. Klöckner Stahlforschung (D).

Coal gasification in the molten iron reactor. The Commission has granted up to 40% of the cost of Phase I of this project. The aim of this first phase was to establish the performance of coal gasification trials according to the principle of the molten iron reactor in a 60 t industrial converter of Eisenwerk-Gesellschaft Maximilianshütte. This Phase I is completed and the results will be diffused.

The process is based on the simultaneous carbonization and decarbonization of liquid iron. In a vessel similar to an O.B.M. converter containing a high-carbon iron bath (hot metal), pulverized coal and oxygen are injected through tuyères installed in the bottom. Lime or converter slag, in ground form, can also be injected through the bottom tuyères. In a simplified way, the following process runs autothermically at a temperature in the range 1450 - 1550°:

- the volatile components of the coal escape and are cracked;
- the carbon is dissolved in the iron;
- the coal ash, together with added materials, such as steel plant slag or lime, forms a basic final slag;
- the sulphur introduced by the coal is bound in the basic reactor slag as well as in the reactor dust;
- the coal dissolved in the molten iron reacts with the oxygen and thus generates the product gas.

In the 60 t converter the coal rate in the trial was 15 - 20 t/h of a 14.4% ash and 29.5% volatile coal. The trials have shown that the molten bath temperature was stable during the gasification. The gas composition and the main pollutants concentration are given in Table 4.

TABLE 4 - Composition and pollutants in the gas produced by the Klöckner process

CO	64%	SO_2+SO_3	0-42 ppm
CO_2	2-3%	Cl	0.2-7.1 ppm
H_2	30-31%	F	0-3.6 ppm
CH_4	730 ppm	HCN	0-4.2 ppm
N_2*	3%	NH_3	0.1-0.9 ppm
H_2S	2-70 ppm	NO_x	0.1-0.5 ppm
COS	5-28 ppm		

* Primarily from carrier gas for coal transportation.

Due to the high process temperature of 1400 up to 1500°C, as well as the function of the molten iron, higher molecular hydrocarbons (tars) were not present. The concentrations of chlorine, fluor, cyanide of hydrogen, ammonia and nitrogen oxides are extremely low. Also the sulphur concentrations are low. This is a particularly interesting feature of this process compared to other gasification processes. The dust quantity was in the range of 20 to 60 g/Nm^3 but this quantity will be easily reduced to about 5 mg/Nm^3 by the gas cleaning system applied in steel making. The recirculation of fines is considered. The results show that the carbon losses can be limited to 1 - 2% and that desulphurisation of the gases occurs partly in the converter stack. In a longer trial (67 t of coal), the converter lining and the tuyères were in an excellent condition. The slag produced - although containing 3% sulphur - is suitable for the production of concrete, as a material for road construction and as a fertilizer.

This Phase I of the project has shown that coal gasification according to the principle of the molten iron reactor can be performed with good results (high gasification capacity, low carbon losses and low pollutants concentration in the coal gas) already with the present state of technology. On the basis of the favourable results, Klöckner-Werke A.G. finally decided to erect a full-size demonstration plant for coal gasification in the molten iron reactor.

2.2.6. Institut pour le Développement de la Gazéification Souterraine (IDGS) (B+D)

Belgo-German experiment on in-situ gasification. The test is under way in Belgium, on the Thulin site, in a coal deposit lying at 860 m depth. It aims at achieving an underground gasifier working at a pressure ranging between 20 and 30 bar. A first trial of reverse combustion linking was carried out from April to November 1982, between two vertical wells, 35 m apart from each other. Even though the combustion of coal was kept continuous for a period of more than six months, with air injections at pressures ranging between 250 and 300 bar, the trial has not led to the creation of a channel, but only to the formation of two zones of increased permeability in the vicinity of the two wells.

The failure of this first trial is to be ascribed to three factors: the self-ignition of the coal in the vicinity of the injection well, the system of ignition and the high level of the back pressure at the recovery well.

A second linking trial started in September 1983 but at the moment no significant results have been achieved.

Taking into account the experience that has been gathered, some elements of the experimental equipment have been modified; these modifications mainly concern:

- replacement of air injection by injection of a cold mixture of air + CO_2;
- use of a new system of ignition by means of a fuel-oil torch;
- reduction of the back pressure and continuous drainage of the water at the bottom of the recovery well.

2.2.7. Groupe d'Etude de la Gazéification Souterraine (GEGS) (F).

Deep underground gasification of coal. The French project is aimed at the gasification of deep lying coal which is not accessible to conventional mining. Since the technology to be applied covers a very wide field, Charbonnages de France, Gaz de France and Institut Français du Pétrole have been pooling their know-how together with Bureau de Recherches Géologiques et Minières and have set up GEGS (Groupe d'Etude de la Gazéification Souterraine).

GEGS's long term objective is to bring into operation an industrial production process for SNG ($36 \cdot 10^6$ J/m^3). To get a gas of this quality, a primary underground produced gas with a medium calorific value ($11 \cdot 10^6$ J/m^3) has to be converted to methane on the surface. This primary gas can be obtained with oxygen and steam as gasifying agents.

A technical economic feasibility study has shown that the method proposed by GEGS could lead to the production of an economically viable gas if injection and production wells required are at least 60 m apart. The method as envisaged consists of three stages:
- pre-linking by hydraulic fracking;
- enlarging the linkage by reverse combustion;
- gasifying with oxygen and steam.

The three aspects are studied mainly by pilot tests and also require some laboratory and theoretical work.

A first main test has been carried out in Bruay-en-Artois in the northern coalfield of France between 1979 and 1981 and has provided some important conclusions:
- adequate well fittings and valves exposed to severe pressure variations;
- possibility to link by hydrofracturing;
- easy ignition of coal with an electric heater;
- difficulties to carry reverse combustion according to spontaneous combustion problems at the injection well;
- possibility to prevent spontaneous combustion with an injection gas of low oxygen content.

A second pilot test is in operation at La Haute-Deule also in the northern coalfield. The results of hydrofracturing have been better than in Bruay and ignition and reverse combustion has begun in 1984. If these operations are successful, gasification with oxygen and steam will be tried in the spring of 1984.

In a parallel spot test in a very shallow seam of the Loire coalfield, another linking method electrolinking has been tried. The results being successful, the coal seam will be shortly open-casted to back-check the results of the linking test.

2.2.8. National Coal Board (UK).

Supercritical gas extraction pilot plant. The aim of this project was to design, construct and operate a pilot plant (25 t/d) for the liquefaction of coal by supercritical gas extraction and the conversion of extracts to high-grade liquids. The design has been completed (February 1982). The project was then merged with the new projects 'Coal liquefaction demonstration facility' in view of a temporary concentration on the liquid solvent extraction technology.

Coal liquefaction facility. The aim of the project was the design, construction and operation of a coal liquefaction plant (25 t/d) using the technique of solvent extraction. The coal extract is converted to distillate oils suitable as feedstock for aromatic petrochemicals, transport fuels, etc. It was provided to include at a later stage, also the supercritical gas extraction process (see above). The design phase has been completed (February 1981).

During 1982, the NCB informed the Commission of its wish to reduce by a factor of 10 the size of the plant. The NCB justified this reduction in size by arguing that, due to the oil market situation, there was no urgent need for larger liquefaction plants and that a 2.5 t/d plant would be more flexible. The Commission did not accept these arguments and the contract was terminated.

2.2.9. Rheinische Braunkohlenwerke A.G. (D).

Hydroliquefaction of lignite. The design, construction and operation of a demonstration plant for the hydrogenation of lignite (360 t/d lignite, 72 t/d products) was foreseen to obtain high-rank products (motor fuels, chemical feedstocks, etc.). The basic engineering phase was completed and was partly financed by the Federal Government whilst the detailed engineering phase started on 1 October 1981. The work was concentrated on the dimensioning of apparatus and machinery, the pretreatment of coal and hydrogenation reactor. However, in 1982, due to the energy and world economic situation, RBW preferred to stop their project.

2.2.10. Haldor Topsøe (DK).

Selective hydrocarbon synthesis. The aim of the project is a mobile plant for the manufacture of gasoline from coal based gas by a novel selective synthesis process. The Danish Haldor Topsøe A/S has developed a process scheme integrating their catalytic technology with Mobil methanol to gasoline technology. This integrated process has potential for significant savings in investments and energy consumption. The process may find application first for treating low-grade natural gas with high CO_2 content or for converting gas from remotely placed wells.

Up to 1982 the contract covered Phase I (engineering and design work), Phase II (construction) and Phase III (operation). Topsøe has built a demonstration pilot plant with a capacity of 1 ton of gasoline product per day which has been transported and installed at its Houston Texas facilities and is now being put into operation. Mobil furnishes experimental quantities of its proprietary zeolite catalyst for use in the pilot plant. Topsøe has completed tests on a smaller pilot scale which have demonstrated the stable operation of the integrated process (Cf. 3.2.1).

3. NEW ACTION (1983)

3.1 Generalities

There is considerable evidence to confirm the effectiveness of the provision of financial support by the Community as regards the commencement of demonstration projects, the considerable response to the invitations to submit projects, the value of the first results, the publication of regular newsletters, meetings between the various parties concerned, and the expectations in relation to the invitations to submit projects aroused in the organizations and firms concerned. However, the experience acquired has also indicated that the application of the arrangements should be improved, and therefore reviewed.

The Commission has estimated that in this particular field the applicants need more support for pilot industrial projects rather than for demonstration projects, the pilot industrial stage being a stage which entails less investment, resources and time, and makes it possible to gather sufficiently precise data in order to progress from the research and development stage either to the industrial stage or to the demonstration stage. In addition, one of the features of the industrial pilot stage is that there is considerable flexibility as regards modifying plant and equipment.

The industrial pilot and/or demonstration stage, which comes after the research and development stage, is the stage permitting the transition to the industrial and commercial exploitation of techniques, processes and products which are innovatory in themselves or in their application.

The industrial pilot stage is characterized by the operation of an installation having an adequate capacity (20 - 25 t/d) and using suitably large components with a view to increasing the reliability of the technical and economic data needed to progress to the demonstration stage, and in certain cases, the industrial and commercial stage.

The demonstration stage is characterized by the operation of an industrial-scale installation which should make it possible to bring together all the technical and economic data, in order to proceed at the least possible risk to the commercial and industrial exploitation of the technology. In most cases, industrial pilot and/or demonstration operations are a vital adjunct to the research and development programmes at Community and national level.

Moreover, in this field it is unlikely that a project will be ready for industrial-scale use in the short term, and an industrial pilot stage (20 - 25 t/d) will usually be needed before the commercial demonstration or industrial application stage. This was, in fact, already evident following the invitations to submit projects - many projects had to be ruled out because they were pilot projects rather than demonstration projects - and during the work on the projects under way.

The Council of Ministers adopted, on 11 July 1983 Regulation (EEC) 1971/83 /4/ on the granting of financial support to pilot industrial projects and demonstration projects relating to the liquefaction and gasification of solid fuels. The publication of an invitation to submit projects has shown that the industrial spheres were awaiting the Community's support.

Table 5 sums up the main data of this first invitation to submit pilot industrial projects and demonstration projects in the field in which we are interested. It must be underlined that the grants to demonstration projects are still repayable at 50% in case of commercial success while the grants to pilot industrial projects are not repayable. In view of the limited amount of funds available, the Commission has continued to support the selected projects only by granting some phases rather than the entire project.

TABLE 5 - Results of the invitation to submit projects (1983)

Proposals submitted	28
Total investment in proposals submitted	805 MioECU
Projects selected	9
Total investment in proposals selected	432 MioECU
Average investment per project	50 MioECU
Financial support granted	28 MioECU

The selected projects, as well as the amount of financial support granted, are reported in Table 6. Among these nine projects, only those of the Rheinische Braunkohlenwerke and British Gas Corporation are projects of demonstration size. The responses to the invitation to submit project proposals in the field of liquefaction and gasification of solid fuels show that the Commission was right in wishing to support pilot industrial projects.

TABLE 6 - "Liquefaction and gasification of solid fuels" Projects (1983)

Beneficiary	Project	Aid MioECU
Haldor Topsøe (DK)	Selective hydrocarbon synthesis (part not yet covered by the previous contract)	1.835
Energy Equipment & Labofina (UK+B)	Industrial fluid bed pilot plant for the gasification of solid fuels	1.031
Krupp-Koppers (D)	48 t/d Prenflo plant	5.138
Charbonnages de France (F)	Fluidized bed coal gasification (U gas process)	5.127
National Coal Board (UK)	Liquid solvent extraction pilot plant	0.166
Rolls-Royce (UK)	Operation of a gas turbine generating set on fuel gas from the British Gas/Lurgi slagging gasifier	1.362
Rheinische Braunkohlenwerke AG (D)	Gasification of browncoal by the HTW process	6.365
National Coal Baord (UK)	Industrial fuel gas fluidized pilot plant	1.779
British Gas Corporation (UK)	Slagging gasifier and HICOM methanation process	4.940

Although the funds are limited, the programme now covers most of the technological fields: fixed bed, fluidized bed, entrained bed and "in-situ" gasification as well as direct and indirect liquefaction. The only field not yet covered is coal devolatilisation under inert atmospheres or under hydrogen but this is probably due to the fact that this technology is only just reaching maturity.

We will now briefly review the nine projects selected for 1983.

3.2 Projects Receiving Support

3.2.1. Haldor Topsøe (DK).

Selective hydrocarbon synthesis. This project is a modified version of the project already in course and that has been mentioned above. The applicant wished to modify the project for the following three reasons:

- the optimisation of the process can increase the utilisation life of the catalyst before regenerating it;
- possibility of carrying out, from natural gas, a simulation of the production of synthesis gas of the same composition as that obtained from coal;
- a new assessment of costs.

The production will be about 900 kg/d of gasoline and 300 kg/d of C_3-C_4 hydrocarbons, equivalent to 2.9 t/d of methanol. The end of the project is expected for December 1986. The total cost of the project is 73.6 MioDKR (9 MioECU).

3.2.2. The Energy Equipment Co. (UK) and Labofina (B).

Industrial fluid bed pilot plant for the gasification of solid fuels. The process is based on the Energy Equipment fluidized bed technology. The plant will gasify a solid fuel to generate a fuel gas at a nominal rate of 18 GJ/h. This gas will have a calorific value in the range of 3700 to 13,000 kJ/m^3. The system will accept all solid fuels. The solid fuel is fludized by air and steam in a sand bed; in a first stage, partial combustion will take place in the bed. In the second stage, the modified installation will produce nitrogen-free gas; the heat of reaction will be provided by the combustion in air of a part of the solid fuel in an interconnected bed. The gases containing nitrogen will not pass into the fuel gas which is generated in the second chamber. This development will make it possible to adapt the low calorific value system to the medium calorific system by the utilisation of the reactor with separated compartments. The project should be finished by February 1986. This project will cost 1.6 Mio Pound Sterling (2.8 MioECU).

3.2.3. Krupp-Koppers (D).

48 t/d PRENFLO plant. The PRENFLO process is the pressurized version of the commercial Koppers-Totzek process. Development of the process under pressure has been carried out in collaboration with Shell. Progression to units of 1000 - 2500 t/d will be possible from the 48 t/d installation. The plant will operate at 30 bar to produce 4000 Nm^3/h of gas from Saar coal. Because of the reaction temperature (1300 - 1600°C) there are no by-products. The gasifying medium is a mixture of oxygen (available on site) and steam. The 150 t/d Shell-Koppers installation

has given good results but some components require modification or development. This cannot be done at reasonable cost on the 150 t/d scale, hence the need for the 48 t/d plant. The plant will be built in the Saar on the site of the Saarberg-Dr.Otto plant. The tests will start in January 1985. The main components will be tested in continuous operation; improvements will be carried out as the tests progress. The end of the project is foreseen for the end of 1987, with a total cost of 86.9 Mio DM (38.8 MioECU).

3.2.4. Charbonnages de France (F).

Fluidized bed coal gasification. The aims of Charbonnages de France in setting up this project are to develop a second generation coal gasification process presenting wide possibilities with regard to the coal used and to be able to sell units worldwide. The technology that has been chosen is that of the U-gas process of the Institute of Gas Technology (USA). The plant will have a capacity of 200 t/d and will operate at 30 bar. This technique should make it possible to gasify a broad range of coals, the gas produced being able to be easily treated and having different uses (chemistry, combustion, combined cycle, methanation). The U-gas process permits a high degree of carbon conversion thanks to the agglomeration of ash in a reactor under pressure, without moving parts and producing a by-product-free gas. A great quantity of fines must be recirculated. The plant will be erected on the "plateforme d'essais de Mazingarbe" (N/PdC); the end of the project is anticipated for the beginning of 1989. The total cost is expected to be 490 Mio FF (71 MioECU).

3.2.5. National Coal Board (UK).

Liquid solvent extraction pilot plant. This project is the same as the one discussed above, but with a reduced capacity of 2.5 t/d. The plant is to be erected on the site of Point of Ayr. The UK Department of Energy is negotiating with private companies a participation in this unique direct liquefaction process. The forecasted price if 30 Mio Pound Sterling (52 MioECU).

3.2.6. Rheinische Braunkohlenwerke AG.

Gasification of browncoal by the HTW process. RBW is building at Ville-Berrenrath a demonstration plant with a capacity of 1500 t/d of raw brown coal using the Winkler process at high temperature (1000°C) and under pressure (10 bar). The aims are to increase productivity and to reduce the compression energy of the process of chemical synthesis down-stream. The high temperature should make it possible to improve the gas quality and to increase the carbon conversion rate; the latter will also be improved by recirculation of the entrained fines. The HTW gasifier is followed by a shift reaction of the CO where CO and H_2O are completely transformed into H_2 and CO_2; The gases are cleaned by the Rectisol process and a Stretford sulphur recuperation unit is planned. The syntehsis gas (37,000 Nm^3/h) will serve for methanol production (15 t/h). The total cost of the project is 453 Mio DM (200 MioECU).

The commencement of the operation phase is expected for the middle of 1985 and the end of the project is planned for 1990. This is the first step of a large gasification plant of 6000 t/d.

3.2.7. National Coal Board (UK).

Industrial fuel gas fluidized bed pilot plant. The main aim of this project is to build and demonstrate a gasifier for the generation of gas for firing furnaces and kilns as used, for example, in the brick, ceramic and glass industries. Such a gasifier would normally be installed close to the furnaces of kilns so that gas can be fed directly from the gasifier to the burner. There is no doubt that there is a need for such a gasifier (the forecasts are that for the Community as a whole the potential market in the heavy clay, refractory and allied industries alone is close to 20 Mt oil equivalent). Studies partially funded by the E.C.S.C. led the NCB to choose a spouted fluidized bed. The characteristics of the technology developed by the NCB are:

- use of a wide range of coals;
- easy to operate and flexible to changes in demand;
- sulphur retention;
- insignificant yield of condensible tar.

The objectives of this plant working at atmospheric pressure and with a capacity of 12 t/d are:

- to investigate different process and plant configurations to improve coal conversion, for example, fines recycle;
- to investigate the direct use of the commercially available grades of coal;
- to demonstrate gas cleaning to reach the environmental standards required by industrial processes;
- to demonstrate the combustion characteristics of the low calorific value gas;
- to demonstrate continuous operation under commercial conditions.

The plant will be built at the Coal Research Establishment at Stoke Orchard, Cheltenham, but will be designed in a modular form so that, when proved, it can be moved to a commercial site, for a demonstration of its performance in a real industrial situation. The project will be finished by the end of 1986 at a total cost estimated to be 2.3 Mio Pound Sterling (4 MioECU).

3.2.8. Rolls-Royce Ltd. (UK).

Operation of an SK30 gas turbine generating set on fuel gas from the British Gas/Lurgi slagging gasifier. This project fits in the British Gas-Lurgi slagging gasifier programme (Cf.2.2.3) and aims at the combination of a Rolls-Royce SK30 gas turbine with the gasifier in order to form a combined cycle power station producing 27 MWe (19 MWe to the public network) and to confirm the adequacy of the gas clean-up and turbine combustor concepts. After completion of Phases 1 and 2 (feasibility study and design), the project consists of the following steps:

Phase 3: (remainder) thermal paint test and combustion rig testing of insurance standard burners (both at Rolls-Royce's Ansty plant);

Phase 4: procurement, delivery and installation of the SK30 generating set (gas, turbine, generator and ancillaries) at Westfield Development Centre;

Phase 5: confirmatory testing, inspections, strip and inspection of the gas turbine on completion of the 90 day run (Phase 4 of the British Gas project, see below).

The end of the project is foreseen for mid 1986. The cost of the project is 2.2 Mio Pound Sterling (3.8 MioECU).

3.2.9. British Gas Corporation (UK).

Demonstration of a commercial scale slagging gasifier and the HICOM methanation process. This is the continuation and extension of the running slagging gasifier project, Phases 1 - 3 of which being covered by a contract. The aims of the present contract are the further development, demonstration and optimisation of the gasifier system and equipment for commercial application. In Phase 4, a 90 day run is planned, during which the extended gasifier (2.3 m diameter, 500 t/d coal) will be run under selected operational conditions. Some of the product gas is to be upgraded to SNG using the British Gas High Carbon Monoxide Methanation Process (HICOM). Phase 5 will be concerned with equipment modifications (new gas offtake, wash cooler and waste heat boiler, automated coal locking system, tar management system), the feeding of fine coal and by-products to the gasifier (run-of-mine-coal), and the final demonstration of the complete commercial configuration. The end of the project is expected for the beginning of 1986. The total cost is around 50 Mio Pound Sterling (86 MioECU).

It must be pointed out that these last two projects have good prospects. Sheffield city intend to carry out a feasibility study for a district heating scheme, fuelled by household refuse and a coal gasification plant of the British Gas type, supplying a series of Rolls-Royce gas turbines with recuperation boiler and steam turbines. This is an original conception not yet utilised in the Community. The proposed technology is able to constitute a solution to the difficult problems raised by the new norms concerning the emissions of sulphur dioxide and nitrogen oxides from district heating plants.

4. CONCLUSIONS

We have seen that the Community programme in the liquefaction and gasification of solid fuels sector covers nearly all the present technologies. The projects are helping to solve the often severe technical problems posed by the construction and operation of large installations.

Considering its energy vulnerability, the Community is duty bound to facilitate the development and demonstration of those technologies. The liquefaction and gasification of solid fuels must not only be regarded from an economic point of view, but also from a strategic point of view. Experience has shown that the Commission was right in wishing to maintain coal research to a reasonable level and this is thanks to ECSC grants among other things. At any rate, now more than ever, efforts must be renewed in the field of solid fuels conversion.

REFERENCES

1. OJ No. L158 of 16.6.1978.
2. OJ No. L93 of 12.4.1979.
3. Resolution of 17.11.1977 on the 2nd draft Commission Regulations; Resolution of 18.02.1980 on the energy objectives for 1990.
4. OJ No. L195 of 19.7.1983.

ENERGY FROM BIOMASS IN THE EUROPEAN COMMUNITY

G. GRASSI
Commission of the European Communities
Directorate General XII for Science, Research and Development
200 rue de la Loi
B - 1049 BRUSSELS

ABSTRACT. In the field of biomass productivity, the main objectives were: the selection of species presenting the highest yield of dry matter/year x ha; and to verify if the new concept of energy-farming could be adopted for the surplus area (good agricultural land) available in the Community (6 mio ha in 1980 - 12 mio ha in 1990).

In the field of biomass harvesting and collection, three machines were developed: one for small trees and forestry wastes (F), two for coppice (UK and IRL) and one study on an integrated harvesting system recovering industrial wood and residues in a one-pass operation (IRL).

In the field of biomass conversion into energy products, good progress was achieved in the two types of process: biological and thermochemical.

DG XII has sponsored, coordinated and directed a programme of pilot plants for the production of synthesis gas (first step in the production of methanol).

1. INTRODUCTION

Biomass is one of the most important among the new and renewable sources of energy. In the medium-term (around the year 2000) biomass may provide around 6% of total E.C. primary energy demand (1.600 MToe).

Therefore, the main objective of the R&D "Energy from Biomass" programme of the Commission is to demonstrate the interest of biomass as an energy source for Europe.

It is also important to remember that this energy contribution is compatible with the average penetration time of new energy technologies which is 20 to 25 years for a 10% market penetration (threshold value for commercial introduction).

G. E. Beghi (ed.), Synthetic Fuels, 455–489.

The following presentation of the European Community's R&D programme on biomass provides a good summary of the work performed over the past few years in European industry, universities and research institutions, particularly the agricultural research centres.

There is a good balance in the programme between biomass cropping, biomass conversion and utilization.

Other side benefits which may result from a larger use of biomass energy are :

- creation of new employment; eventually 0.5 million new jobs could be created within the Community;

- development of rural areas by decentralization of energy production;

- new export markets, giving industry the means to enter these markets;

- improvement of environmental conditions by the recycling of biomass residues;

- in the agricultural sector, there is a future prospect of replacing some of the present surpluses by using biomass for the energy and other non-agricultural markets;

- as is generally recognized, biomass energy will play an increasing role in the Community's policy for aid to developing countries.

2. TOTAL ENERGY POTENTIAL OF BIOMASS IN THE COMMUNITY

The medium-term potential of biomass is the biggest of all the new and renewable sources of energy. Within the European Community, biomass has the real potential to produce an amount of usable net* energy approximating 90 Mtoe by the year 2000 (Table I). This is around 6% of the estimated energy consumption of the 10 Member States in the year 2000.

About half the estimated total energy production can be derived from agricultural and crop wastes, municipal wastes and sewage and wood with little or no change in agricultural, forestry or other land use. The other half can be derived from small changes in agricultural or forestry procedures which would result in increased yields of wood, the use of catch crops and the use of agricultural surpluses (as an alternative to export at subsidized low prices). Again this would not require significant change in land use.

---

* Values are computed for material after conversion (i.e. conversion losses, typically 50%).

TABLE I : Total net energy potential (after conversion)

Biomass raw material	net* energy (M.Toe)
Wastes and residues	
1. Animal wastes	11.5
2. Crop residues	12.5
3. Forestry/wood residues	7.9
4. Municipal solid wastes	8.9
5. Sewage	1.8
Firewood	6.6
Energy forestry and crops	
Dry (forestry and short rotation plantation	28.1
Wet (Catch crops, algae and aquatic plants)	8.5
Total	85.8

In addition agricultural surpluses could be available, for example :

	1980/81 figures (10^6 t.)
Grains (total)	7
Wheat	11
Sugar	4

The rapid increase of costs for the agrarian market, a number of 12 mio unemployed persons in the EC, the need for development of European energy sources led the Commission to carry out several research projects to see if it were possible to reduce surpluses and to replace agricultural production by energy farming.

* Values are computed for material after conversion (i.e. conversion losses, typically 50%).

Energy farming is a new field which has not been exploited in the past because agricultural production has been totally food and material oriented, and has never been optimized in terms of energy production. Even short-rotation forestry, a promising approach to biomass production has always given low yields because energy production has not been the primary objective.

At present, in the Community there is a surplus area of at least 5-6 mio hectares, which could be used for energy farming. It can be expected that this surplus area will increase to about 9 mio hectares in 1986 and 12 mio hectares in 1990. In Table II details on the potential surplus area for energy farming have been presented.

TABLE II

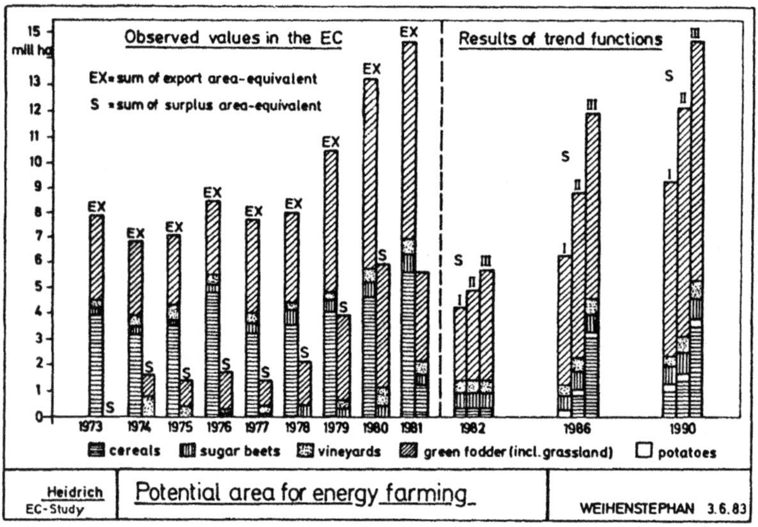

Although it is generally accepted that biomass offers good prospects for competitive energy production, under today's conditions, the subsidies necessary for competitive energy farming exceed the possible savings from a reduction of the land use for food production. If more land is set free, more costs for intervention can be saved (see Table III). For example,

- to make the production of ethanol instead of sugar from sugarbeets interesting for the farmer, a subsidy of 1600 - 1800 ECU/ha would be necessary; the savings, in case of not producing sugar, are about 800 ECU/ha.

- the necessary subsidy to use rape for energetic purposes (combustion of the oil in Diesel engines) is about 800 ECU/ha larger than the subsidy for rape when used in the conventional way (500 ECU/ha).

- In the case of short rotation forestry, the amount of subsidies to be given to this type of land does not produce any net savings for the agricultural budget of the EC.

TABLE - III

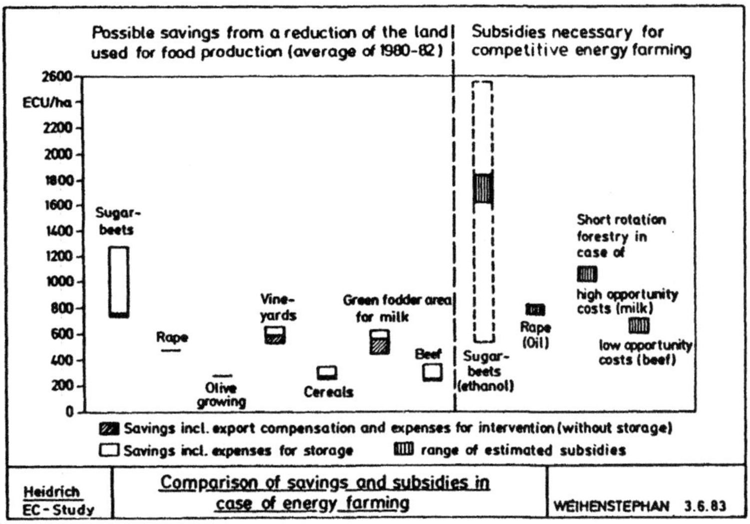

Nevertheless, rising energy prices and/or declining prices for food products, together with technical improvements in the field of energy farming and conversion, might well modify this situation. In any case, more research work is necessary to increase yields of energy crops by breeding better plant varieties, improving farm methods and decreasing costs for energy farming by construction of new powerful and better adapted machines for harvesting, processing and energy conversion.

3. PRODUCTION, HARVESTING, COLLECTION OF BIOMASS

There are two types of biomass :

- terrestrial biomass;
- aquatic biomass.

3.1. Review of research activities on terrestrial biomass wastes, energy crops and forestry

We should remember here that the main objective is to collect or to produce as high a yield as possible of dry matter per hectare, per annum per a given energy input. The components of the system which must be considered are: the total yield per unit area (in dry weight), the energy content and the cultural energy ratio (energy output/input). As shown in Table IV, these vary over a large range for different crops and cultural systems.

TABLE IV

Yields and cultural energy requirements

Cultural system	Max.Yield (t/ha) (dry weight)	Energy Content of cultures* Toe/ha	Cultural Energy ratio output/input
Forestry	10	3.6	10 - 20
Short rotation forestry	15	5.3	5 - 15
Algae	60	21.4	8 - high
Aquatics	15	5.3	5 - 15
Catch crops	8	2.9	3 - 4
Grass	15	5.3	2.4 - 5.6
Natural vegetation	5	1.8	High
Wheat	5	1.8	3.4
Sugar beet	10	3.6	3.6
* 1 Tonne oil equivalent = 42 GJ			

The level of information available on biomass arising from crop residues, catch crops and natural vegetation is summarized in Table V.

TABLE V - State of the Art

Resource	Research Areas			
	Size of Resource (potential)	Yield	Production Technology	Conversion Technology
Crop residues	Adequately determined	Adequately estimated	Well-tried	Well-developed
Catch crops	Being refined	Being collated analyzed	Well-established	Remains to become economic
Existing natural vegetation	Well-known in some cases	Well-known but stability unknown	Relatively unknown	Many options being developed
New energy crops	Much known about a very small number of species	Initial yields known for several species	Almost unknown	Many options being developed; new methods will be required

In general, the extent of the resource and order of yield is well-known while the production technology is well-known for agro-ecosystems but relatively unknown for natural vegetation. The constraints on the commercial viability of the resulting biofuels depend upon the conversion technology as well as the price of feedstock. However, current costings of feedstock and conversion processes show that economic viability has almost been achieved in some cases.

Although crop residues, catch crops and natural vegetation show many promising characteristics for a significant contribution to future energy needs, there are obviously a number of problems to be overcome by research and development (Table VI). From the producers' standpoint, the most critical factor is probably the maintenance of yield stability over long periods of time. The removal of nutrients from site, mechanical effects on the soil and the physiological responses of perennial plants to harvesting may lead to instability, but inputs of nutrients in rain may reduce the nutrient depletion. Also, a new technology may be necessary to harvest natural vegetation, but the impressive developments of forest machinery suggest that this problem is mainly one of commercial incentives.

TABLE VI : Characteristics of the agricultural wastes and energy crop (non-forest ecosystems) projects

	NON-FOREST ECOSYSTEMS		
	RESIDUES AND CATCH CROPS	EXISTING VEGETATION	NEW ENERGY
POSITIVE ASPECTS	a) Traditional land use slightly modified	Traditional land use almost unchanged	-
	b) Should not detract from main use of land	Should not compete with other uses of land	-
	c) Energy inputs low; existing equipment used	Energy inputs low; no planting	There is a vast genetic library available
	d) Catch crops should increase stability of main crop	Species diversity should stabilize energy crops	
	e) The size of the resource is very large	The size of the resource is vast	-
NEGATIVE ASPECTS	a) Extra nutrients will be removed	Nutrients will be removed for the first time	-
	b) Biomass from catch crops will be limited in quality and mainly restricted to anaerobic digestion	Biomass may be variable in quality and will grow in mixtures	-
	c) Soil could be damaged by extra traffic	Erosion could be increased by harvesting	-
	d) -	Often difficult terrain to work	
	e) -	New technologies need to be developed	New management methods need to be developed

The 2nd R&D programme of the Commission of the European Communities (DG XII) included 13 research projects which can be subdivided into three main areas: natural systems, agricultural wastes and energy crops, forestry production and harvesting.

The different projects follow a gradient of innovation: crop residues are already available and their use is almost economically justified. The use of catch crops and existing natural vegetation is a little further from implementation while the development of new energy crops from wild species must be a long term project.

The group of projects also shares an important characteristic in that their proposed exploitation of biomass will have only small effects on current land use. There are often strong limitations on the designation of land to new crops owing to conservation interests, amenity factors, landscape quality and land ownership.

In many cases, therefore, there will be relatively little conflict over competing land uses. For example, residues already exist, catch crops increase the efficiency with which existing arable land is used while some forms of natural vegetation can be harvested from land which cannot be used for conventional crops.

a) Natural systems

This research programme included the investigation of biofuel production from natural vegetation in Great Britain (contract ESE-R-015-UK), a new method of cultivation of Arundo donax in France (ESE-R-042-F), harvesting Turkey oak coppice in Italy (ESE-R-035-I), harvesting Mediterranean-type ecosystems in Greece (ESE-R-063-GR) and the quality and quantity of latex which can be produced from the Euphorbiaceae family in Greece.(ESE-R-064-GR)

The important point to note about all these projects is that there are relatively large resources technically available which could be utilized with only small energy inputs. The use of natural vegetation for energy is an important management alternative for such ecosystems. Stability of production following continued harvests may be difficult and must be verified.

In Britain Callaghan has identified and studied 3 species with a high potential of use for energy. These are bracken, cordgrass and Japanese knotweed. Of these, the latter should more properly be considered as an energy crop. Bracken and cordgrass occur naturally and can be harvested annually with yields of 7 and 11 dt/ha/yr respectively. Continued harvesting of bracken may lead to its eradication and consequently its use for energy could be coupled with an eradication programme.

The 2 projects in Greece started later than the other projects and so full results are not available. However, it seems likely that yields of

50 dt/ha on a 10 year rotation are possible for the maquis ecosystem. Management of this system for energy will have an additional advantage in preventing the ravages of fire.

The work on Arundo donax is particularly interesting with productivities of up to 20 dt/ha/yr over 10 years. The ease with which it can be integrated into agricultural systems must make it an attractive proposition in southern France. Problems still remain concerning the best management system and the optimum frequency of harvests to maintain productivity.

Harvesting of natural Turkey oak coppice using mechanization is the first time that it has been done in Italy. There is a large potential resource and preliminary indications show that it can be done without damaging the cut stools and so allowing coppice re-growth. Here productivities of 2 dt/ha/yr might be expected on rotations of 20 years. Cost figures are needed before it can be concluded that such a practice is economically viable.

A possible problem with the use of natural vegetation for energy is the low density of supply and often the remoteness of the resource from the point of end use making transport costs the limiting factor. Furthermore, it is felt that there is a strong need to identify possible uses and conversion processes for which natural vegetation is suited as this may effect management and harvesting regimes. That natural vegetation can be harvested (technically and/or economically) is as yet unproved.

Future work

It was felt that there was a need for more fundamental work on plant productivity and its limiting factors along with a continued search for new wild species suitable for energy cropping.

Large scale trials are required of systems already identified to establish reliability of yield, harvesting and management. Such trials should also be used to produce cost estimates as these so far have been generally lacking from the programme.

Multiple cropping such as agroforestry as well as crop fractionation deserve more attention.

b) Agricultural wastes and energy crops

This group contains projects investigating the use of straw and animal wastes (contract ESE-R-014-F) and growing catch crops for fuel in Great Britain (ESE-R-034-UK). Straw appears to be an economically viable fuel for combustion on farms for use in space heating and grain drying but its use off-farm will be limited due to unfavourable competition with coal.

The study on catch crops indicates that there is a technical potential production of about 1 to 7 dt/ha/yr but the economic potential has not been examined in detail. Catch crops are considered suitable for anaerobic digestion.

The economics of the use of catch crops and straw need to be further refined. This might best be accomplished by using a farm case study approach. This should also encompass a survey of farmer attitude to catch crops, availability of land and possible end uses for the energy crops.

c) Forestry

There are 6 projects in this group. Four are primarily concerned with production and 2 with harvesting.

c.1. Forestry Production

There are three sources of forest biomass for energy; existing scrub woodland (coppice and single stem), thinnings and residues from conventional forestry, energy plantations (coppice and single stem). Scrub woodland, although of low productivity, is an important resource by virtue of its large extent particularly in France, Greece and Italy.

For production, coppice receives the most attention with work in Great Britain (ESE-R-017-UK), France (ESE-R-018-F), Ireland (ESE-R-036-EIR). Single stemmed short rotation forestry is studied in Great Britain (ESE-R-016-UK). The French project also involves work on the production of inter-specific crosses within the genus Alnus

Productivity data is beginning to emerge from these projects. For coppice which is limited to the more fertile and sheltered sites productivities of 12 to 15 dt/ha/yr appear possible. With good species selection to fit the site and environment coupled with good management, these production levels could be maintained.

Single stem plantations, which are likely to be on poorer ground than coppice and of which more will be available, have productivities of between 6 and 12 dt/ha/yr. The results in Ireland of 8.5 dt/ha/yr after 5 years growth with Lodgepole pine on virgin peat appear particularly promising for such a difficult site.

Residues can be estimated at about 20% of yearly productivity.

The management of forest biomass energy plantations is becoming better understood and the importance of good weed control in the two years following planting has been recognized. Information on costs and logistics of establishing plantations is beginning to emerge although this is, in the main, limited to small areas.

The potential of short rotation forestry to supply wood for fuel appears promising particularly if good quality land will be released from agriculture following revision of the Community's agricultural policy.

c.2 Forest Harvesting

In this programme 3 harvesting machines are being developed; one for forestry wastes and small trees (ESE-R-049-F) and 2 for coppice (ESE-R-016-UK and ESE-R-019-EIR). In addition a desk study has been undertaken to examine ways of harvesting forest biomass for energy (ESE-R-016-UK).

The French forestry wastes harvester appears to be a promising design but it is too early to be confident of its performance.

Of the two coppice harvesters that have been built the Irish Bord na Mona machine was designed for harvesting the produce of large plantations on cut-over peat bogs and the Loughry machines for harvesting coppice on small areas of disadvantaged farm land. The Bord na Mona machine produces billets and the Loughry machine bundles. Both machines have suffered through not having sufficient material available for extensive harvesting trials.

The desk study has identified the potential importance of integrated Harvesting Systems to recover forest residues in a 'one pass' operation. This is particularly important considering the vast resource of small roundwood available in the short to medium term in north-west Europe.

c.3 Future Work on Forestry

Here below a Forest Energy Research and Development Programme is listed indicating those areas covered by the present programme of the Commission and those requiring further work and their priorities.

A. MANAGEMENT

I Complete Systems (2) * : both strategic (3) and specific (2) studies are required. They should include consideration of economics, logistics, institutional constraints, land availability, energy balance and ecological impact.

B. BIOLOGICAL

I Production Models (2) : of existing single stem (1) and coppice (2) stands and new energy plantations of single stem (3) and

---

* Key
1 - current input, satisfactory
2 - current input, further work required
3 - no current input, new work required
4 - no current input, long term outwith biomass programme
5 - no current input, high priority for further work

coppice (2) are required. Therefore the trials laid down in the present programme should be monitored in future.

II Silviculture (2) : there is a need for further trials to test species, spacing, harvesting cycle, site and cultural techniques

III Selection (2) / Breeding (4) : is still required for species of *Alnus*, *Eucalyptus*, *Salix* and to a lesser extent *Populus*.

C. MECHANISATION

I Site Preparation (3)

II Planting (3)

III Tending (3)

IV Harvesting Existing Forests (2) : more work is needed on harvesting scrub single stem (3) and coppice (2); forest residues ('one and two pass' systems) (2) and whole-tree harvesting of thinnings (2).

V Harvesting Energy Forests (2) : the two coppice harvesters (Loughry and Bord na Mona) should be tested extensively and developed further. The Bord na Mona machine should be made self-propelled.

D. MATCHING

I The interrelation of biomass production, harvesting and conversion should be studied (5).

- Further small scale trials are needed in Europe to gain more information for species and clone selection.

- The potential importance of Integrated Harvesting Systems to produce both an energy and a conventional forest product in a 'one pass' operation should be studied further.

- Field trials are needed to assess the production rates and economic viability of the systems across a range of crop and terrain types.

---

* Key
1 - current input, satisfactory
2 - current input, further work required
3 - no current input, new work required
4 - no current input, long term outwith biomass programme
5 - no current input, high priority for further work

CONCLUSION ON FUTURE R & D ACTIVITY FOR TERRESTRIAL BIOMASS

To conclude our review on terrestrial biomass the main research requirements are summarised in Table VII, where the need for future research and development once again reflects the gradient of innovation from residues to new energy crops (Table VII). At one extreme where the extent, yield and management of the resource are well known; at the other extreme, so little is known about the vast genetic library of wild species which may be useful as new energy crops that some screening programme is necessary.

There is a strong need to establish large trials so that management and maintenance of productivity can be studied on a scale that might be met in practice.

TABLE VII - FUTURE DEVELOPMENTS

RESIDUES	1. The efficiencies of recovery and conversion should be improved. 2. Can straw yields be increased without a decrease in grain yield ?
CATCH CROPS	1. The constraints on implementation should be established 2. The understanding of yield needs needs to be refined 3. An economic conversion technology should be developed
EXISTING NATURAL VEGETATION	1. The stability of yield after harvesting should be investigated further 2. Harvesting technologies should be explored
NEW ENERGY CROPS	1. A genotype survey is essential

3.2 Review of research activity on aquatic biomass.

The various R & D projects of the European Community (DG XII), concerning the use of algae as biomass, can be divided into two groups utilising either micro-algae or macro-algae.

Micro Algae have been studied in the context of :

(a) closed systems or photoreactors (Paris and Cadarache)
(b) open systems consisting of fresh water ponds (Liege)
(c) open systems consisting of sea water ponds (Calabria)

Macro Algae have been evaluated in the context of

(d) growth on artificial substrates (North East Atlantic)
(e) direct harvesting under natural conditions (Venice Lagoon).

The objective of micro-algae closed systems is to study the feasibility of industrial production of hydrocarbons using algae.

To this end factors affecting hydrocarbon production of the green unicellular algae "Botryococcus braunii" such as the morphology and metabolism and the effects of the biological and physico chemical parameters of its culture have been investigated.

Productivity was 21 $mg/m^2/day$ as average of 365 days/year corresponding to 4 % of total sunlight photoconversion.

It has been shown that at least two species of Botryococcus exist, one producing linear hydrocarbons, the other producing branched hydrocarbons. The proportion of hydrocarbons is high in both species. It is considered that productivity may be increased further from 4 % up to 6 % sunlight photoconversion by strains selection, by culture parameter optimisation and through the use of a bacteria associated culture.

A first step towards the realisation of industrial production of hydrocarbons by Botryococcus cultivation has been taken by the setting up of a tubular photoreactor from which initial data have been obtained. The production cost is around 750 to 900 ECU/ton of dried Biomass.

Porphyridium Cruentum has been grown producing 50 % of cellular biomass which could be used for CH_4 or alcoholic fermentation, and 50 % of exocellular biomass (polysaccharide related to a carraghenan) with a value of 3.400 to 7600 ECU/ton.

Micro Algae in open systems have been grown successfully in large artificial lagoons (ponds of 2200 m^2), supplied by water from a river, heated up from cooling water of a power plant. The culture produces about 10 tonnes dry biomass $hectare^{-1}$ $year^{-1}$.

The production of biogas with a methane content of 65 % has been demonstrated.

The changes which occur during the course of the year have been studied also.

In Southern Italy seawater ponds have been used for the culture of algae, the growth of which is facilitated by the high ambient temperature.

Yields of 60 tonnes of dry biomass hectare^{-1} year^{-1} have been obtained and yields as high as 90 tonnes hectare^{-1} year^{-1} are expected in appropriate locations. Algae can be converted efficiently into methane (0,35 m^3 of CH_4 per Kg of volatile solids).

In open sea mariculture studies have been made of the growth of various types of macro algae on horizontal and vertical polypropylene ropes moored offshore.

Growth rates approaching 15 Kg per meter (fresh weight) have been measured.

Macro algae growing naturally in the lagoon in Venice have been harvested using boats fitted with suitable equipment. Harvesting is facilitated by tidal currents which remove macroalgae from the bed of the lagoon and produce concentrations in particular areas. The annual crop is estimated to be approximately 10^6 tonnes fresh weight of macroalgae.

Future activity

In the future further work on Botryococcus cultivation in a tubular photoreactor will eventually include the selection of high productivity strains and genetic engineering and a search for suitable biological and chemical agents to protect Botryococcus cultures against contamination.

Concerning the open systems it is envisaged to extend the present lagoons up to a size of about one hectare and to submit the harvested algae to a plant of biomethanisation.

The micro-algae cultivation in seawater will also be tested on a larger scale in order to demonstrate the industrial potential of the technology. For the macro-algae in open sea systems the maximum growth rates achievable and the factors which affect it will be established.

4. CONVERSION TECHNOLOGIES FOR BIOMASS

Biomass conversion technologies are now available in a commercial form. There is no single conversion technology which can be used with all biomass raw materials. Due to variations in water content and chemical composition a number of technologies have to be developed. These can be classified in terms of the conversion technology as thermal (direct combustion, gasification, pyrolysis, catalytic synthesis) or biological (fermentation or anaerobic digestion). In general, thermal processes are favoured for processing of dry ligno-cellulose materials and can result in production of heat, process steam, electricity, gas, charcoal or liquid fuels - either as the result of catalytic synthesis from gas (e.g. methanol or hydrocarbons) or as the crude mixed oxyhydrocarbons produced by pyrolysis processes. More sophisticated catalytic hydrogenation processes for direct production of liquid fuel are being developed.

Biological conversion processes are applicable to wet (over 80 % moisture) substrates. Easily fermentable substrates (sugar, starch, molasses, milk whey, etc.) are the starting point for the production of ethanol. Polluting liquid wastes, manures etc., may be used for the generation of methane by anaerobic digestion. Solid wastes may also be used as a source of methane using land-fill techniques. Algae and catch crops can be used as substrates in anaerobic digestion. The relationship between raw materials conversion technology and products is shown in Table VIII.

Table VIII - Raw materials : conversion technologies

Wastes/Residues	Conversion technology	Products
a) wet -manures, wine (grape pulp), olive wastes abattoir wastes	Anaerobic digestion	Methane gas
b) dry - straw, husks, peel	Thermal	Heat charcoal, gas liquid fuel
c) milk whey/ molasses	Fermentation	Ethanol, butanol, chemicals
d) forestry/wood	Thermal	Heat, charcoal, gas, liquid fuel (methanol, etc.)
e) municipal solid waste	Thermal/landfill	Heat/methane
f) sewage	Anaerobic digestion	Methane

Wastes/Residues	Conversion technology	Products
Energy crops and plantations		
a) wet - catch crops, algae, aquatic plants	Anaerobic digestion	Methane
b) dry - forestry	Thermal	Methanol, charcoal, gas
Surpluses		
Grains	Fermentation	Ethanol, charcoal, gas
Sugar	Fermentation	Ethanol, charcoal, gas

We shall present in the following pages more details concerning the different conversion technologies. In Table IX we present the average total energy ratio concerning complete biomass systems (production, harvesting, conversion)

Table IX - Total net energy ratio

Production plus Conversion energy ratio	Conversion efficiency %	Total net energy ratio (*)
Anaerobic digestion (gas)	50 - 60	Wastes: high** catch crops: 2-3
Fermentation/distillation (ethanol)	20 - 30	0.8 - 3.5
Thermal conversion (methanol: from gasification/pyrolytic products)	30 - 60	2 - 5
Direct combustion (heat)	60 - 80	5 - 15

$$\text{* Total net energy ratio} = \frac{\text{energy output in product}}{\text{non-renewable energy inputs (total energy inputs for production and conversion of biomass)}}$$

** Wastes used on site with minimal transport have very low energy requirements - this also applies to landfill

4.1 Biological Conversion of Biomass

Four main research areas were included in the E.C. Energy from Biomass Programme : (i) biogas production by anaerobic digestion of purpose grown terrestrial biomass and effluents and residues from the food industry and from agriculture (5 contracts); (ii) biogas from marine and freshwater algae (1 contract); (iii) enzymic liquefaction and saccharification of cellulosic wastes (2 contracts) and (iv) fermentation alcohol from agricultural biomass. The progress achieved during the last four years of the programme is :

a) Anaerobic Digestion

Important progress was achieved on laboratory and pilot scale studies of a variety of reactor designs treating soluble and solid substrates at mesophilic and thermophilic temperatures and involving one-step or two-step digestion techniques.

Two-phase anaerobic digestion of solid substrates was shown to be technically feasible by the independent research of Rijkens and Voetberg in Wageningen, Colleran and co-workers in Galway and Nyns, Naveau and co-workers in Louvain-la-Neuve. The Wageningen group demonstrated the feasibility of the process at laboratory and pilot scale for a variety of solid residues currently causing considerable disposal problems in the Netherlands. In Galway, the process was regarded as a viable means of generating additional energy at farm-level by the digestion of crop wastes in season through linking a hydrolysis reactor to an anaerobic filter operating on a year-round basis for animal slurries and other liquid wastes. Two-phase digestion of algal substrates was also shown to be feasible and to operate with a high methane conversion efficiency at Louvain-la-Neuve. Considerable differences with respect to the design and mode of operation of the first phase, liquefaction reactor were reported.

The results of comparative pilot scale studies on the digestion of farm wastes and energy crops using a high rate contact digester, a plug flow digester, a hydraulic digester and an anaerobic filter were reported by Stafford and co-workers. High methane yields at a retention time of 3.3 days were obtained with the hydraulic digester treating pig waste. Naveau and co-workers presented data on a continuous single stage system for algal digestion. High CH_4 yields at a loading rate of 0.4 kg $VS_o.m^{-3}.d^{-1}$ were reported and efficient integration of the various steps from cultivation to digestion was achieved.

In summary, two-stage digestion presents a novel attractive approach to biomass methanation.

Nevertheless, the development of stable and reliable ancillary equipment for anaerobic digesters was seen to be an urgent prerequisite for full scale adoption of anaerobic digestion technology. In particular, suitable pumps, flow meters, heat exchangers and sensors to measure COD and VFA levels in feed, effluent and within the reactor are required. Further study on modelling and automation of anaerobic digestion is needed.

b) Hydrolysis and Liquefaction of cellulosic wastes.

The utilisation of commercially available cellulolytic and pectolytic enzyme preparations for the liquefaction and saccharification of beet pulp, potato fibre, tomato plant and wood wastes and a turnip catch crop was investigated by Beldman, Voragen and co-workers in Wageningen. A very high degree of conversion was obtained for beet pulp, potato fibre and stubble turnips. Tomato plant and wood wastes proved to be more resistant to hydrolysis due to their lignin content and pretreatment was found to be necessary. Extrusion increased the enzymic digestibility of both substrates but may be non-viable economically.

Future work

A clear need has been identified for additional fundamental research on the polymer degrading enzymes involved in biomass saccharification.

Research on reactor design is seen to be of importance with respect to removal of fermentation byproducts and optimisation of saccharification rates. Efficient and low-cost pretreatment methods must be developed in order to achieve saccharification of lignocellulosic biomass. Priority was also given to the need to develop efficient solid state fermentation systems which would allow liquefaction and saccharification to be combined with subsequent fermentation processes in a single reactor. A need was also identified for further research on the fermentability of pentose sugars and uronic acids formed during biomass saccharification.

c) Fermentation Ethanol

Goma and co-workers of INSA - Toulouse have reported on a method for solid state fermentation of Jerusalem artichoke which yields 10 % ethanol after 12 hours fermentation. A synergistic effect between ethanol and sugar inhibition was demonstrated. Detailed studies on ethanol inhibition revealed that a dual effect was involved - (i) inhibition as a result of ethanol accumulation within the cell and (ii) inhibition by an unknown excreted co-metabolite which is more inhibitory than ethanol. Inhibition was shown to be correlated with a decrease in cell-bound water. Technological studies on the use of immobilised cells; the inclusion of various additives to enhance cellulose hydrolysis and the development of an extractive fermentation system which couples solvent extraction of ethanol with fermentation in the same reactor were also carried out. The results obtained are also applicable to acetonobutylic and acidogenic fermentations for liquid fuel production.

Future work.

It was agreed that a critical reappraisal of the ethanol by fermentation route for energy production from biomass should be carried out. Further laboratory studies are necessary on inhibition effects in ethanol fermentation, particularly with a view to ascertaining the identity and mode of action of the co-metabolite inhibitor. Technologically innovative concepts need to be explored with regard to continuous fermentation at high cell density. In this regard, cell recycling systems and the coupling of ultrafiltration and fermentation and flocculating devices require further investigation. Computer control of fermentation processes also merits additional study.

4.2 Thermochemical Conversion of Biomass.

The objectives of this presentation are :

- first to identify the state of art and achievements of the last four years in the various thermochemical conversion areas sponsored by the EEC;

- second to identify the remaining problems and constraints inhibiting implementation of the technologies developed;

- third to suggest promising future research and development areas which deserve support in a future programme.

a) Direct combustion

Extensive work has been carried out on straw combustion in Denmark, Germany and the U.K. with agreement that continuous feeding is preferred to batch operation; most technical problems have now been solved, with the major disincentive to wider adoption being the high initial cost of such units. The primary requirement now is for low cost, high efficiency furnaces with automatic charging systems and adequate safety controls. A compromise between quality and cost may be necessary. The differing environmental regulations of member countries is an inhibitory factor in some locations.

A current problem is that many furnaces now in use are of poor design with a low efficiency and with consequent environmental problems. These factors tend to discourage wider adoption. However, over 500,000 combustors were claimed to have been sold in Germany and 1 in 6 Danish farmers have a unit. Systems of up to 10 MW supplying district heating schemes were reported in four locations in Denmark, although generally 1000 kW was believed to be the limiting capacity. A further limiting factor might be feedstock in some locations, for example Holland, where straw production has virtually disappeared in 5 years. Denmark, however, still has a straw surplus of 2 million t/y straw.

Alternative feedstocks : wood and bark can be satisfactorily burned in high quality, low cost systems; and a comparative study of gasification and combustion of urban waste has recently been commissioned by the CEC.

It was generally felt that biomass combustion systems were sufficiently developed and available commercially to remove the need for further substantial R & D support. There are, however, improvements in efficiency and pollution levels that are desirable; and interesting possibilities exist for developing fluidised bed combustor systems.

For electricity production large scale biomass production systems are generally needed.

For a 20 MW el. power station an area of 23.600 ha, producing 5 t/ha x year (dry matter), is needed assuming for the generator an efficiency of 25 % and an operating time of 6000 hr/year.

b) Pyrolysis

Strictly speaking pyrolysis is understood as the thermal decomposition of organic matter into char, pyrolytic oil and gas. This technology well known a long time ago in processing coal, was noticed again during the seventies as a process to handle municipal residues and wastes. But the heterogeneous constitution of these above residues showed up several

›cessing problems. As biomass has rather homogeneous chemical ›perties pyrolysis is discussed at the moment to be suitable to convert ›mass into useful energy holders. Here below, several schematic ‑outs of Pyrolytic converters are shown :

Schematic layout of a rotating converter with charging and discharging appliance and flue gas duct (Kiener)

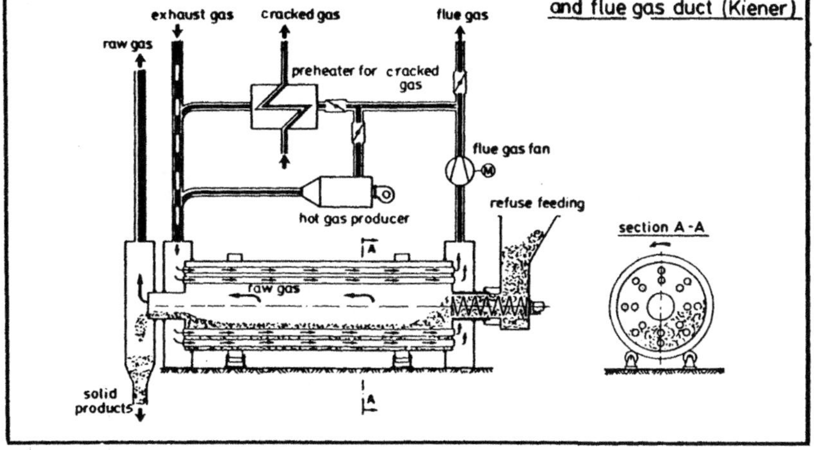

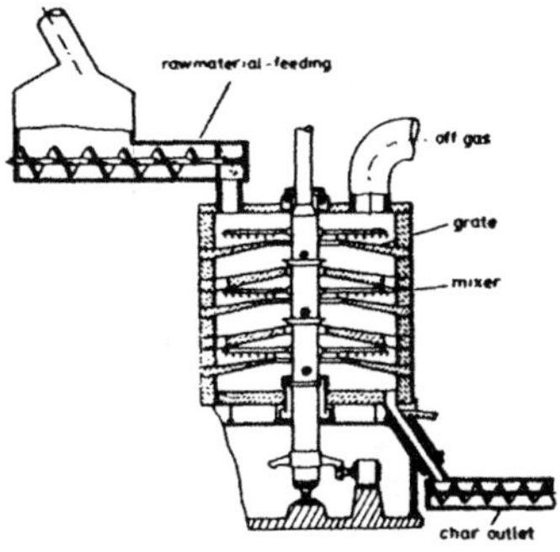

Rackfurnace for Charproducing
(Herreshofofen)

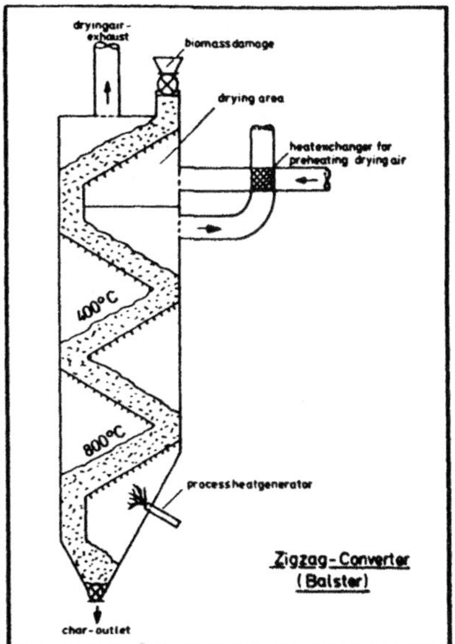

Zigzag-Converter
(Balster)

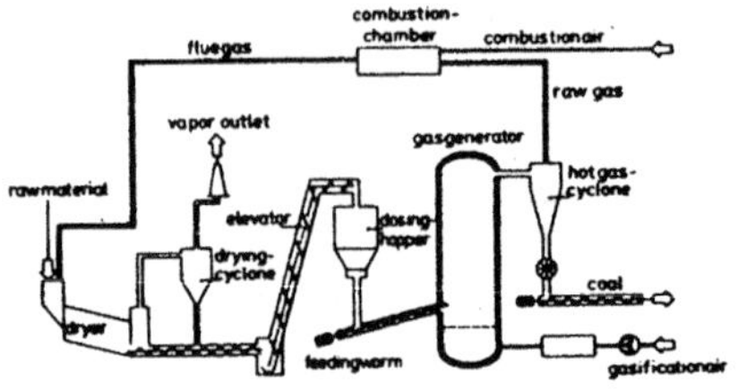

Fluidized bed plant for carbonisation of biomass (TAG, ERCO)

Cross flow pyrolysis (Foster Wheeler)

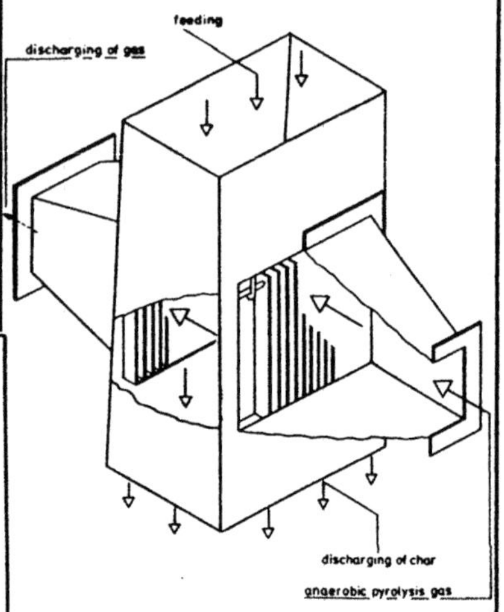

Fluidized bed - Pyrolysis Reactor (GARRETT)

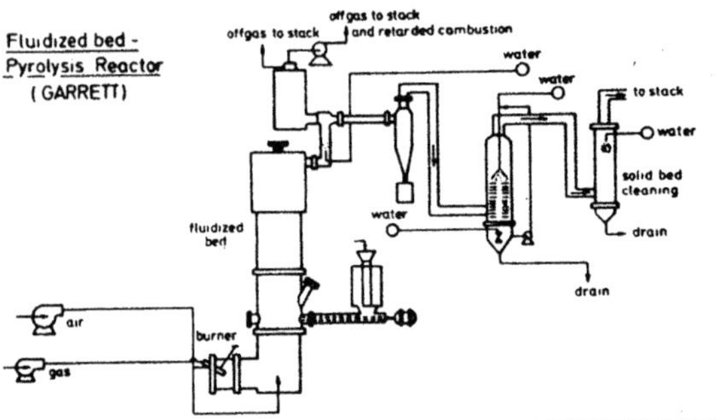

Converterstation, System TECH AIR

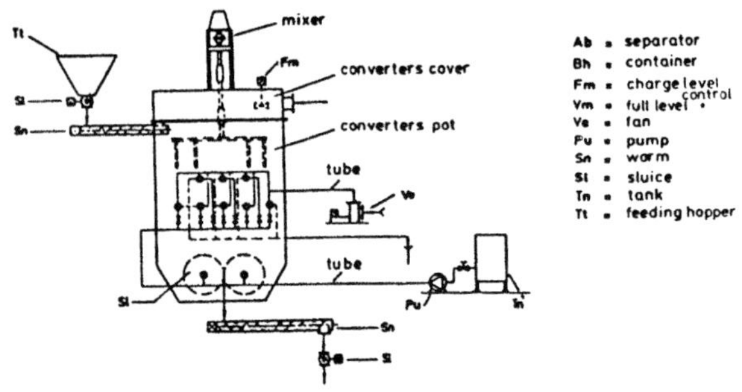

Pyrolysis in Pressure-Autoclaves (PPT-Process-Pyrolysis)

(Heyner)

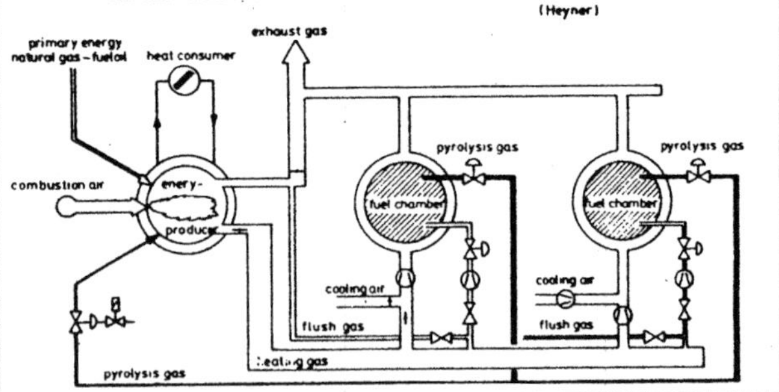

In Table X experimental data obtained by a pyrolysis installation (Italenergie SPA - Italy) are presented in relation with different types of agricultural residues and operating temperature.

TABLE X: Pyrolysis experimental results

Operatg Temp. (°C)	Conversion Product output (% by weight)	Type of agricultural residues				
		Olive oil residues	Bunch of grapes	Straw	Lucerne	Algae
450	Charcoal	37	35	31	30	45
	Bio-oil	8)		23	10)	
	Water	22)	31	16	37)	20
	Gas	33	34	30	23	35
550	Charcoal	32	-	30	27	-
	Gas	38	-	47	42	-
	Bio-oil	7	-	7)		
	Water	23	-	16)	31	-

The produced charcoal has a calorific value of about 28 MJ/kg and represents a high valuable energy holder which is commonly sold as barbecue coal or basic material for chemical industries for rather high prices of 360-400 ECU/t. If the raw material is available at low prices - e.g. as wastes from agroindustry or forestry - even under European conditions the production of charcoal may be economic as long as the high prices for charcoal will be paid. The value of pyrolytic oil depends mainly on the degree of gas condensation. If water is still remaining as vapor, pyrolytic oil has a calorific value of about 20-25 MJ/kg.

This oil can be added to fuel oil with a percentage of 20 % for the use of conventional oil burners. Short duration successful tests on diesel engine, utilising 100 % of pyrolytic oil have been reported. If the condensable components of the gas will be condensed totally, pyrolytic oil consists more or less of contaminated water and is therefore useless. Generally pyrolysis can be considered as a technology which may be applied in rather smaller units than other conversion systems like e.g. methanol synthesis.

Some work has been carried out on maximising the liquid fraction, for example by flash pyrolysis, and an overview of the status and progress of work in this area would be valuable. There is a definite incentive for R and D on upgrading the product to improve stability, reduce corrosive constituents and produce a more uniform and compatible fuel.

c) Gasification

The CEC programme has, over the last eight years, seen projects develop from fundamental and laboratory investigations to substantial demonstration units. It is clear that the European market for such systems will not develop rapidly, but the world market has a much greater potential to assimilate the technology being developed. What, then, has been learned; what are the problems; and what should subsequent work programmes consider ?

A general point concerns feedstocks. The conclusion is that, especially for moving beds, shorter term development should be orientated to plants dedicated to a specific feedstock. This requirement would place greater emphasis on a closer interface between biomass production and conversion. Feedstock preparation is a fertile area for further investigation in terms of size control, drying, handling and storage. The nature of biomass being essentially fibrous is likely to cause more problems in feeding than, for example, coal; and there is thus scope for more work in this area particularly for pressurised reactor systems.

<u>Small scale gasifiers (50-500 kg/hr).</u> Most work in this area has been completed; table XI gives an overview.

Table XI : Small scale gasification.

Reactor type	Capacity kg/hr	Reference
Co-current moving bed gasifier	50	Twente University of Technology (2)
	50-250	Fritz Werner Industrie-Ausrüstungen GmbH (3)
Crosscurrent gasifier	40	Foster Wheeler * (4a)
Fluidized bed gasifier	50-100	Twente University of Technology (4b)
	100-150	Creusot Loire (4c)
	100	Fritz Werner Industrie Ausrüstungen GmbH (3)
Cyclonic gasifier	100	CEMAGREF (5)

The CEC programme has been very stimulating and successful in this field. In fact the programme resulted in commercialization of small scale air gasification plants in Europe and it is felt that after the present project period there will not be much need for additional R & D. An exception might be gasification of "difficult feedstocks" with high ash content or low ash melting point (e.g. municipal waste, rice hulls).

The European market is now seen to be at a critical stage, and, although there are at least ten companies offering commercial gasifiers, most market growth potential is seen to be in developing countries.

Medium scale gasifiers (500-5000 kg/hr).

Four pilot plants aiming at synthesis gas have been completed. These projects had be sponsored by the CEC in the "Methanol from wood" programme. Two other systems aiming at synthesis gas had been tested in the past CEC programme. Table XII gives an overview.

Table XII : Gasifiers and pilot plants for synthesis gas from wood.

Reactor type	Capacity kg/hr	Pressure bar	Consortium
1. Fluidized bed gasifier	2500	10-15	Creusot-Loire
2. Double fluidized bed gasifier	1000	atmospheric	John Brown/ Wellmann
3. Circulating fluidized bed gasifier	400	atmospheric	Lurgi Kohle und Mineralöltechnik GmbH
4. Co-current moving bed	2000	10-15	Noverlerg
5. Cross current gasifier	1000 2500	atmospheric 10	Creusot-Loire Entreprise
6. Fluidized bed gasifier	1000	atmospheric	Italenergie

The construction of 4 pilot plants has just been completed. It is however, too early to comment on these activities.

The technical survey will identify remaining problem areas that still ask for attention. The economical evaluation will probably give reliable data on the feasibility of medium scale synthesis gas production from wood. Despite the large numger of economic data already published sufficient accurate figures are not yet available due to uncertainties in both real costs of feedstock and cost of medium scale gasification.

A possible attractive follow-up may be found in integrating a fast rotation forest production area with a methanol synthesis plant, provided the economics will not be too prohibitive.

Though the programme concentrates on synthesis gas production, it will automatically also result in the commercialization of medium scale air gasification.

Of great importance is a comparative evaluation of gasification and combustion with respect to plant capacity and the mix of energy products needed. It would be of great help as a decision making guide for those who have a choice in installing either a gasifier or a combuster.

A market survey is believed to be of great importance to identify and differentiate opportunities for different systems.

In tables XIII/XIV/XV we have summarized the most important achievements, constraints, possible future R&D subjects relative to thermochemical conversion of biomass.

TABLE XIII COMBUSTION : Summary of Achievements, Constraints and R&D Opportunities for Thermochemical Processing of Biomass

Achievements to date	Constraints against implementation	Opportunities for further R&D
Commercialisation realised depending on location	High capital cost below 1MW	Further design improvements
Some design improvement	Current models poorly designed	Retrofitting applications
Combustion of wet material with heat recovery	Temperature needs to be high	Efficiency improvements
	Residence time needs to be high	Environmental improvements with gas and liquid effluents
Low grade heat recovery for domestic applications	Excess air needs to be low	Application of fluid bed technology
Material handling improvements e.g. pelletising	Environmental requirements (site specific)	Comparison of combustion and gasification; technically and economically
500 000 units sold in 5 years in Germany	Feedstock supply problems in some locations (e.g. Holland)	

TABLE XIV : SMALL SCALE SYSTEMS FOR PYROLYSIS AND GASIFICIATION

Summary of Achievements, Constraints and R&D Opportunities for Thermochemical Processing of Biomass

Achievements to date	Constraints against implementation	Opportunities for further R&D	Further R&D
Pyrolysis			
Available commercially Some design problems solved	Condensibles inhibit any application except direct burning Multiple products (gas liquid and char) Environmental requirements Novelty precludes wider acceptance	Fundamental studies Maximise liquid yields Modelling Effective utilisation of all products	Liquid stability engines optimization mobile units development monitoring
Low Heating Value Gasification			
Available commercially up to 1 MW. More than 10 companies offering systems in Europe Retrofitting possibility Some design problems solved Developments in design procedures and modelling	Feedstock flexibility and difficult feedstocks particularly with moving beds Capital costs on small scale (typically up to 1kWe for moving beds and up to 1MWe for fluidised beds) Environmental requirements from wet cleaning systems	Monitoring performance of units already sold Feedstock preparation and requirements (including drying) Interface between biomass production and conversion Hot gas cleaning	Adapatation of moving beds to small particles and high ash solids Gas quality specification Ash behaviour including fusion for wastes above 5% ash optimization Improved design procedures

Table XIV continued

Achievements to date	Constraints against implementation	Opportunities for further R&D	Further R&D
	Gas cleaning necessary for internal combustion engine applications Novelty precludes wider implementation	Pressurisation alone, and with turbines for power generation Secondary gasification/reforming for control of tars etc.	Export potential (adaptation to feedstock and environment) Flexibility-feedstock, operation, turndown System evaluation and application assessment Comparison of gasification and combustion

TABLE XV : SMALL SCALE SYSTEMS FOR PYROLYSIS AND GASIFICATION

Summary of Achievements, Constraints and R&D Operations for Thermochemical Processing of Biomass

Achievements to date	Constraints against implementation	Opportunities for further R&D	Further R&D
Indirect Liquefaction			
Four pilot plants in progress sponsored by EEC plus others. Too early for assessment	Imbalance of airings and economic fuel synthesis Capital cost	Small scale economical fuel synthesis operation up to 300 t/d Pressurisation Catalysis (with fundamental work) synthesis gas upgrading	Technico-economic feasibility studies and optimisation Case studies of complete process from biomass production to fuel end-product Reactor modelling
Direct Liquefaction (Pressure hydrogenation)			
Possible but too early to judge	Yield Selectivity Feasibility	Fundamental investigations Monitoring activity (Later)-Feasibility studies Characterisation of products	

Table XV continued

Achievements to date	Constraints against implementation	Opportunities for further R&D	Further R&D
Pyrolysis			
Some recent progress on increasing liquid yield	Stability, corrosivity, and carcino-genicity of liquids Product yield Feasibility and viability	Upgrading of condensibles to stable non-corrosive products Maximising liquid yield Scaling up, modelling and feasibility studies	

5. ECONOMIC CONSIDERATION ABOUT THE UTILSIATION OF BIOMASS-DERIVED PRODUCTS

Unfortunately, general economic data concerning complete biomass sytems (production, harvesting, collection, conversion, products marketing and distribution) do not exist.

Furthermore, it is not possible to draw final economic conclusions because there is a continuous improvement of technology and process consequent to the large R&D effort.

The level of the specific investment required in the different conversion technologies is approxiately as follows (based on 20 % water content):

- pyrolysis: \$ 90/m^3 biomass x year
- Ethanol plant: \$ 140-230/m^3 biomass x year
- Wood sawing plant (for beating): \$ 140/m^3 biomass x year
- Wood thermoelectric plant: \$ 200/m^3 wood x year
- Wood gasifier (1MW electricity generator: \$ 190/m^3 wood x year
- Biogas plant: \$150-500/m^3 solid organic matter x year

and some data about the market value of some biomass fuels:

-	Ethanol	:	300-1000 ECU/t
-	Methanol	:	350- 750 ECU/t
-	Pyrolytic Oil	:	100- 150 ECU/t
-	Charcoal	:	60- 100 ECU/t

(1 ECU = 0,85 $ - June 1984)

6. CONCLUSION.

I. Biomass is one of the most important renewable sources of energy. Within the European Community biomass, by the year 2000, may provide 90 MToe net energy;

II. Biomass productivity shows quite large variation :

natural systems	:	2 to 20 t/ha/y (dry matter)
forestry	:	6 to 15 t/ha/y
catch-crops	:	2 to 8 t/ha/y
algae	:	40 to 80 t/ha/y

III Harvesting technologies need intensive further development;

IV Biological conversion of biomass is very promising for the production of methane, production of ethanol, production of additives in substitution of lead in petrol;

V. Thermochemical conversion of biomass is an important fast process for the production of liquid fuels like methanol, pyrolytic-oil;

VI Energy-crops may become a new economic activity replacing agricultural production by energy farming;

VII The full exploitation of biomass could offer the opportunity of creating 500.000 new jobs in the E.C.

REFERENCES :

- Solar Energy R&D in the EC, Energy from Biomass, Vol.3 Reidel Publ.Co
- Solar Energy R&D in the EC, Energy from Biomass, Vol.5 " " "
- Proceedings of 2nd Conference "Energy from Biomass" - Berlin
- various documents of C.E.C.

INDEX

GPSR Compliance

The European Union's (EU) General Product Safety Regulation (GPSR) is a set of rules that requires consumer products to be safe and our obligations to ensure this.

If you have any concerns about our products, you can contact us on ProductSafety@springernature.com

In case Publisher is established outside the EU, the EU authorized representative is:

Springer Nature Customer Service Center GmbH
Europaplatz 3
69115 Heidelberg, Germany

Batch number: 10755284

Printed by Printforce, the Netherlands